Flow Analysis

Flow Analysis

A Practical Guide

Víctor Cerdà

Laboratory of Environmental and Analytical Chemistry (LQA2)
University of the Balearic Islands

Laura Ferrer

Laboratory of Environmental Radioactivity (LaboRA)
University of the Balearic Islands

Jessica Avivar

Laboratory of Environmental Radioactivity (LaboRA)
University of the Balearic Islands

Amalia Cerdà

Head of the Social Responsibility Unit of TIRME, S.A.

ELSEVIER

AMSTERDAM • BOSTON • HEIDELBERG • LONDON
NEW YORK • OXFORD • PARIS • SAN DIEGO
SAN FRANCISCO • SINGAPORE • SYDNEY • TOKYO

Elsevier
Radarweg 29, PO Box 211, 1000 AE Amsterdam, The Netherlands
The Boulevard, Langford Lane, Kidlington, Oxford OX5 1GB, UK

First edition 2014

British Library Cataloguing in Publication Data
A catalogue record for this book is available from the British Library

Library of Congress Cataloging-in-Publication Data
A catalog record for this book is available from the Library of Congress

For information on all Elsevier publications
visit our web site at store.elsevier.com

Printed and bound in Poland
14 15 16 17 18 10 9 8 7 6 5 4 3 2 1

ISBN: 978-0-444-59596-6

Contents

COMPANION WEB SITE
In the website of Elsevier we provide more information which will help you to build your flow systems, such as suppliers, meetings, literature resources, databases, figures and more.
http://store.elsevier.com/product.jsp?isbn=9780444595966.

Preface

Flow techniques have aroused especial interest in relation to many other automatic methodologies of analysis. Ever since segmented flow analysis was developed by Skeggs in 1957, flow techniques have been in continuous evolution towards new developments from flow injection analysis (FIA) by J. Ruzicka and E.H. Hansen in 1975 up to multisyringe flow injection analysis by our group in 1999 and, more recently, multipumping flow systems in 2002 by Lapa et al.

The fast evolution of flow techniques has led to a scarcity of literature on some. In fact, only FIA has been the subject of several books and SIA has been of only one, dealing briefly with it. This prompted the authors to write a book intended to fill the existing gap by describing each technique individually and comparing it to its alternatives in terms of advantages and disadvantages. As shown in this book, there is no solid argument in favor of using any specific flow technique in particular; rather, one can derive substantial advantages from the combined use of two or more.

Computers are virtually indispensable in all modern flow techniques. This initially hindered further development owing to the unavailability of suitable commercial software and a general lack of experience in coupling personal computers to instruments. For this reason, nearly a whole chapter is devoted to AutoAnalysis, a general-purpose software suite that facilitates the implementation of all flow techniques, chromatographic ones included. AutoAnalysis revolves around a paradigm that affords the continual incorporation of new instruments and apparatuses. This has facilitated the integration of new techniques, which have emerged ever since the software was originally developed and is bound to be the case with others that may arise in the future.

This book is not a one-person work, but rather the product of a group of researchers who have invested many hours in developing some of the configurations described in it.

We wish to express our gratitude to all those who have made this book possible and thank them for the pleasant time we have enjoyed together and not only while working. Thus, we are indebted to the firm Sciware Systems for its ceaseless support to our work, and for its willingness to cooperate in the development of some prototypes, which took long hours of discussion.

As William Smellie wrote in the preface to the first edition of Encyclopedia Britannica (1768–1771), "Utility ought to be the principal intention of every publication. Wherever this intention does not plainly appear, neither the books nor their authors have the smallest claim to the approbation of mankind." It is therefore our humble intention to be useful to those who want or need to introduce automation in their laboratories.

We hope the book will be successful enough to warrant a second edition, so we are open to any comments or suggestions readers may wish to make towards an expanded, updated version.

The authors

Palma de Mallorca, October 2013

Author Biographies

Prof Dr Víctor Cerdà Martín was born in Palma de Mallorca (Spain), and graduated from the University of Barcelona (Spain) in chemistry, where he also developed his PhD in chemical sciences. He leads the Laboratory of Environmental and Analytical Chemistry in the University of the Balearic Islands, and the group of Analytical Chemistry, Automation and Environment since 1984. His main line of research has been focused in the development of automated methodologies of analysis based on flow techniques and their application to the determination of parameters of environmental interest. Editor, author, and coauthor of more than ten books, more than 15 chapters and approximately 500 research articles in international journals. He has supervised more than 30 doctoral theses. He has been the principal researcher of more than 25 research projects since 1985, among them is the COMETT II Project of the EU that lead to the creation of the "European School on Environmental Sciences and Techniques". He has been vice-chancellor of Scientific Policy and Innovation in the University of the Balearic Islands. He is also the head and founder of the university spin-off Sciware Systems, S.L., devoted to the development of new automated analytical solutions, and the President of the Association of Environmental Sciences and Techniques (AEST).

Dr Laura Ferrer Trovato was born in La Plata (Argentina), and graduated from the National University of Mar del Plata (Argentina) in Biology. Then she obtained a PhD in biological sciences from the Southern National University (Argentina). She moved to Spain in 2001, where she completed her second PhD in chemical science and technology from the University of the Balearic Islands. She has written four book chapters related to environmental monitoring. Dr Ferrer has published more than 60 research articles and presented more than 70 contributions at international conferences related to the environment, automation, and radioactivity. She is the head of the Laboratory of Environmental Radioactivity (LaboRA) and member of the university spin-off Sciware Systems, S.L. Dr Ferrer is responsible for supervising a number of PhD and master theses candidates in the field of automation and environmental monitoring.

Dr Jessica Avivar Cerezo born in Palma de Mallorca (Spain), and graduated from the University of the Balearic Islands (Spain) in chemistry, where she also developed her PhD in chemical science and technology. She is the coauthor of a book chapter related to environmental monitoring. Dr Avivar has published approximately 20 research articles and has presented more than 20 contributions at international conferences related with the environment, automation, and radioactivity. Researcher at the University of the Balearic Islands, secretary of the Laboratory of Environmental Radioactivity (LaboRA) and member of the university spin-off Sciware Systems, S.L. Her research has been mainly focused in the development of automated approaches for radionuclides environmental monitoring.

Dr Amalia Cerdà Lacaci was born in Palma de Mallorca (Spain), and graduated from the University of the Balearic Islands (Spain) in chemistry, where she also developed her PhD in chemistry. She is coauthor of two books and two book chapters related with the automation of analytical methods. She has published more than 10 research articles and presented approximately 30 contributions at international conferences related to the environment. She has participated in several research projects of the European Union, Spanish, and Autonomic Governments. She has been a professor lector at the University of the Balearic Islands in the Chemistry Department and nowadays is the head of the Environmental and Quality Department of TIRME S.A., the company responsible for the urban solid waste management of Mallorca. She is in charge of the design, development, and implementation of integrated management systems in the company.

CHAPTER 1

Evolution and Description of the Principal Flow Techniques

CHAPTER OUTLINE

Advances in science and technology have resulted in an increasing demand for control analyses and posed various challenges to analytical chemists such as the need to develop new methods exhibiting as much selectivity, sensitivity, sample and reagent economy, throughput, cost-effectiveness, simplicity and environmental friendliness as possible. Automation and miniaturization of solution-based analysis are essential to make them fast and efficient for routine and research tasks [1].

Flow techniques have undoubtedly aroused especial interest in relation to many other automatic methodologies of analysis. Ever since, segmented flow analysis (SFA) was developed by Skeggs in 1957, flow techniques have been in continuous evolution toward new developments. They have

Flow Analysis. http://dx.doi.org/10.1016/B978-0-444-59596-6.00001-2

gained importance for clinical, industrial and environmental purposes as they allow highly reproducible fast determinations. Since the beginnings of automation of analytical methods, various different flow techniques have been developed and used for analytical or monitoring applications. There is no solid argument in favor of using any particular flow technique separately; rather, substantial advantages can be derived from their combination. Since flow-based methods are nonseparative tools, advantages of combining flow techniques with separation techniques are also noteworthy (see Chapter 3). Among the benefits of automation of analytical procedures, the increase of sample frequency, minimization of sample contamination or alteration, miniaturization of the analytical system, analyst security improvement and lower reagent and sample consumption, implying less personal and consumable costs should be highlighted.

Thus, a detailed description of flow techniques, their evolution, their hyphenation advantages and a critical comparison between them are presented in this chapter.

1.1 The need for analysis automation

The need to seriously consider the development of automatic methods of analysis arose in the 1950s, when clinical tests started to be increasingly used for diagnostic purposes in medicine. This led to a rapid increase in the demand for laboratory tests which, for obvious economic reasons, could not be met simply by hiring more laboratory staff.

The large number of samples with which analysts can be confronted imposes the use of expeditious analytical methods, such as automatic ones. Among the benefits of automation of analytical procedures, the increase of sample frequency, minimization of sample contamination or alteration, miniaturization of the analytical system, and lower reagent and sample consumption, implying less personal and consumable costs should be highlighted. Furthermore, automation improves methods' repeatability and reproducibility, since human error is avoided and the analytical system is more isolated from the environment. Automation is of great value also, when dealing with samples or chemicals which are harmful, e.g. radioactive samples or cancerogenic chemicals. Thus, analyst safety is also improved by automation, e.g. radiochemical methods automation is of high interest since this reduces analyst exposure to radiation.

1.1.1 Automatic and automated methods

Despite the major conceptual and operational differences between automated and automatic, these terms are frequently confused. Automated is referred to partly automated methods and automatic to fully automated ones.

A fully automated method should allow the whole analytical process to be completed with no intervention from the analyst; that is to say it should also be capable of making decisions by itself, including operating conditions alteration in response to the analytical results. All too frequently, methods are deemed automatic simply because one or several steps of the analytical process are performed in an automated manner. However, an automatic method should be capable of completing all steps including sampling, sample preparation and dissolution, interference removal, aliquot withdrawal, analyte measurement, data processing, results evaluation and decision making, and also of restarting the whole process in order to adapt it to the particular needs of a new sample if necessary.

Table 1.1 Steps of the Analytical Process and Equipment Used to Automate the Operations Involved

Step	Operations	Equipment	Difficulty	Controller
Sampling	Solids	Robot	Unusual	Dedicated microprocessor computers hosts
	Liquids	Dispensers	Low	
	Gases	Pumps	Low	
	Mixtures	Robot	Unusual	
Physical conditioning	Size reduction	Dispenser, robot and mill	Low	
	Drying	Robot and stove or lamp	Unusual	
	Dissolution	Robot and vortex, heater	Unusual	
	Filtration	Robot and filtering unit	Unusual	
	Centrifugation	Robot and centrifuge	Unusual	
	Weighing	Robot and balance	Unusual	
	Aliquot withdrawal	Dispenser, robot and balance	Unusual	
	Injection	Dispenser and valve	Low	
		Robot and syringe	Usual	
Chemical conditioning	Extraction	Robot	Unusual	
	pH adjustment	Dispenser	Low	
	Masking	Dispenser	Low	
	Derivatization	Special sampler	Usual	
Measurement	Transducing	Transducer	Low	
	Signal conditioning	Noise filter	Low	
	Amplification	Amplifier	Low	
	Data storage	Graph recorder	Low	
Calibration	Gases	Valves	Low	
	Solids	Valves, dispensers	Low	
Data processing results evaluation and decision making		Microprocessors and dedicated computers	Low	

In Table 1.1 are shown different ways of automating individual steps of the analytical process [2].

Obviously, a fully automatic method is very difficult to develop, especially for solid samples, first steps in the analysis of which can rarely be performed in an inexpensive manner. Usually, the operations posing the greatest difficulties among those involved in such steps are those requiring some mechanical handling, automation of which is only possible in most cases by using a robot arm adapted to the particular chemical operations to be performed. Because this equipment is too expensive for most analytical applications, fully automated methods for the analysis of solid samples are very scant and largely restricted to the control of manufacturing processes in practice.

Automation of analyses involving liquid samples is facilitated by their usually adequate homogeneity and easy mechanical handling by use of peristaltic or piston pumps, or some other liquid management devices, i.e. a liquid driver. Nonetheless, this is not the case when dealing with solid samples, analysis of which frequently involves their prior conversion into liquids by dissolution. The dissolution step is the bottleneck of analytical processes involving solid samples, as it is frequently slow and must be performed manually.

Earliest automatic methods used dedicated devices suited to particular applications. This restricted their scope to very specific uses such as the control of manufacturing processes or in those cases where the number of samples to be analyzed was large enough to justify the initial effort and investment required. Ideally, analytical equipment should be versatile, capable of accommodating a wide variety of assays without the need for system reconfiguration, and compatible with a wide range of detectors.

Moreover, the power of computers has grown so dramatically that they now allow an automatic method to be reset for adaptation to quite different needs; such a high flexibility simplifies the development of effective solutions at substantially reduced costs. The advantages of current flow methods are also in part the result of the inception of computers; in fact, the flexibility of computers allows the same equipment (hardware) to be used with little or no alteration in order to implement the same analytical method on different types of samples simply by altering the software.

1.1.2 Flow techniques contribution to greener analytical chemistry

We may classify flow methods in nonresolutive (SFA, flow injection analysis (FIA), sequential injection analysis (SIA), multicommutated flow injection analysis (MCFIA), multipumping flow systems (MPFS), etc.) and resolutive ones (chromatographic techniques, capillary electrophoresis, etc.). In this book we will mainly focus on nonresolutive flow methods of analysis.

Nowadays, the amount and toxicity of the generated wastes is as important as any other analytical feature when developing a new analytical method.

Ideally a green method should not produce chemical wastes, if so these should not be toxic and their amount should be minimized. Moreover toxic wastes should be recycled and if possible reused and/or the analytical method should include waste treatment and disposal.

Although not all flow-based methods can be considered green methods, since most of them are reagent based, these contribute to reduce environmental impact by reducing reagent consumption and so waste generation. Especially multicommutated flow techniques which inject the amounts of reagent strictly required. Moreover, the progress of flow techniques toward miniaturization and reagents saving has contributed to greener analytical chemistry [3]. As an example, a method for carbaryl pesticide determination reduced the amount of the main reagent (*p*-aminophenol) employed from 140 to 5 μg exploiting a multicommutated flow system [4].

Nonresolutive flow techniques can be classified in two main groups. That is to say, initially flow systems were operated exclusively by hand, e.g. in SFA and FIA. Subsequently, however, computers facilitated the development of computerized controlled flow techniques such as SIA, MCFIA, multisyringe flow injection analysis (MSFIA) and MPFS, based on multicommutation operation. Multicommutated flow techniques have shown great potential in comparison with previous flow techniques in minimizing reagents consumption and waste production, providing more environmentally friendly methodologies, since liquids are only propelled to the system when required and returned to their reservoirs when not. All of them have common components such as impulsion pumps, which act as liquid drivers and a series of plastic tubes or manifold intended to carry liquid streams to the detector. Finally, in a search for miniaturization and integration of the whole analytical procedure have appeared miniaturized systems such as lab on valve technique (LOV), chip on valve (ChOV) and lab in a syringe (LIS).

1.2 Skeggs' solution: SFA

As cited previously, the need to seriously consider the development of automatic methods of analysis arose in the 1950s, when clinical tests started to be increasingly used for diagnostic purposes in Medicine. In order to fulfill these increasing analysis demands, Skeggs developed SFA [5], which afforded not only substantially increased throughput, but also substantial savings in samples and reagents. SFA laid the foundations for modern nonresolutive flow techniques. It should be noted that SFA was developed as a mechanical tool for automating a number of analytical methods. In many other areas, automation had to wait until the advent of microprocessors and computers.

SFA is an automatic continuous methodology developed by Skeggs in 1957.

Its associated equipment (Figure 1.1) usually includes a peristaltic pump for continuous aspiration of sample and reagents, and a series of plastic tubes (manifold) intended to carry liquid streams to the detector.

SFA systems usually consist of several channels and use one assembly such as that schematically depicted in Figure 1.1 per analytical parameter to be determined.

Once aspirated, samples are segmented by inserting air bubbles in the liquid streams (Figure 1.2) that are subsequently removed before they reach the detector.

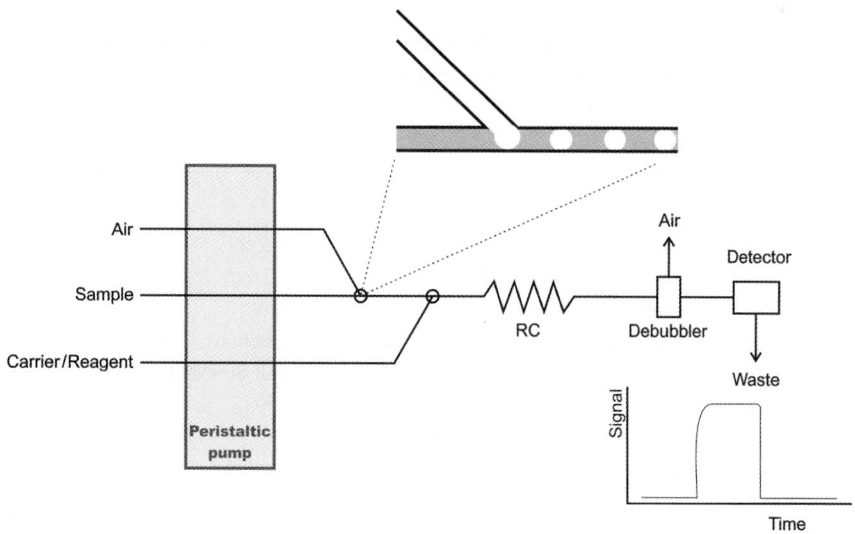

FIGURE 1.1

Scheme of a segmented flow analysis (SFA) system. RC: reaction coil.

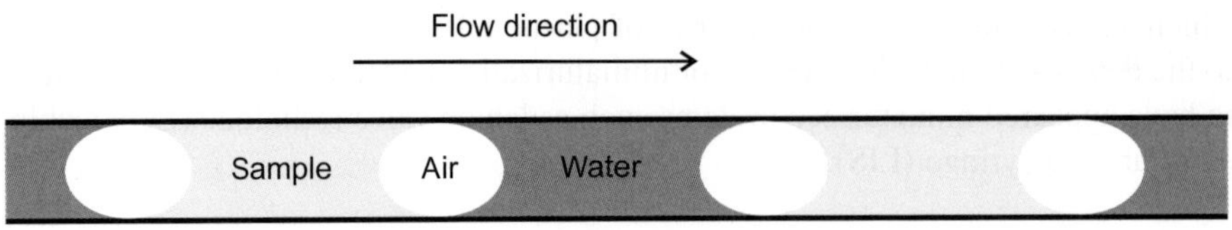

FIGURE 1.2

Injection sequence in SFA.

Air bubbles serve various purposes, namely:

1. Carryover avoidance between samples, by inserting a segment of flushing water between individual samples in order to remove any residues of the previous sample potentially remaining on tubing walls.
2. Prevention of the reaction plug dispersion.
3. Facilitation of the formation of a turbulent flow in order to homogenize the sample/reagent mixture in the plug sandwiched between each pair of bubbles.

However, the use of air bubbles has some disadvantages. Thus, their high compressibility results in pulsation; their injection and subsequent removal complicates the operational design; and their presence reduces the efficiency of separation techniques, e.g. dialysis, liquid–liquid extraction, hinders the implementation of stopped-flow methods and precludes miniaturization in many cases.

Since each individual segment is isolated from its neighboring segments of flushing water, the recording provided by the detector is roughly a rectangle the height of which is proportional to the analyte concentration, as long as the reagents are permanently present in overstoichiometric amounts (see Figure 1.1).

In its day, SFA provided an effective solution for laboratories engaged in large numbers of repetitive determinations on a daily basis. However, its high costs hindered its expansion (Table 1.2).

Table 1.2 Characteristics of Segmented Flow Analysis

Sample volume	0.2–2 ml
Response time	2–30 min
Tubing diameter	2 mm
Detection conditions	Equilibrium
Throughput	About 80 injections/h
Precision	1–2%
Reagent consumption	High
Flushing cycle	Essential
Kinetic methods	Unfeasible
Titrations	Unfeasible
Response type	Rectangle
Measured parameter	Peak height

1.3 Flow injection analysis

FIA technique was developed by J. Ruzicka and E.H. Hansen [6] in 1975. Although at first sight FIA may resemble SFA, FIA is rather different from it in both conceptual and practical terms. Basic components of FIA are virtually the same as those of SFA which include a peristaltic pump to propel the sample and reagents, and a series of plastic tubes (manifold) carrying the liquids to the detector (see Figure 1.3).

However, unlike SFA, the sample is not inserted by continuous aspiration; rather, a constant volume of sample is inserted into a stream of carrier liquid via an injection valve (Figure 1.4) for merging with the reagents used by the analytical method applied.

Tube lengths and the rotation speed of the peristaltic pump are dictated by the reaction time. Thus, if a long time is required for kinetic reasons, then a long piece of tubing is inserted—usually in coiled form—in order to increase residence times of the sample and reagents in the reactor.

Another difference between SFA and FIA, is that unlike SFA which operates under turbulent flow regime, FIA works in laminar flow, which reduces the likelihood of carryover between successive

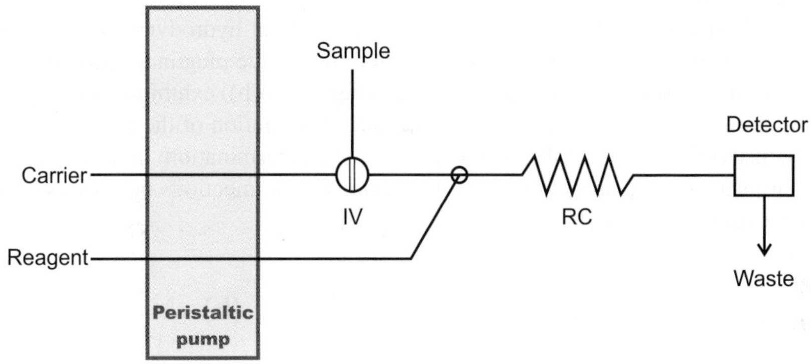

FIGURE 1.3

Schematic depiction of a typical FIA system. IV: injection valve, RC: reaction coil.

FIGURE 1.4

Manual injection valve typically used in FIA. (For color version of this figure, the reader is referred to the online version of this book.)

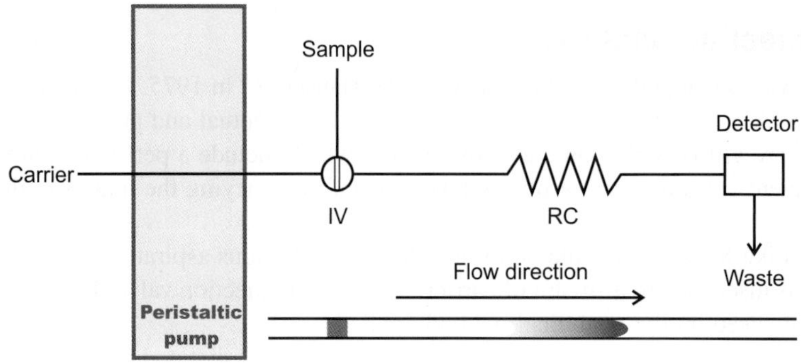

FIGURE 1.5

Single-channel FIA manifold. IV: injection valve, RC: reaction coil.

samples. Also, FIA does not require separation of samples with intervening bubbles, that is to say it uses unsegmented flow.

In Figure 1.5 is shown an FIA manifold used in a theoretical hydrodynamic study involving the injection of a dye. At the bottom is the profile exhibited by the dye plug in its way from the injection point to the detector. The result is an asymmetric peak (Figure 1.6(b)) exhibiting a χ^2 type distribution. The height and area of the peak are proportional to the concentration of the target species, which facilitates the construction of a calibration curve for its determination in unknown samples. In Figure 1.6(a) are shown the peaks obtained from quadruplicate injections of a series of standards of increasing concentration of analyte.

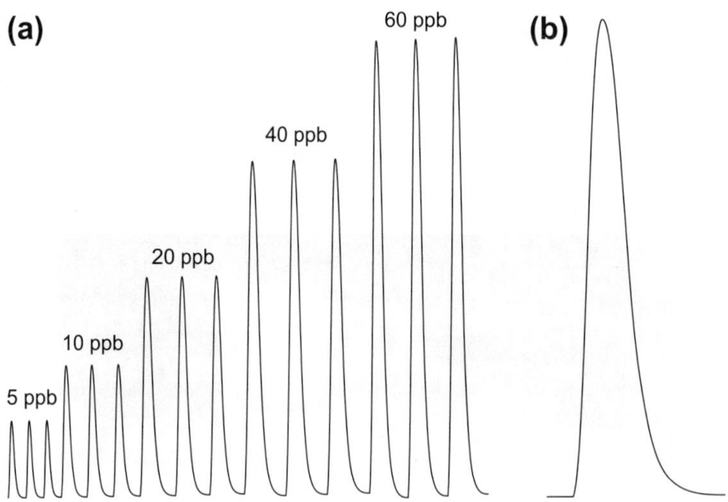

FIGURE 1.6

(a) Peaks used to construct a calibration curve. (b) Typical FIA recording.

Table 1.3 Characteristics of Flow Injection Analysis

Injected sample volume	50–150 μL
Response time	3–60 s
Tubing diameter	0.5 mm
Detection conditions	Equilibrium not needed
Throughput	About 100 injections/h
Precision	1–2%
Reagent consumption	Low
Flushing cycle	Unnecessary
Kinetic methods	Feasible

Table 1.3 summarizes the most salient features of FIA. As can be seen, operating conditions are rather different from those of SFA. One of the most relevant differences is the reduction of sample volume which is reduced from milliliters to microliters. Response times are also substantially shorter and tubing diameters smaller in FIA than in SFA.

While SFA usually requires that the analytical reaction reaches chemical equilibrium, FIA does not. In fact, FIA only requires the extent of reaction to be constant and reproducible, which is facilitated by the high reproducibility provided by the hydrodynamic behavior of the system. Moreover, since FIA uses much thinner tubing and lower flow rates, it consumes samples and reagents much more sparingly than does SFA.

In addition, FIA is much more flexible than SFA allowing the implementation of analytical methodologies unaffordable to the latter, e.g. kinetic methods, stopped-flow methods.

Another major advantage of FIA over SFA is its ease of implementation. In fact, a dedicated manifold can be readily assembled from fairly inexpensive parts, viz. a peristaltic pump, injection valves, flow-cells, Teflon tubing, connectors and available measuring instruments, e.g. spectrophotometers, potentiometers, ammeters, atomic absorption spectrometry equipment. This propitiated a vast expansion of FIA among research laboratories and led to the development of a large number of applications relative to other, more recent techniques within a few years after its inception.

1.4 Sequential injection analysis

SIA [7] was developed by J. Ruzicka and G. Marshall as an alternative to FIA. SIA has proved that its scope departs markedly from that of the earlier technique. Figure 1.7 shows a schematic depiction of a typical SIA system.

This technique is based on the use of selection valves. In Figure 1.8 is shown a typical selection valve of low pressure design and made of chemical resistant polymers PEEK (stator) and poly(tetrafluoroethylene) (PTFE) (rotor).

The central port of the valve is connected to a two-way piston pump via a holding coil, to address the peripheral ports of the unit, for sequential aspiration of the various constituents, and also to the detector, via the central communication channel. Side ports can also be used for other purposes such as discharging waste or connection to other devices, e.g. a microwave oven, photooxidation system or mixing chamber.

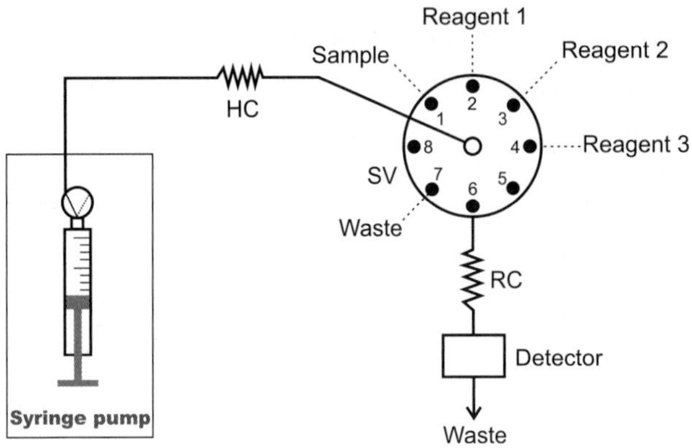

FIGURE 1.7

Schematic depiction of a typical SIA system. HC: holding coil, RC: reaction coil, SV: selection valve.

FIGURE 1.8

(a) Switching valve used in SIA. (b) Inner parts of the valve. (For color version of this figure, the reader is referred to the online version of this book.)

One of the essential features of SIA is its computerized control. The computer selects how the central port of the valve is connected to its side ports, starts and stops the pump in order to aspirate or dispense liquids, selects their volume and adjusts the flow rate. Also, the fact of implementing a computer makes possible data acquisition and processing.

In a theoretical SIA system using a single reagent, the central port of the switching valve would connect to the sample channel and the pump would be set to aspirate a preset volume of sample at a fairly low flow rate in order to avoid bubbles formation. Then, the central port would connect to the appropriate side port in order to aspirate a preset volume of reagent. Next, the valve would be actuated to connect its central port with the channel leading to the detector and an appropriate volume of carrier would be dispensed to drive the sample and reagent to the detector in flow reversal mode.

SIA works under laminar flow regime, exhibiting the typical asymmetric shape of FIA peaks.

As can be seen in Figure 1.9, dispersion in an SIA system leads to the sample and reagent plugs overlapping and the reaction product (P) formation to be detected.

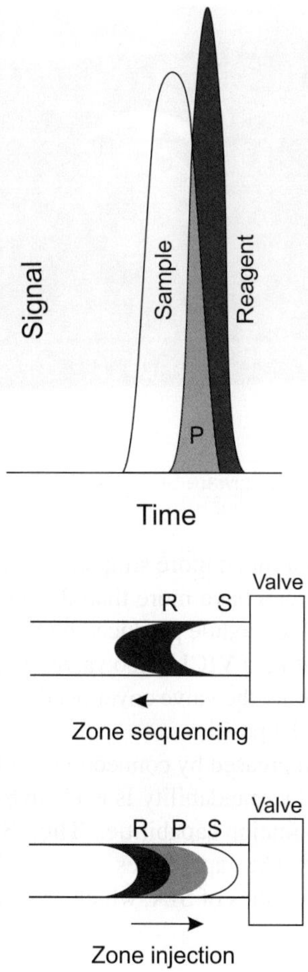

FIGURE 1.9

Overlapping sample and reagent plugs in an SIA system. P: product, R: reagent, S: sample.

A typical SIA manifold includes two types of coil, namely: a holding coil inserted in the channel connecting the piston pump to the central port of the switching valve which prevents the sample and reagents from reaching the piston pump—cleaning of which otherwise would be labor-intensive and time-consuming—and a reaction coil in the channel leading to the detector that is intended to ensure an adequate overlap between the sample and reagent plugs in order to allow a detectable amount of reaction product to form.

Unlike FIA, SIA can be turned into a true multiparametric analysis system simply by using a switching valve with an appropriate number of channels to hold the different analytical reagents, delivery of which can be precisely programmed via the associated computer. In this respect, SIA is much closer to SFA in multiparametric capabilities, being able to determine up to 20 parameters per

FIGURE 1.10

Programmable selection valve module from Sciware Systems. (For color version of this figure, the reader is referred to the online version of this book.)

sample—except that SIA operates in a much more simple, green and, especially, economical manner. Currently available switching valves can have more than 20 side ports.

Figure 1.10 shows a switching valve module produced by Sciware Systems (Sciware Systems S.L., Palma de Mallorca, Spain) implementing a VICI Valco valve, which allows the number of lateral ports to be programmed. This implies that only the valve cover needs to be changed according to the selected number of lateral ports, e.g. 6, 8, or 10 positions.

Also, the number of ports can be increased by connecting a side port in a valve to the central port of several others. Such a high degree of expandability is exclusive of SIA, no other flow technique can match SIA in multiparametric determination capabilities. Thus, SIA is an excellent tool for developing automatic analyzers with multiparametric capabilities.

Table 1.4 summarizes the characteristics of SIA, which are seemingly quite similar to those of FIA (see Table 1.3).

Table 1.4 Characteristics of Sequential Injection Analysis

Injected sample volume	50–150 μL
Response time	3–60 s
Tubing diameter	0.8 mm
Detection conditions	Equilibrium not needed
Throughput	About 70 injections/h
Precision	1–2%
Reagent consumption	Very low
Flushing cycle	Unnecessary
Kinetic methods	Feasible
Titrations	Feasible
Response type	Peak
Measured parameter	Peak height or area

Careful scrutiny, however, reveals some differences between the two. One is the dramatically reduced consumption of sample and reagents in SIA; this, however, is not a result of using smaller injected volumes, but rather of the way an SIA system operates. In FIA operation, sample and reagent consumption are virtually independent of the analysis frequency as the peristaltic pump continuously propels sample and reagents at a constant flow rate throughout. In SIA, however, the piston pump only works during the time strictly needed to aspirate the amount of sample and reagent needed for a given determination. Aspirating an additional amount of sample is only required when the sample is replaced with a different one as the old sample should be completely flushed out of the aspiration tube in order to avoid carryover. As an example, an SIA monitor for determining ammonium ion in waste water uses 10 times less reagents than does a comparable FIA monitor [8]; this is of a high economic and practical significance, especially with equipment that is intended to operate unattended over long periods, e.g. an automatic analytical monitor.

Also, due to the fact of employing piston pumps, SIA is more robust than other flow techniques using peristaltic pumps. Peristaltic pumps use tubing of materials that are relatively easily damaged by some fluids, viz. acids, bases and, especially, solvents; by contrast, piston pumps use glass and Teflon tubing, which are highly inert materials and ensure a long service life. Also, it is important to remind that in SIA, the sample, reagents and solvents seldom reach the propulsion system, which just holds the carrier solution at most.

One difficulty of SIA operation arises from the way plugs are stacked; this hinders mixing of the sample and reagents, especially with more than two, which require using the sandwich technique. Another feasible solution in determinations involving many reagents is using a mixing chamber in one of the side ports to homogenize the different sample/reagent mixtures with the aid of a magnetic stirrer for withdrawal of small aliquots as required [9].

One of the greatest initial hindrances to SIA development was the need to use a computer in order to govern the system. This resulted in the development of barely a few tens of methods during its first year of existence. The scarcity of commercially available software and the lack of experience in interfacing computers to analytical instruments caused the slow development of SIA initially, despite its proven advantages. Only during the past decade, with the inception of commercial software, has SIA gained ground in the field of routine analyses.

Nonetheless, what first was a hindrance to SIA development has resulted one of its greatest advantages in front of earlier flow techniques. Thus, the fact of being a computerized technique results in that residence times need no longer be controlled via the length of the manifold tubes and the flow-rates of a peristaltic pump; rather, they are controlled in a highly reproducible manner by the computer. Also, the ability to adjust the flow rate required in each step of the process and to change it at will at any time make SIA a highly flexible analytical tool. Nevertheless, whereas in FIA in order to implement a different method very frequently entails altering the configuration of the manifold, switching to another method in SIA seldom requires more than using a different computer file containing the operational settings to be used with each procedure. Obviously, changing the reagents will also be necessary; however, a switching valve with an adequate number of ports can be used to hold the reagents needed for several determinations in different ports, so simply choosing the appropriate settings file will usually suffice to determine another analytical parameter.

The incorporation of computers into SIA systems has also facilitated the implementation of stopped-flow methods. It suffices to calculate the volume of carrier to be delivered and stop the system when a peak is obtained at the detector in order to readily implement various analytical methodologies including classical kinetic, spectrophotometric, polarographic, voltammetric and anodic stripping methods. Stopped-flow methods are also very useful for radiometric detections, where longer counting times are required.

In addition, computers have increased the flexibility of analytical systems by allowing a number of operations mimicking those performed manually. Samples can be aspirated and mixed with a reagent; their mixture driven to a photoreactor; and an aliquot withdrawn, sent to a solid-phase preconcentration unit and eluted from it with a view to sequentially detecting the analytes by using several detectors that can be arranged serially at the same port or radially at different ports or even in a serial/radial mixed configuration.

1.5 Multicommutated flow analysis

The MCFIA technique, devised by B.F. Reis et al. [10], exploits fast-switching three-way solenoid valves (see Figure 1.11). Earliest MCFIA systems used a single-channel propulsion system to aspirate the liquids via individual valves. Since aspiration devices tend to insert air bubbles or degas liquids in the system, it is preferable to use liquid propulsion devices such as peristaltic or multipiston pumps instead.

As an example an MCFIA system has been schematically depicted in Figure 1.12. This could be used for a number of purposes by rapidly switching the valves. It uses a peristaltic pump and has four solenoid valves V1, V2, V3 and V4 arranged in such a way that the propelled liquid is returned to the reagent reservoir while the valves are Off, but inserted into the system while they are On. By alternately switching V1 and V2 On, one could dilute the sample to a preset extent with carrier. Given that solenoid valves can be switched very rapidly, variably thick slices of carrier and sample could be alternately inserted, which will interpenetrate on their way through RC1. Valve V5 would allow the flow to be directed either to a copperized cadmium column in order to reduce nitrates to nitrites, or to the upper channel avoiding this reduction reaction. Finally, valves V1 and V4 could be used to

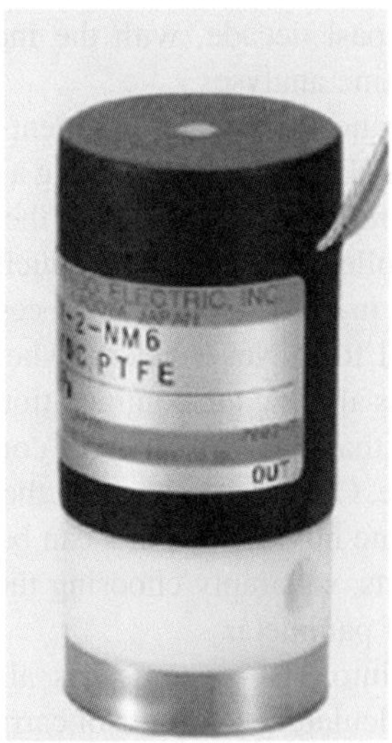

FIGURE 1.11

Three-way solenoid valve. (For color version of this figure, the reader is referred to the online version of this book.)

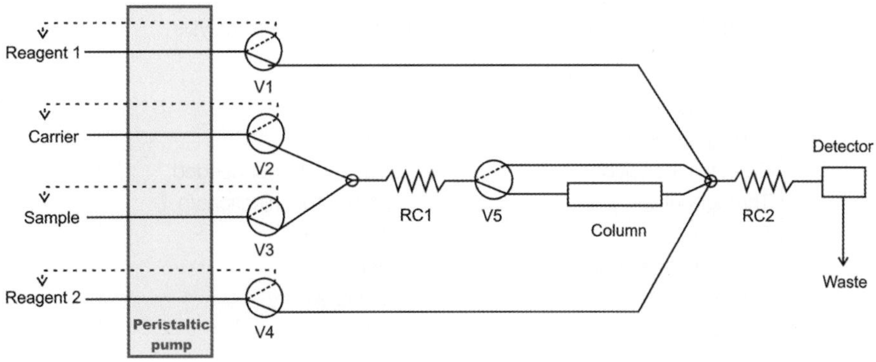

FIGURE 1.12

Schematic depiction of a typical MCFIA system. RC: reaction coil, V: valve.

inject preset volumes of reagents by switching them On over an appropriate interval. RC2 is intended to facilitate homogenization of the diluted unknown sample with the reagents added in the last step.

One major shortcoming of solenoid valves is the unfavorable effect of the heat released by the solenoid coil when the valves are On for a long time. The resulting increase in temperature can deform the Teflon inner membranes of the valves and render them unusable. Overheating can be avoided by using an effective electronic protection system. Figure 1.13 shows a system where the valve is switched On by applying a direct current (DC) voltage and then continues to operate under a pulse train of the same nominal voltage but of a much lower efficient voltage, which substantially reduces the amount of dissipated power.

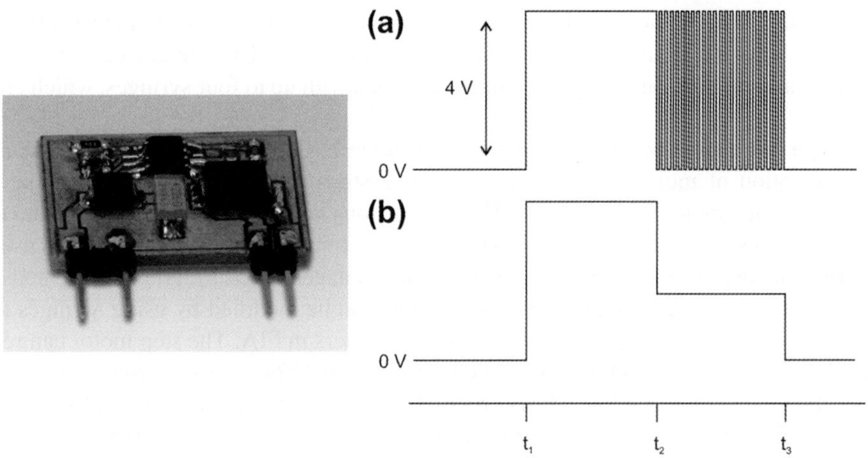

FIGURE 1.13

Left: photograph of a valve protector. (a) Voltages used to switch and keep the valve On. (b) Equivalent DC voltages. (For color version of this figure, the reader is referred to the online version of this book.)

Table 1.5 Characteristics of Multicommutated Flow Analysis

Injected sample volume	50–150 μL
Response time	3–60 s
Tubing diameter	0.8 mm
Detection conditions	Equilibrium not needed
Throughput	About 100 injections/h
Precision	1–2%
Reagent consumption	Very low
Flushing cycle	Unnecessary
Kinetic methods	Feasible
Titrations	Feasible
Response type	Peak
Measured parameter	Peak height or area

Table 1.5 summarizes main characteristics of MCFIA. As can be seen, it shares some advantages of FIA such as a high throughput (higher than that for SIA) and a reduced consumption of reagents as a result of the return of unused sample and reagents to their respective reservoirs. However, MCFIA also shares one disadvantage of FIA, namely: the vulnerability of peristaltic pump tubing against aggressive reagents and, especially, solvents.

1.6 Multisyringe flow injection analysis

MSFIA [11,12] was developed in 1999 by our research group in cooperation with the firm Crison (Alella, Barcelona, Spain). The aim was to combine the advantages of previous flow techniques while avoiding their disadvantages. It was designed as a novel multichannel technique combining the multichannel operation and high injection throughput of FIA with the robustness and the versatility of SIA.

Figure 1.14 shows a typical multisyringe burette for use in MSFIA. The device consists of a conventional automatic titration burette which can be equipped with up to four syringes, which are used as liquid drivers.

Each syringe has a three-way solenoid valve (N-Research, Caldwell, NJ) at the head, which facilitates the application of multicommutation schemes. Pistons are mounted on a common steel bar driven by a single step motor (40000 steps). Thus, all pistons are moved simultaneously and unidirectionally for either liquid delivering (dispense) or aspirating (pick up); this is equivalent to use a multichannel peristaltic pump in FIA but avoids the disadvantages of its fragile tubing, and frequent recalibrations. The ratio of flow-rates between channels can be modified by using syringes of appropriate cross-sectional dimensions similarly to tubing diameters in FIA. The step motor can reach total displacement, corresponding to 40000 steps, between 20 and 1024 seconds, providing a wide speed range. Thus, the multisyringe module allows precise handling of microliters and a wide flow rate range (0.057−30 ml/min, depending on the syringe volume 1−10 ml). Syringes of 0.5, 1, 2.5, 5, 10 and 25 ml are available, enabling a wide flow rate range and a great combination. High chemical robustness is provided by the use of resistant polymers poly(ethylene-*co*-tetrafluoroethylene) (head valves) and PTFE (piston heads, poppet flaps).

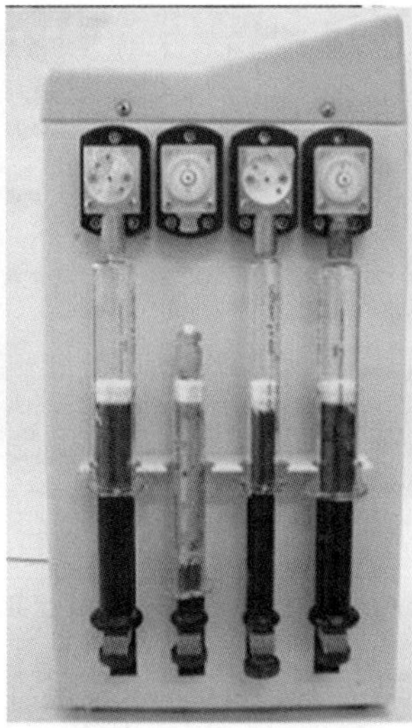

FIGURE 1.14

Frontal view of an MSFIA burette (from Crison Instruments). (For color version of this figure, the reader is referred to the online version of this book.)

Solenoid valves located at the heads of the syringes (On: to the system; Off: to the reservoir) allow four kinds of liquid displacement: On-dispense, Off-dispense, On-pick-up, and Off-pick-up (Figure 1.15).

Four backside ports (1–4, Figure 1.16) enable the power of additional external multicommutation valves, micropumps or other instruments either directly or via a relay allowing remote software control (e.g. Inductively Coupled Plasma Mass Spectrometer (ICP-MS)). This amplifies the possibilities to construct sophisticated flow networks (see Figure 1.17).

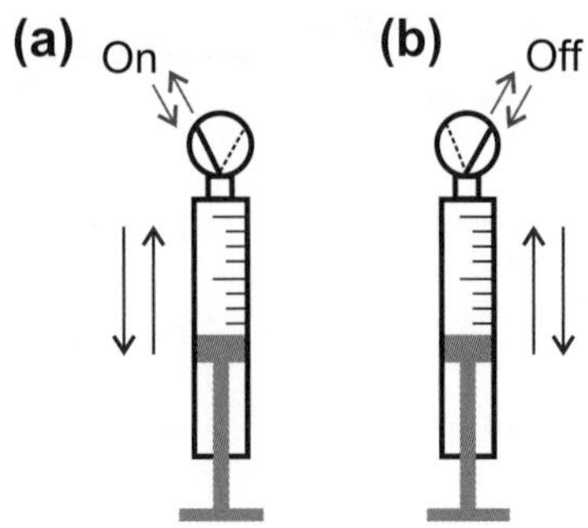

FIGURE 1.15

Schematic depiction of the solenoid valves placed at the head of each syringe. (a) Activated solenoid: "On" position and (b) Deactivated solenoid: "Off" position. (For color version of this figure, the reader is referred to the online version of this book.)

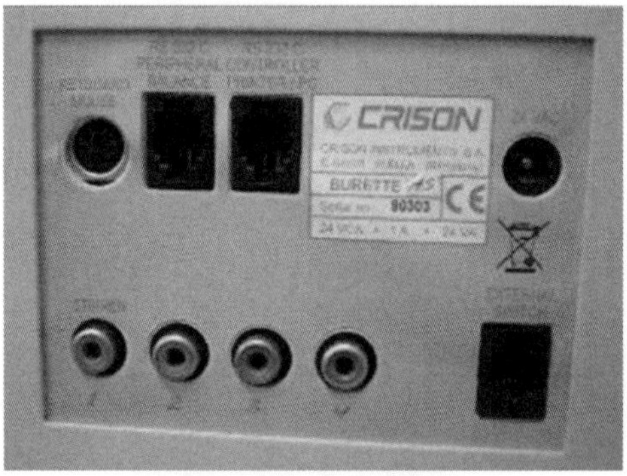

FIGURE 1.16

Backside connections of a multisyringe burette. (For color version of this figure, the reader is referred to the online version of this book.)

External solenoid valves are commonly used to drive the liquids in flow based systems (Figure 1.11). The DC voltage of each port may be computer controlled, to a maximum value of 12 V and a maximal current of 0.5 A, enabling the connection of up to three external solenoid valves (V) to each port.

As said before, MSFIA systems combine some of the advantages of the above-described flow techniques, namely:

1. The high throughput of FIA, which is a result of the sample and reagents being incorporated in parallel. This in turn leads to an improved mixing efficiency in relation to SIA.
2. The robustness of SIA. In fact, the liquids only come into contact with the walls of the glass syringes and Teflon tubing, as no peristaltic pump tubes are used.

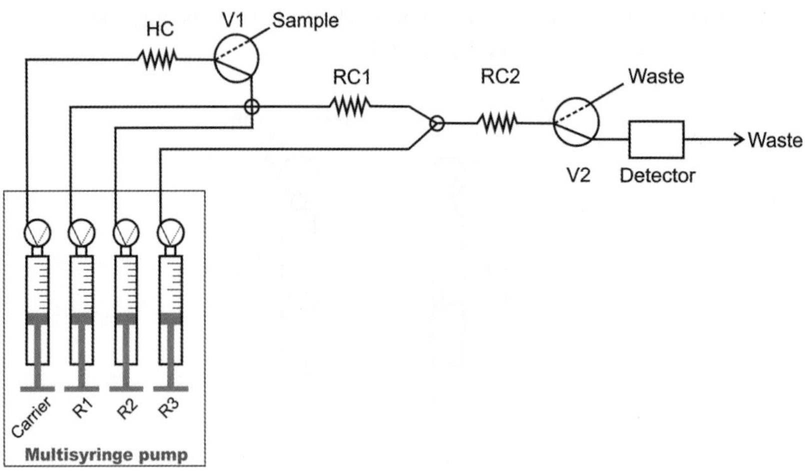

FIGURE 1.17

Schematic depiction of a typical MSFIA system. HC: holding coil, R: reagent, RC: reaction coil, V: valve.

Table 1.6 Characteristics of Multisyringe Flow Injection Analysis

Injected sample volume	50–150 μL
Response time	3–60 s
Diameter of manifold and sample tubing	0.8 mm
Diameter of reagent aspiration tubing	1.5 mm
Detection conditions	Equilibrium not needed
Throughput	About 100 injections/h[1]
Precision	1–2%
Reagent consumption	Very low
Flushing cycle	Unnecessary
Kinetic methods	Feasible
Titrations	Feasible
Response type	Peak
Measured parameter	Peak height or area

[1] *Up to 200 injections/h if an additional burette is used.*

3. The low sample and reagent consumption of SIA by implementation of the multicommutation concept, since liquids are delivered to the system only when required and so the reagents and sample consumption are low.
4. The high flexibility of SIA manifolds. Flow rates and propelled volumes are precisely known and defined by software-based remote control of the multisyringe device, as in SIA. Also, switching to a different analytical method simply requires loading the file containing the appropriate settings for the new method and changing the reagent vessels, as in SIA.
5. The ability to use MCFIA solenoid valves, which can be actuated without the need to stop their pistons. Switching between valves is so fast that no overpressure arises in the operation.

By use of parallel moving syringes as liquid drivers, it also overcomes the shortcomings of peristaltic pumping such as pulsation, required recalibration of flow rates and limitations regarding applicable reagents. Thus, MSFIA is an ideal multichannel technique for challenging analytical procedures, which require high and precise flow rates, and high pressure stability such as those with sorbent columns implementation [13], enabling at the same time, the handling of aggressive and volatile solutions. The only disadvantage of MSFIA in front of other flow techniques is the periodical syringe refilling which causes a lower injection frequency than using an FIA approach. However, by alternate use of two burettes allows the throughput to be doubled (up to 200 injections/h, which can hardly be matched by any other flow technique) (Table 1.6). Unlike FIA, MSFIA requires the use of a computer to control the system. This, however, poses no special problem as a variety of affordable software for implementing any flow technique is by now available.

1.7 Multipumping flow systems

MPFS, which were developed in 2002 by two research groups at the Pharmacy Faculty of the University of Porto (Portugal) and the Piracicaba CENA (Brazil) [14], are based on the use of solenoid piston pumps (Figure 1.18) where each stroke propels a preset volume of liquid (3, 8, 20, 25, 50 μL) the flow rate of which is determined by the stroke frequency.

A micropump (Figure 1.19) is a solenoid operated device designed to provide a discrete dispensed volume of fluid. The flow path is isolated from the operating mechanism by a flexible diaphragm. When the solenoid is energized, the diaphragm is retracted creating a partial vacuum within the pump body. This pulls liquid through the inlet check valve (A) and simultaneously closes the outlet check valve (B). When the solenoid is de-energized a spring pushes the diaphragm down, expelling a discrete volume of liquid through check valve B while simultaneously closing check valve A. Micropumps require a complete

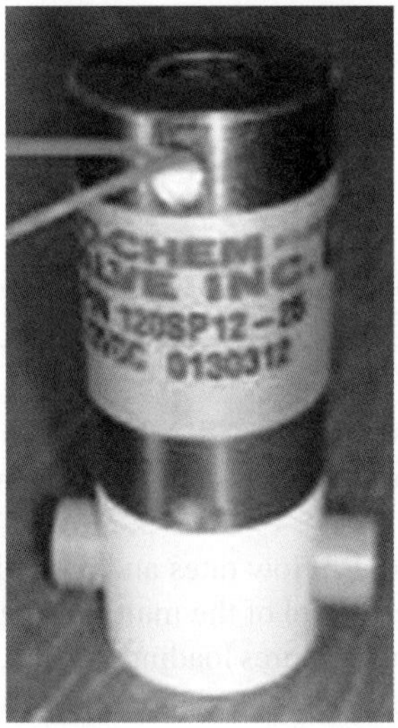

FIGURE 1.18

Solenoid piston pump. (For color version of this figure, the reader is referred to the online version of this book.)

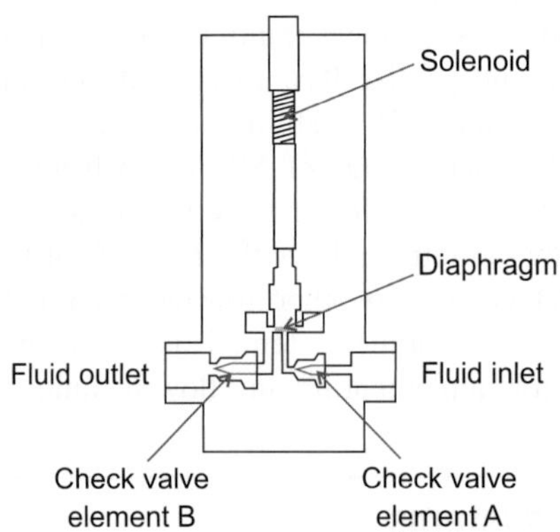

FIGURE 1.19

Scheme of a micropump (from Biochem valve). (For color version of this figure, the reader is referred to the online version of this book.)

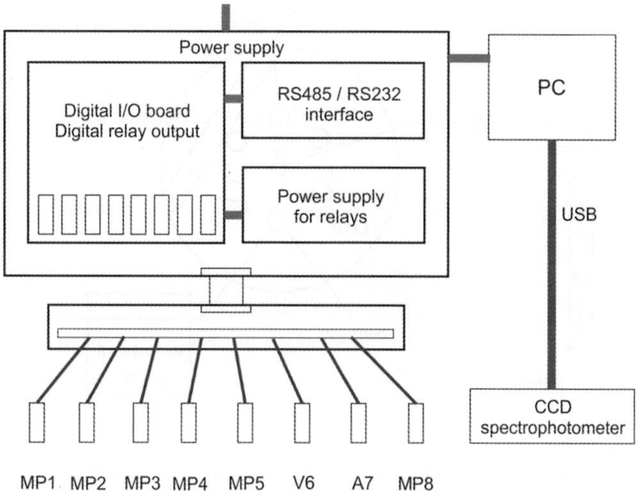

FIGURE 1.20

Schematic depiction of an MPFS controller system. A: agitator, MP: micropump, V: solenoid valve.

on–off cycle for each discrete dispense. Repeatedly cycling the solenoid creates a pulsed flow. The body of the micropumps is made of steel at the top (solenoid part) and teflon at the bottom where is the pumping system (diaphragm). Micropumps when activated use an internal source of 12 V.

Principal advantages of these systems are their high flexibility, easy configuration, robustness and low cost as a result of the pump operating as both liquid propeller and valve. Minimal reagent consumption is achieved, since each micropump is operated individually in inserting the solutions. In comparison with other flow techniques, the pulsed flow of the micropumps is better and faster at homogenizing the reaction zone [15], which leads to higher analyte peaks. Main features to be highlighted are the simplicity and very low costs of the controlling circuits, favoring economic, portable and miniaturized flow analyzers [16,17], which facilitate field measurements. Further advantages of MPFS are a high versatility and flexibility of the flow system networking, especially in combination with multicommutation solenoid valves (V). However, some disadvantages of micropumps are the susceptibility to blockage by particles and to backpressure, requiring recalibration of the dispensed volume.

A schematic depiction of a controller system (Sciware Systems S.L.) is shown in Figure 1.20. This module can be controlled through the interface RS232.

A typical multipumping system is similar to an MCFIA system (Figure 1.12). In fact, the former can also be used to implement MCFIA as it affords controlling any combination of valves and solenoid micropumps. The primary difference between the two is that multipumping systems require controlling not only valve switching, but also the stroke frequency, in order to ensure reproducible flow-rates.

1.8 Lab on valve

LOV [18,19] significantly facilitates integration of various analytical units in the valve and provides great potential for miniaturization of the entire instrumentation.

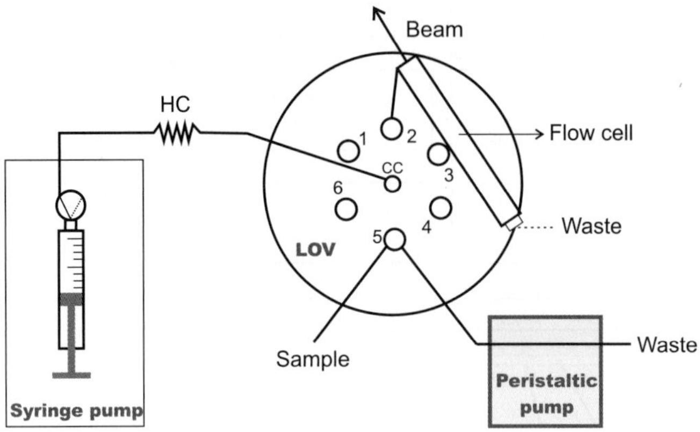

FIGURE 1.21

First LOV system including a flow through cell. CC: central conduit, HC: holding coil.

The first LOV system was a single, monolithic fabricated Poly(methyl methacrylate) (PMMA) component mounted atop a six-port selection valve, including a flow through cell furnished with optical fibers (Figure 1.21). Nowadays, there are different designs of LOV systems, fabricated from methylmethacrylate (Perspex), ultem (polyetherimide) or kel-F as monolithic structures encompassing integrated microchannels.

The central port of the integrated LOV sample processing unit, is usually connected to a piston pump, via a holding coil, which is made to address the peripheral ports of the unit, for sequential aspiration of the various constituents, via the central conduit (CC) in the selection valve (see Figure 1.8). Precise volumes of sample and reagents are stacked in a holding coil by sequential aspiration from the microfluidic device mounted atop a rotary selection valve and propelled afterwards by flow reversal. It is shown that sample handling in the sequential injection mode, which employs forward, reversed and stopped flow, can be programmed to accommodate a wide variety of assays, such as solution metering, mixing, dilution, incubation and monitoring. One of the LOV channels can serve as microcolumn position to carry out the bead injection (BI) process (Figure 1.22), as a flow cell for optical measurements (see Figure 1.21) [20,21] or as a "jet-ring-cell", to carry out BI and real time monitoring of the optical changes of the beads [22].

In addition to compactness, other advantages of these "lab-on-valve" systems is the permanent rigid position of the sample processing channels that ensures repeatability of microfluidic manipulations and a large volume to surface ratio, which minimizes the unwanted adsorption on channels walls that may result in carryover. This provides proven robustness and reliability of operation, and makes the microfluidic system compatible with real-life samples and peripheral instruments.

It is noteworthy that LOV-based techniques have not only been extensively employed in homogeneous solution-based assays, but have also shown promise in heterogeneous assays because flexible fluid manipulation is also suitable for delivering beads in flow-based manifolds, i.e. precise fluid manipulation by the LOV system and the channel configuration also make it a powerful platform for BI [22,23]. In combination with the renewable surface concept, BI has been widely exploited for separation and preconcentration of analytes in the presence of complex matrix components. Most importantly, the automated transport of solid materials in such a system allows their automatic renewal at will and thus provides measurement, packing and perfusion of beads with samples and

FIGURE 1.22

LOV with a microcolumn integrated. (For color version of this figure, the reader is referred to the online version of this book.)

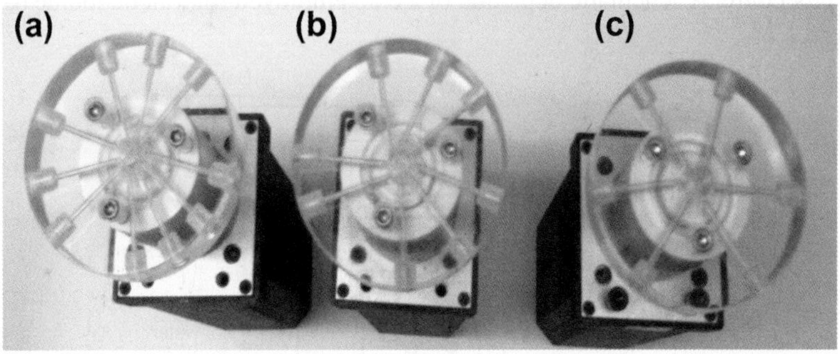

FIGURE 1.23

With the same VICI Valco motor, the number of positions may be programmed allowing different LOV designs. (a) 10 channels, (b) eight channels and (c) six channels. (For color version of this figure, the reader is referred to the online version of this book.)

reagents with a high degree of repeatability. Furthermore, being operated in a closed system, LOV systems are characterized by low consumption of sample and reagent, reduced analysis time, high reproducibility and minimal sample contamination. Moreover, with the same VICI Valco motor, the number of positions may be programmed allowing different LOV designs with different number of ports simply by changing the chip, e.g. for 6, 8 or 10 channels (see Figure 1.23).

LOV has also proved to be an effective front end to various detection techniques such as ICP-MS [24] and electrothermal atomic absorption spectrometry [23] and an useful tool in biochemical analysis [25,26].

Shortcomings of LOV are quite similar to those of classical SIA, such as a worse zone penetration and longer time of analysis than in FIA due to the use of wider tubes, the refilling of the syringe and the sequential injection mode. Another disadvantage is the impossibility of confluence mixing. Some of these disadvantages are overcome by coupling it to other flow techniques, such as MSFIA [27].

Summarizing, true advantages of using microfluidics are the compactness, the automation of all the steps of the experimental protocol and, last but not least, the integration of all manifold components into a permanent rigid structure that enhances the repeatability of sample processing operations.

1.9 Chips

Chips are monolithic flow devices, integrating different functions in a reduced size such as confluent mixing, reaction coil, and thermostating, which allow minimizing the dimensions of the entire analyzer. These are a step forward in automation and miniaturization of all the laboratory protocols. Strictly speaking LOV could be included in this category, however given its widespread use we decided to dedicate a whole section to LOV, separately from other chips.

Some of these chips have been applied in automating kinetic methods [28], some incorporating the sensor in the chip. This is because reproducible metering and mixing of all solutions, precise timing until data readout, and shielding of the reaction mixture from outer contamination are imperative requirements, which can be perfectly fulfilled using miniaturized chips integrating most of the steps of the analytical procedure. A schematic depiction of a chip employed for kinetic iodide determination is shown in Figure 1.24.

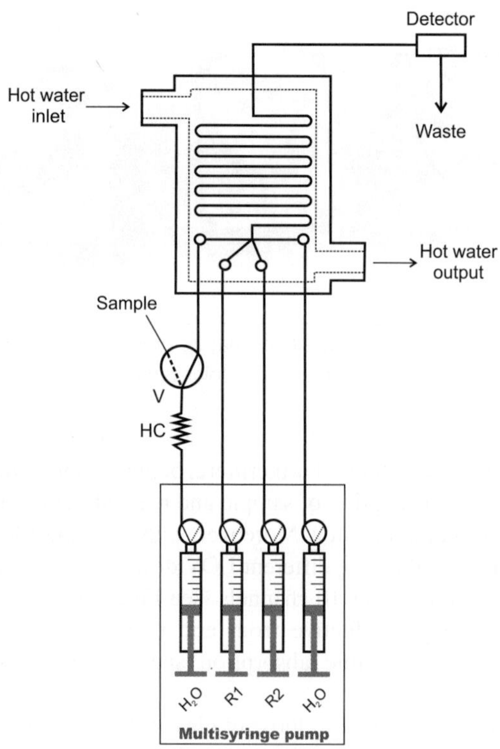

FIGURE 1.24

Schematic depiction of a chip employed in kinetic methods. HC: holding coil, R: reagents, V: solenoid valve.

One example of application of these chips is Chip on valve (ChOV). ChOV is a microfluidic methodology that consists on a chip mounted atop a selection valve. A schematic depiction of one of the first prototypes is shown in Figure 1.25.

Chips have also been applied to cross-injection analysis (CIA) [29] (see Figure 1.26).

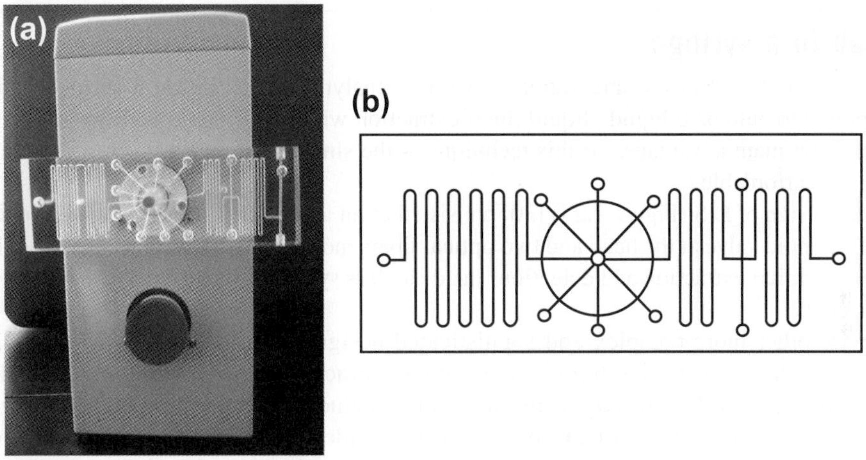

FIGURE 1.25

Chip prototype for implementing in ChOV. (a) ChOV mounted atop a selection valve. (b) Schematic depiction of the chip design. (For color version of this figure, the reader is referred to the online version of this book.)

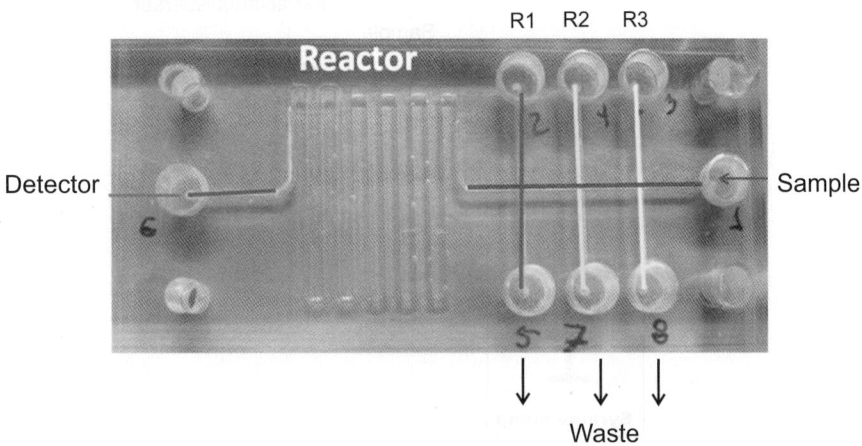

FIGURE 1.26

Chip designed for cross injection analysis (CIA). R: reagents. (For color version of this figure, the reader is referred to the online version of this book.)

These techniques (ChOV, CIA) share some characteristics with their predecessor LOV such as compactness, the permanent rigid position of the sample processing channels that ensure repeatability of microfluidic manipulations, controlled by conventional-sized peripherals, being also compatible with real-life samples and peripheral instruments.

1.10 Lab in a syringe

This technique is based in the integration of various analytical steps inside a syringe. It has been mainly applied to automate liquid–liquid microextraction with in-syringe spectrophotometric detection. One of the main advantages of this technique is the simple instrumental setup required, which makes it very affordable.

The first design of LIS (Figure 1.27) [30] consisted of an automatic syringe pump equipped with a three-way solenoid valve at the head and two optical fibers mounted to the syringe, enabling the whole procedure (i.e. microextraction and detection) inside a glass syringe, connected to the central port of a selection valve.

There are other more complex and sophisticated designs such as lab in a syringe with magnetic stirring-assisted dispersive liquid–liquid microextraction [31]. This setup allows the entire analytical procedure to be carried out in the syringe including sample mixing with reagents and extraction. A magnetic micro stirring bar is placed inside the syringe to achieve homogeneous and rapid mixing. This in combination with an adaptor, which is placed onto the barrel of a glass syringe and swirling around the longitudinal axis of the syringe, holding two strong neodymium magnets, provides a magnetic field in the syringe along its whole length making possible a magnetic stirring in the syringe. The sample is mixed with all required reagents homogeneously, nearly-instantaneously, and within the syringe. In contrast to batch-wise

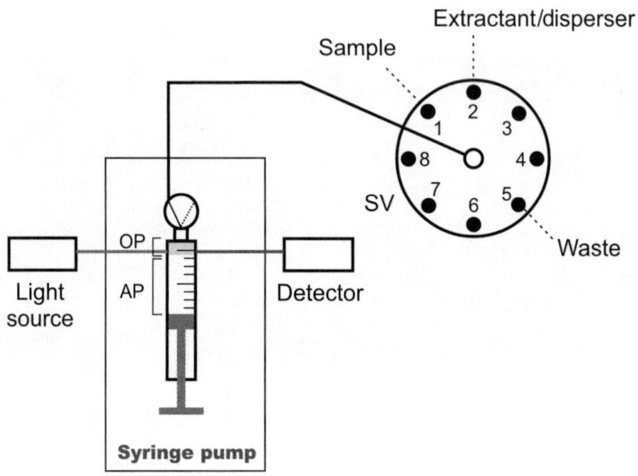

FIGURE 1.27

Lab in a syringe setup. AP: aqueous phase, OP: organic phase, SV: selection valve.

automation using atmospherically open reaction chambers, an adaptable but simultaneously sealed reaction volume is obtained, which allows simple and rapid cleaning procedures and new analytical methodologies.

1.11 Combined use of flow techniques

Obviously, when developing a specific application more than one flow technique can be used in order to maximize the advantages of each individual choice.

One potentially effective combination can be that of SIA and MSFIA (Figure 1.28). In fact, the latter allows all preliminary operations on the sample (e. g. preconcentration on a solid support, photo-oxidation) to be conducted by mimicking the work of manual systems while the latter can contribute with increased precision in injecting the reagents in parallel via a multisyringe and hence more efficient mixing with the sample.

Another attractive combination is that of MSFIA with a module comprising two switching valves; this allows each valve to control two syringes and facilitates the simultaneous implementation of two improved SIA processes at no substantially increased cost (see Figure 1.29).

Finally, MSFIA/MPFS combinations are also interesting (Figure 1.30). In fact, the DLL for the multisyringe burette has been altered so that it can control the solenoid valves via the connecting strip at the back.

Finally, MSFIA/LOV (Figure 1.31) is also a great combination, providing the full automation of all the steps including the column replacement. MSFIA is a very versatile flow technique that allows its easy hyphenation with LOV. Both flow techniques LOV-MSFIA complement each other, improving their individual advantages according to the required analytical needs. This allows drastic reduction of reagents consumption, waste generation, and time saving [24,27].

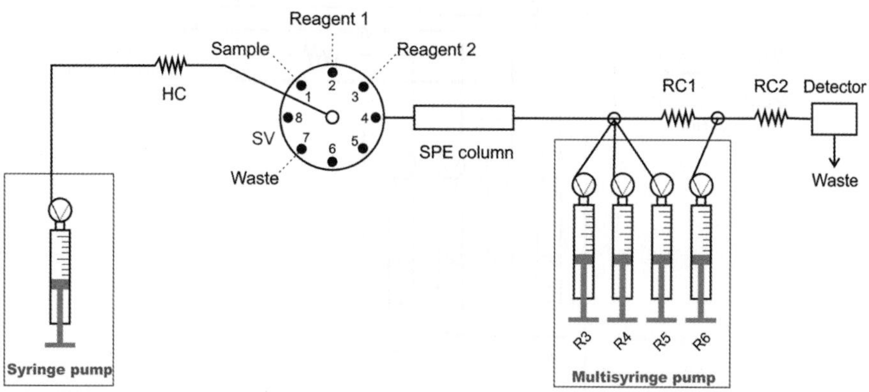

FIGURE 1.28

SIA-MSFIA system. HC: holding coil, R: reagents, RC: reaction coil, SV: selection valve.

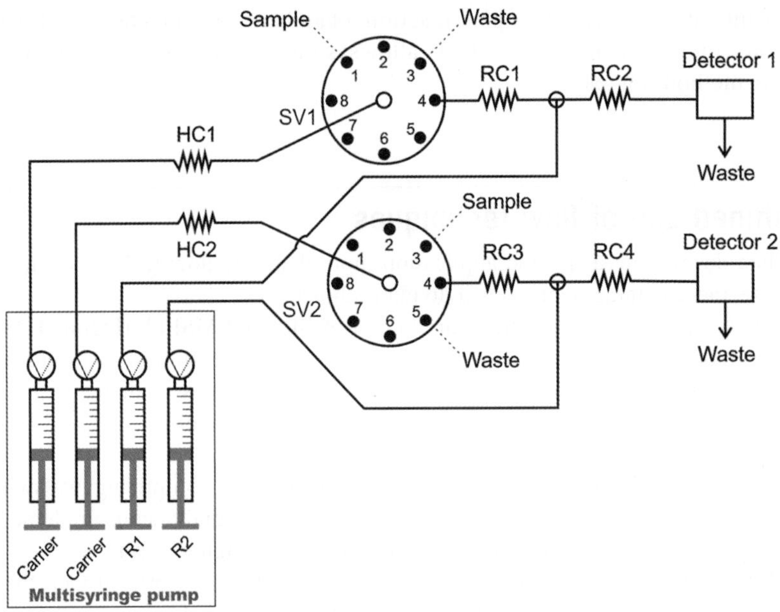

FIGURE 1.29

Schematic depiction of a combined MSFIA-SIA system used for dual SIA. HC: holding coil, R: reagents, RC: reaction coil, SV: selection valve.

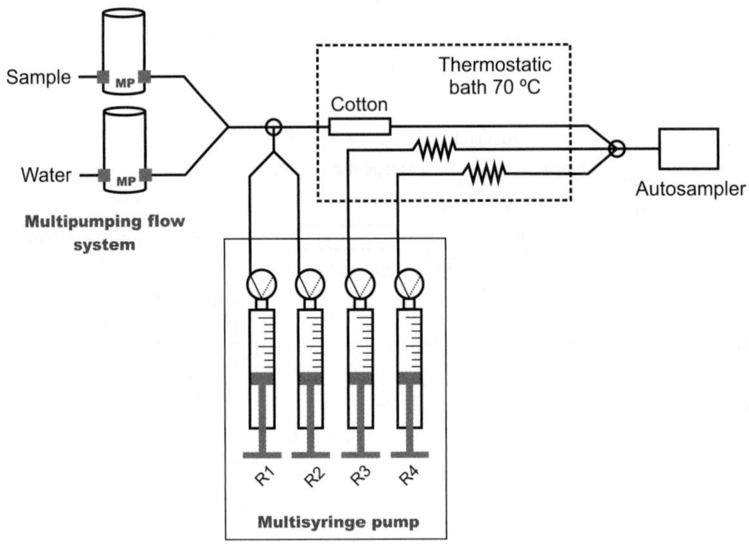

FIGURE 1.30

MSFIA-MPFS combination for radium determination in waters. MP: micropump, R: reagent.

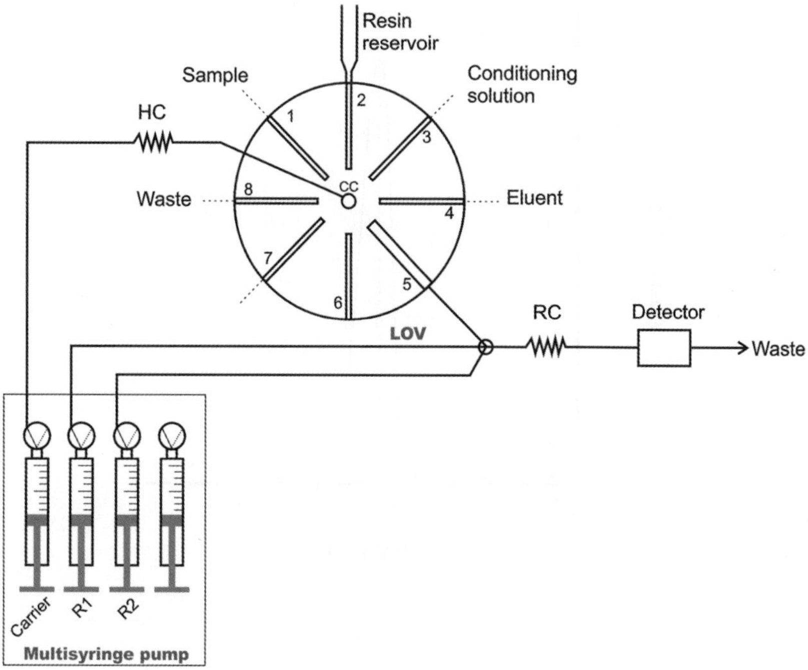

FIGURE 1.31

LOV-MSFIA system. CC: central conduit, HC: holding coil, R: reagent, RC: reaction coil.

1.12 Theoretical foundations of flow techniques

1.12.1 Graphical representation of signals provided by flow systems

Analytical response of most flow techniques, but SFA, is a markedly asymmetric peak the asymmetry of which can vary depending on the particular operating conditions. Figure 1.32 shows a typical peak.

The recording is defined by the following parameters:

I = injection point
h = peak height
A = peak area
$t_{residence}$ = residence time
t_{start} = start time
W = peak width at a given signal level
$W_{1/2}$ = peak width at half-height

The shape of a typical analyte peak can be defined in mathematical terms by using a χ^2 like function such as

$$Y = \frac{\left(\frac{X-L}{\alpha}\right)^S e^{\frac{-(X-L)}{2\alpha}}}{K\alpha} \tag{1.1}$$

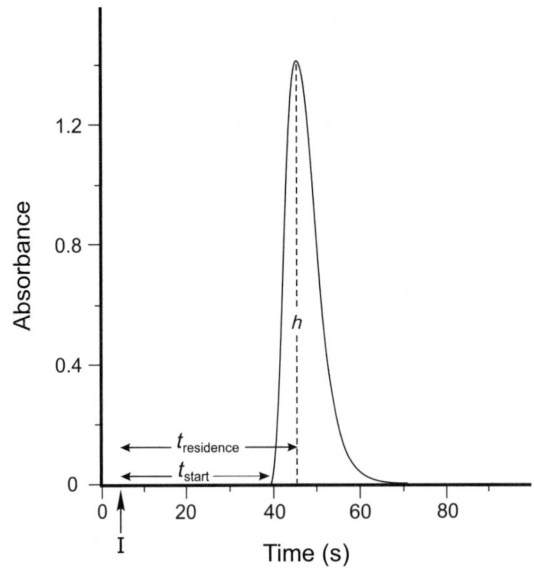

FIGURE 1.32

Typical analytical signal (peak) obtained with flow techniques.

Where,

L = tubing length
S = curve asymmetry, which increases with decreasing S
d = tubing inner diameter
q = flow rate
α = peak-base width, $\Delta t = 57\ d^{0.3}\ L^{0.4}\ q^{-1}$
K = a proportionality constant

1.12.2 Variation of the sample profile along its travel: convective and diffusive phenomena

Shape and symmetry of analyte peaks in flow techniques are dictated by convection under a laminar flow regime and diffusion by effect of concentration gradients. In the absence of convective and diffusive phenomena, an injected sample plug reaching the detector would provide a rectangular profile (see Figure 1.33(a)) similar to that obtained at the injection point. Rather, the flow motion causes the sample plug to adopt a parabolic profile typical of a laminar flow regime (see Figure 1.33(b)) by effect of the velocity of the liquid in the center of the tubing which is higher than at the walls. The peaks in Figure 1.33 were obtained from a χ^2 function, with varying S values at different points, namely: 1 (b), 5 (c) and 15 (d).

As time elapses, the concentration gradient produces diffusion in a direction normal to the parabolic profile (see Figure 1.34, top); simplifying, such diffusion is split into a radial component and

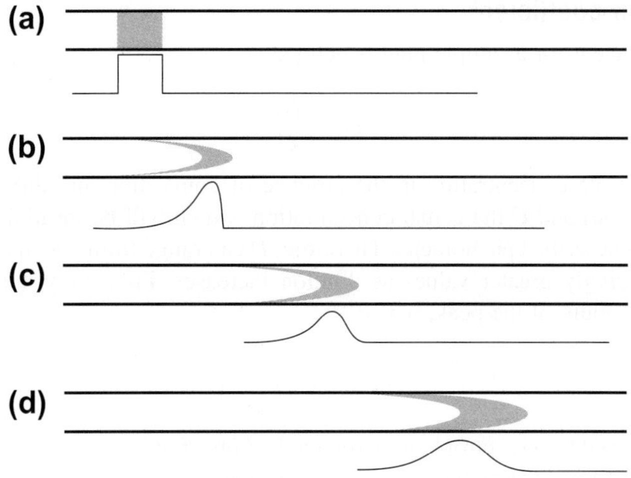

FIGURE 1.33

Dispersion of the sample peak at the injection point (a) and under predominantly convective (b), convective and diffusive (c) and predominantly diffusive flow (d).

an axial one that run normal and parallel, respectively, to the tubing axis. This diffusive phenomenon tends to smooth peak fronts and tails, thereby resulting in more symmetric peaks. After the plug has traveled a long way, diffusion prevails and peaks are increasingly symmetric.

Figure 1.33(b) illustrates the high asymmetry of a peak obtained under predominantly convective flow. The peak, however, grows more symmetric by effect of the increasing prevalence of diffusive flow as the sample plug approaches the detector (see Figure 1.33(c)). Also, the peak shrinks as the distance traveled by the sample plug increases.

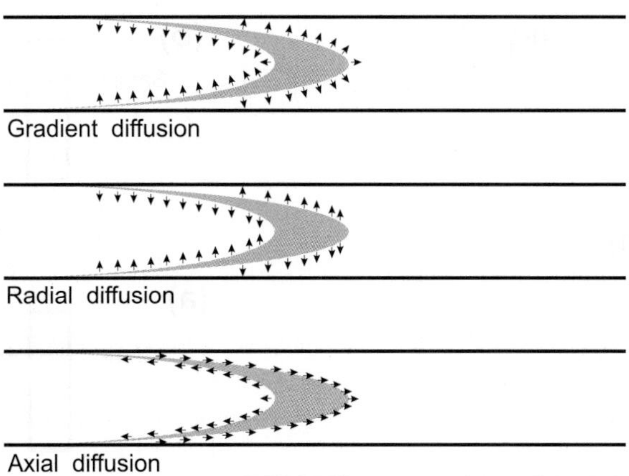

FIGURE 1.34

Diffusive phenomena caused by the concentration gradient.

1.12.3 Dispersion coefficient

The dispersion coefficient for a sample plug is defined as

$$D = \frac{C_0}{C} \tag{1.2}$$

Where, C_0 is the sample concentration in the absence of convection and diffusion (i.e. the initial concentration of sample) and C the actual concentration, which will be invariably lower than C_0 by effect of the previous described phenomena. Therefore, D can range from one in the complete absence of dilution to increasingly greater values as dilution increases. Frequently, this parameter is only calculated at the maximum of the peak, where

$$D_{max} = \frac{C_0}{C_{max}} \tag{1.3}$$

If the peak is obtained by a spectrophotometric method obeying Beer's law, then the concentrations in this equation can be replaced by the respective absorbances

$$D_{max} = \frac{A_0}{A_{max}} \tag{1.4}$$

In practice, D can be determined in various ways. With flow methods involving the injection of a fixed sample volume, the sample, e.g. a dye, can be injected into the carrier and the peak height can be measured in order to obtain A_{max} and then the sample can be aspirated through the carrier channel to obtain A_0. These operations are depicted in Figure 1.35.

With more flexible injection systems allowing variable injection volumes without altering the experimental setup, e.g. SIA, and also with time-based injection systems, it may be more practical to measure h_0 (Figure 1.36) by injecting a very large sample volume in order to avoid dilution by diffusion of the central portion of the plug.

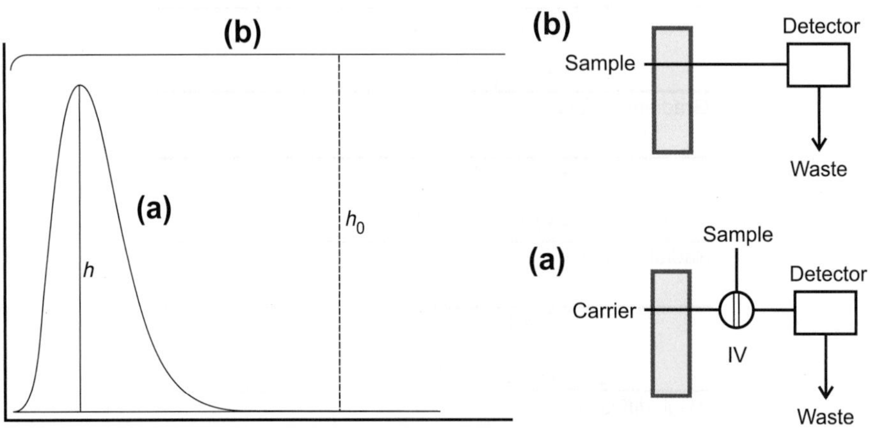

FIGURE 1.35

Determination of the dispersion coefficient in FIA: (a) injection of a known sample volume and (b) subsequent aspiration of the sample through the carrier channel. IV: injection valve.

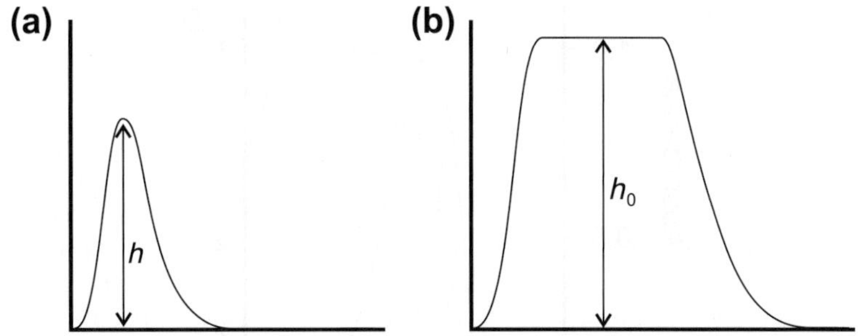

FIGURE 1.36

Determination of the dispersion coefficient by injecting variable volumes of sample. (a) Small sample volume. (b) Large sample volume.

Thus, the dispersion coefficient is a function of variables including the injected sample volume, tubing length and flow-rates, among others. Optimizing a flow system for a specific application requires a prior knowledge of the way the dispersion coefficient is influenced by these variables. This is examined in the following sections.

1.12.4 Influence of the injected sample volume

As noted earlier, the sample volume influences the height of the analyte peak and hence the dispersion coefficient of a flow system. Figure 1.37 illustrates this influence in FIA, allowing several conclusions to be drawn in this respect, namely:

1. The starting time is independent of the injected sample volume.
2. The maximum of the peak shifts as a function of the injected volume.
3. Dispersion decreases with increasing sample volume, and also increases the peak width.
4. The signal eventually saturates by effect of the central portion of the plug undergoing no further dilution by the carrier.

Although large injected sample volumes provide high sensitivity, this also increases peak width. Therefore, it is usually preferable to use small sample volumes in order to increase throughput while ensuring adequate sensitivity. The parameter $S_{1/2}$ is used to denote the sample volume required to obtain a dispersion coefficient $D=2$ (i.e. the peak height corresponding to one-half the signal for the undiluted sample). As shown below, $S_{1/2}$ provides an interesting reference for application of several criteria. Thus, with injection of volumes smaller than $S_{1/2}$, the dispersion is inversely proportional to the injected sample volume.

1.12.5 Relationship between the dispersion coefficient and injected sample volume

The dispersion coefficient is related to the injected sample volume by the following equation [32]

$$\frac{1}{D_{\max}} = \frac{C_{\max}}{C_0} = 1 - e^{-0.693\frac{S_V}{S_{1/2}}} \tag{1.5}$$

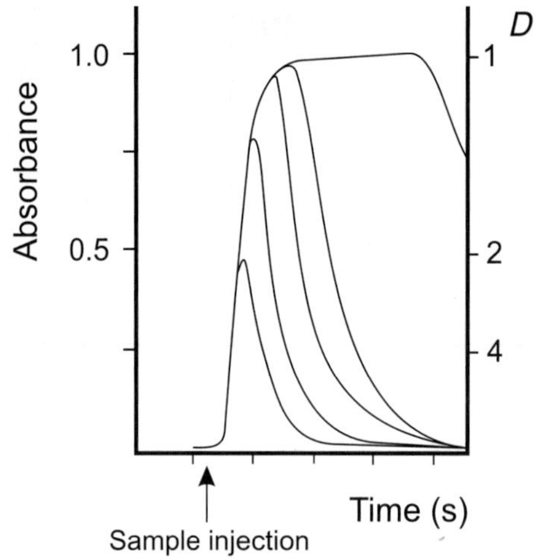

FIGURE 1.37

Influence of the sample volume on peak shape.

When using spectrophotometric measurements obeying Beer's law, the concentrations can be replaced with absorbances, so

$$\frac{A_{max}}{A_0} = 1 - e^{-0.693\frac{S_V}{S_{1/2}}} \tag{1.6}$$

which can be expressed in logarithmic, rearranged form, as

$$\ln\left(1 - \frac{A_{max}}{A_0}\right) = -0.693\frac{S_V}{S_{1/2}} \tag{1.7}$$

Therefore:

$$-\log\left(1 - \frac{A_{max}}{A_0}\right) = \frac{-0.693}{2.303\ S_{1/2}} S_V \tag{1.8}$$

A plot of the term on the left-hand side, which can be readily measured experimentally, against the injected sample volume should be a straight line from which the $S_{1/2}$ value can be easily calculated.

Three SIA experiments were conducted to calculate $S_{1/2}$ in different ways [33]. Figure 1.38 shows a series of peaks obtained in the first test, where the center of each aspirated sample plug remained at a constant distance from the switching valve throughout. To this end, a variable volume of water was aspirated after the dye in such a way that the distance between the center of the dye plug and the valve would always be 250 μL. As can be seen, peak height increased by increasing sample volume.

Based on theory, $S_{1/2}$ can be obtained in two ways, namely: (a) by plotting the variation of peak height against the injected volume and identifying the volume resulting in $A_{max} = A_0/2$ (A_{max} would be 0.633 in this case, so the sought volume would be 236 μL); or (b) by plotting $-\log[1 - (A_{max}/A_0)]$ against sample volume (S_V) and calculating $S_{1/2}$ from the slope by this a volume

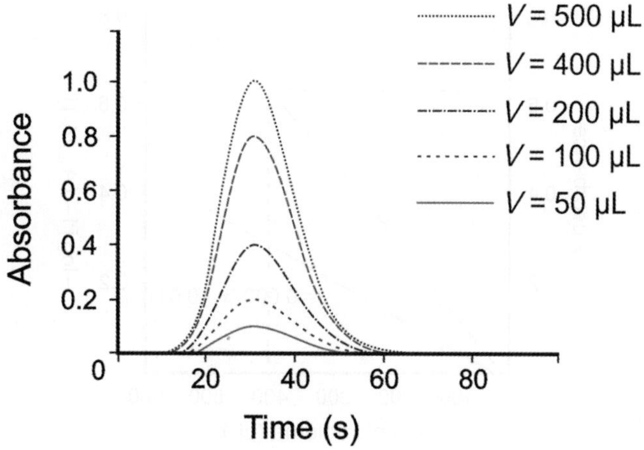

FIGURE 1.38

SIA peaks obtained from injections of variable volumes of dye ranging from 50 to 500 μL. (For color version of this figure, the reader is referred to the online version of this book.)

of 230 μL would be obtained, which is consistent with the previous one. As can be seen from Figure 1.39, the linear relationship between A_{max} and the injected volume is especially strong when the latter is less than $S_{1/2}$. Also, linearity is good throughout the S_V range.

In a second experiment to determine $S_{1/2}$ we changed the inner diameter of the tubing (0.5 mm). As can be seen from Figure 1.40, the theory still holds. However, dispersion was higher by effect of the thinner tubing used, because the sample needs to travel a longer way. Burette addition rate and

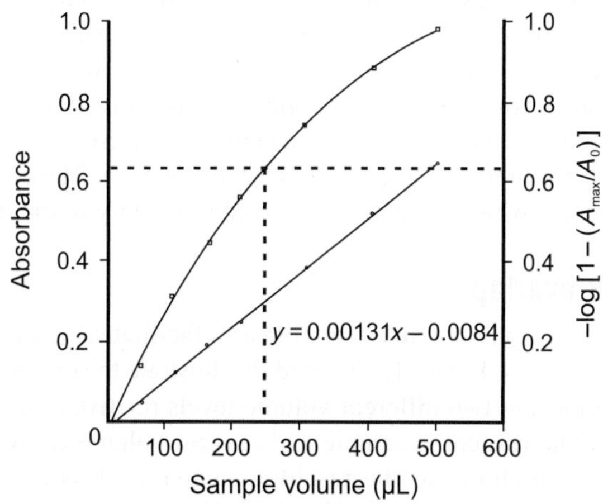

FIGURE 1.39

Variation of peak height with the sample volume injected using tubing of 0.8 mm internal diameter (i.d.).

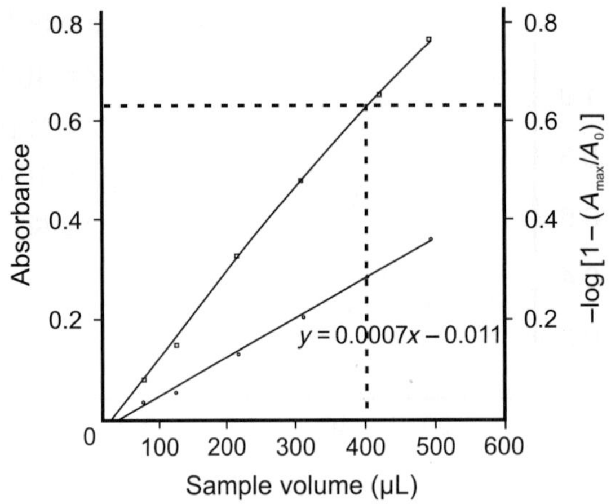

FIGURE 1.40

Plot of $-\log [1 - (A_{max}/A_0)]$ against S_V used to calculate $S_{1/2}$ with the experimental setup of Figure 1.37 but tubing of 0.5 mm i.d.

volumes were the same as in the previous experiment, thus the residence time was roughly identical. However, $S_{1/2}$ was rather different: 405 μL. In any case, the plot of $-\log [1 - (A_{max}/A_0)]$ against S_V was also quite linear.

In a third experiment, $S_{1/2}$ was calculated without centering the sample plug (i.e. by aspirating increased volumes of dye followed by no water). The results thus obtained with the sample near the switching valve differed markedly from those of the previous experiments.

Thus, the distance traveled by the center of the plug increased as the sample volume increased. Also, given that the sample was near the valve, the distance to the detector was shorter and the dispersion was lower than in the previous experiments. Figure 1.41 exposes the poor linearity of the fit. Thus, the $S_{1/2}$ value obtained from a plot of A_{max} against the injected volume was 108 μL, whereas that provided by a plot of $-\log [1 - (A_{max}/A_0)]$ vs S_V including all points—which were poorly aligned—was 223 μL and if it was obtained from the first two points it resulted in 104 μL.

For convenience, the third method is the most widely used in SIA despite the poor linearity between A_{max} and the injected sample volume (especially above $S_{1/2}$). The choice was dictated by the decreased dispersion obtained, which results in improved sensitivity in the determination.

1.12.6 $S_{1/2}$ and peak overlap

$S_{1/2}$ can be especially useful to optimize SIA operation as it facilitates the calculation of the most suitable sample and reagents volumes. Figure 1.42(a) and (b) illustrate the extent of overlap between sample and reagent peaks obtained at two different volume levels relative to $S_{1/2}$. As can be seen, if the sample is aspirated before the reagent, a sample and reagent volume equivalent to $0.5S_{1/2}$ and $S_{1/2}$, respectively, constitute a good choice as the resulting sample peak is completely overlapped with the reagent peak. With two reagents, however, it is preferable to use the sandwich technique and intercalate the sample between the two reagents in such a way that $V_{r1} = V_{r2} = S_{1/2}$ and $V_s = 0.5S_{1/2}$ (see Figure 1.42(c)).

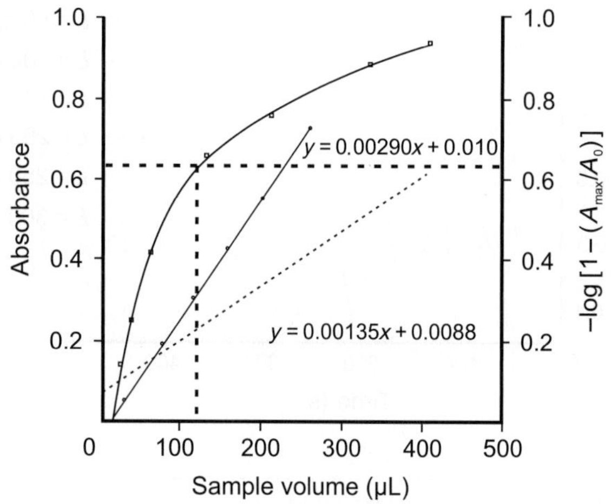

FIGURE 1.41

Plot of $-\log[1-(A_{max}/A_0)]$ against S_V used to calculate $S_{1/2}$ with the experimental setup of Figure 2.7, tubing of 0.8 mm i.d. and the sample near the switching valve.

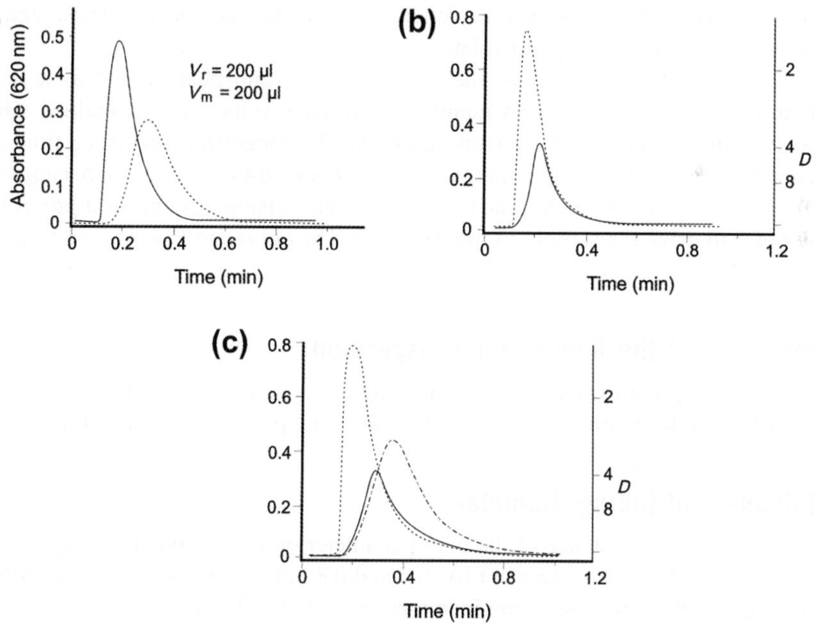

FIGURE 1.42

(a) Overlapped reagent and sample peaks obtained at $V_r = V_s = 2S_{1/2}$. (b) Overlapped reagent and sample peaks obtained at $V_r = V_s = 0.5S_{1/2}$. (c) Overlap of the sample reagent peak obtained between two sample peaks by using $V_{r1} = V_{r2} = S_{1/2}$ and $V_s = 0.5S_{1/2}$.

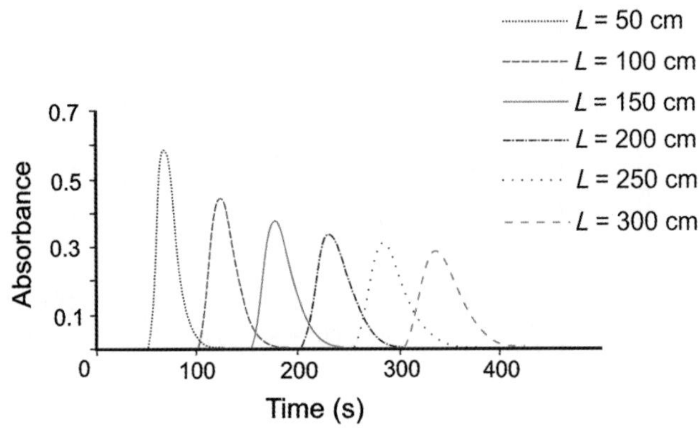

FIGURE 1.43

Variation of peak characteristics as a function of tubing length in a single-channel flow system. $q = 2$ ml/min; $d = 0.8$ mm; $S = 4$; $K = 57$. (For color version of this figure, the reader is referred to the online version of this book.)

1.12.7 Influence of tubing length on dispersion

Although the influence of the channel tubing length on the dispersion coefficient has been extensively studied since the inception of FIA [34] the conclusions reached cannot be extrapolated to other flow techniques for reasons that are explained later on.

Figure 1.43 shows the variation of the residence time and peak height as a function of the tubing length of a single-channel FIA manifold. Similar results can be obtained with other flow techniques. As can be seen, the longer the tubing is, the longer the residence time and higher is the dispersion as a result. In order to increase the residence time without increasing the dispersion one must use the stopped-flow technique (i.e. stop the propulsion system for a preset time and then restart it in order to drive the sample to the detector). The curves in Figure 1.43 were simulated by using equation 1.1.

1.12.8 Influence of the flow rate on dispersion

The flow rate has the opposite effect of the tubing length. Thus, with a given length, increasing the flow rate decreases the residence time and also dispersion in the peaks (see Figure 1.44).

1.12.9 Influence of tubing diameter

Figure 1.45 illustrates the influence of the tubing diameter on dispersion. The graph was obtained by using the most usual tubing diameters in FIA (0.5 and 0.8 mm), and also in SIA and MSFIA (0.8, 1.0 and 1.5 mm). As can be seen, dispersion increases as diameter does.

1.12.10 Influence of coil diameter

When relatively long flow channels must be used in order to ensure thorough mixing or adequate reaction development, the tubing is usually coiled to save space. A coiled piece of tubing is usually

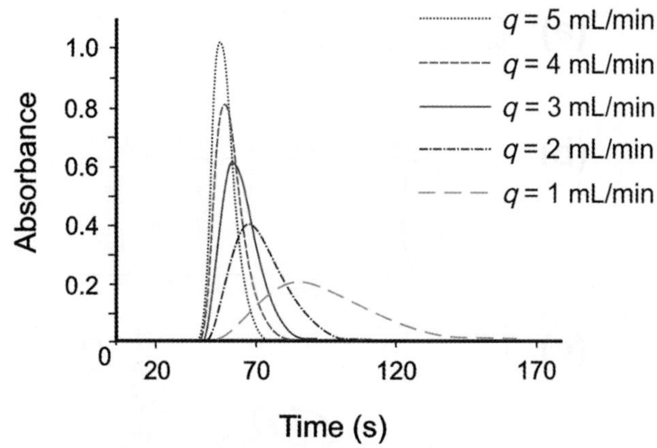

FIGURE 1.44

Variation of peak characteristics as a function of the flow rate in a single-channel flow system. $L = 50$ cm; $d = 0.8$ mm; $S = 4$; $K = 57$. (For color version of this figure, the reader is referred to the online version of this book.)

referred to as a "loop", "coil" or "reactor" in this context. The fact that the diameter of a coil influences dispersion inside it has been used to develop the so-called "knotted reactors". Figure 1.46 shows three typical tubing configurations, namely: straight tubing, which is the preferred choice for short channels; coiled tubing, which is more practical and saves space when longer channels are to be used; and knotted tubing.

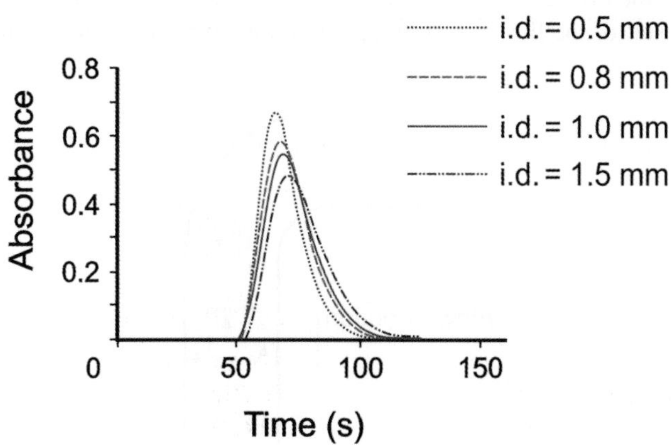

FIGURE 1.45

Variation of peak characteristics as a function of the tubing inner diameter in a single-channel flow system. $L = 50$ cm; $q = 2$ ml/min; $S = 4$; $K = 57$. (For color version of this figure, the reader is referred to the online version of this book.)

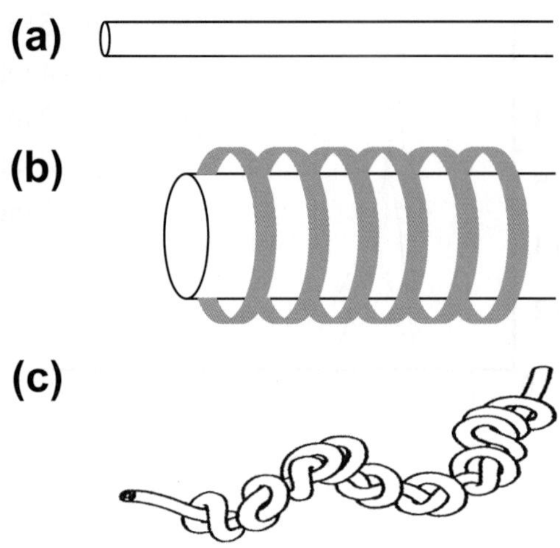

FIGURE 1.46

Typical tubing configurations. (a) Linear. (b) Coiled. (c) Knotted.

The dispersion coefficient decreases (viz. peaks grow narrower and taller) with decreasing the coil diameter. This can be understood by assuming the centrifugal force to generate radial flow (Figure 1.47).

At low flow-rates, the force hardly distorts the parabolic profile; as the flow rate is raised, however, distortion increases and secondary flows normal to the principal one are produced increasing radial mass transfer. At low flow-rates, the centrifugal force is small and dispersion is similar to that in a straight tube; at higher values however, dispersion is much less substantial and decreases with increasing bending of the tubing. In knotted tubing, where bending is exacerbated, the previous effect is very strong as a result of the flow direction being continually altered by the tubing walls.

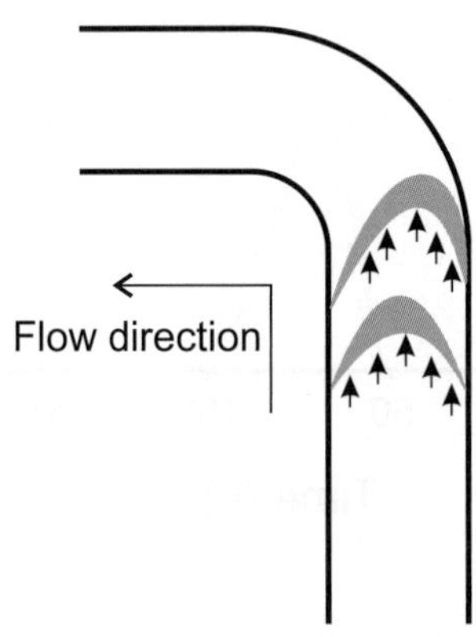

FIGURE 1.47

Distortion in the flow profile around bends reduces dispersion.

References

[1] M. Trojanowicz, Advances in Flow Analysis, Wiley-VHC, Weinheim, Germany, 2008.

[2] V. Cerdà, G. Ramis, An Introduction to Laboratory Automation, J. Wiley, New York, 1990.

[3] F.R.P. Rocha, J.A. Nóbrega, O.F. Filho, Flow strategies to greener analytical chemistry. An overview, Green Chem. 3 (2001) 216–220.

[4] B.F. Reis, A. Morales-Rubio, M. de la Guardia, Environmentally friendly analytical chemistry through automation: comparative study of strategies for carbaryl determination with p-aminophenol, Anal. Chim. Acta 392 (1999) 265–272.

[5] L. Skeggs, Automatic method for colorimetric analysis, Am. J. Clin. Path. 28 (1957) 311–322.

[6] J. Ruzicka, E.H. Hansen, Flow injection analyses. I. New concept of fast continuous flow analysis, Anal. Chim. Acta 78 (1975) 145–157.

[7] J. Ruzicka, G.D. Marshall, Sequential injection: a new concept for chemical sensors, process analysis and laboratory assays, Anal. Chim. Acta 237 (1990) 329–343.

[8] O. Thomas, F. Theraulaz, V. Cerdà, D. Constant, Ph. Quevauviller, Wastewater quality monitoring, Trends Anal. Chem. 16 (1997) 419–424.

[9] P.C.A.G. Pinto, M.L.M.F.S. Saraiva, J.L.F.C. Lima, A flow sampling strategy for the analysis of oil samples without pretreatment in a sequential injection analysis system, Anal. Chim. Acta 555 (2006) 377–383.

[10] B.F. Reis, M.F. Giné, E.A.G. Zagatto, J.L.F.C. Lima, R.A. Lapa, Multicommutation in flow analysis. Part 1. Binary sampling: concepts, instrumentation and spectrophotometric determination of iron in plant digests, Anal. Chim. Acta 293 (1994) 129–138.

[11] V. Cerdà, J.M. Estela, R. Forteza, A. Cladera, E. Becerra, P. Altimira, P. Sitjar, Flow techniques in water analysis, Talanta 50 (1999) 695–705.

[12] B. Horstkotte, O. Elsholz, V. Cerdà, Review on automation using multisyringe flow injection analysis, J. Flow Injection Anal. 22 (2005) 99–109.

[13] M.I.G.S. Almeida, J.M. Estela, V. Cerdà, Multisyringe flow injection potentialities for hyphenation with different types of separation techniques, Anal. Lett. 44 (2011) 360–373.

[14] R.A.S. Lapa, J.L.F.C. Lima, B.F. Reis, J.L.M. Santos, E.A.G. Zagatto, Multi-pumping in flow analysis: concepts, instrumentation, potentialities, Anal. Chim. Acta 466 (2002) 125–132.

[15] C. Pons, R. Forteza, A.O.S.S. Rangel, V. Cerdà, The application of multicommutated flow techniques to the determination of iron, Trends Anal. Chem. 25 (2006) 583–588.

[16] B. Horstkotte, C.M. Duarte, V. Cerdà, A miniature and field-applicable multipumping flow analyzer for ammonium monitoring in seawater with fluorescence detection, Talanta 85 (2011) 380–385.

[17] B. Horstkotte, C.M. Duarte, V. Cerdà, Multipumping flow systems devoid of computer control for process and environmental monitoring, Int. J. Environ. Anal. Chem. (2011), http://dx.doi.org/10.1080/03067319.2010.548601.

[18] J. Ruzicka, Lab-on-valve: universal microflow analyzer based on sequential and bead injection, Analyst 125 (2000) 1053–1060.

[19] J. Wang, E.H. Hansen, Sequential injection lab-on-valve: the third generation of flow injection analysis, Trends Anal. Chem. 22 (2003) 225–231.

[20] M. Grand, H.M. Oliveira, J. Ruzicka, C. Measures, Determination of dissolved zinc in seawater using micro-sequential injection lab-on-valve with fluorescence detection, Analyst 136 (2011) 2747–2755.

[21] I. Lähdesmäki, P. Chocholous, A.D. Carroll, J. Anderson, P.S. Rabinovitch, J. Ruzicka, Two-parameter monitoring in a lab-on-valve manifold, applied to intracellular H_2O_2 measurements, Analyst 134 (2009) 1498–1504.

[22] S.S.M.P. Vidigal, I.V. Tóth, A.O.S.S. Rangel, Exploiting the bead injection LOV approach to carry out spectrophotometric assays in wine: application to the determination of iron, Talanta 84 (2011) 1298–1303.

[23] Y. Yu, Y. Jiang, M. Chen, J. Wang, Lan-on-valve in the miniaturization of analytical systems and sample processing for metal analysis, Trends Anal. Chem. 30 (2011) 1649–1658.

[24] J. Avivar, L. Ferrer, M. Casas, V. Cerdà, Fully automated lab-on-valve-multisyringe flow injection analysis- ICP-MS system: an effective tool for fast, sensitive and selective determination of thorium and uranium at environmental levels exploiting solid phase extraction, J. Anal. At. Spectrom. 27 (2012) 327–334.

[25] M.D. Luque de Castro, J. Ruiz-Jiménez, J.A. Pérez-Serradilla, Lab on valve: a useful tool in biochemical analysis, Trends Anal. Chem. 27 (2008) 118–126.

[26] X.W. Chen, M.L. Chen, S. Chen, J.H. Wang, Flow-based analysis: a versatile, powerful platform for DNA assays, Trends Anal. Chem. 27 (2008) 762–770.

[27] J. Avivar, L. Ferrer, M. Casas, V. Cerdà, Smart thorium and uranium determination exploiting renewable solid phase extraction applied to environmental samples in a wide concentration range, Anal. Bioanal. Chem. 400 (2011) 3585–3594.

[28] F.Z. Abouhiat, C. Henríquez, B. Horstkotte, F. El Yousfi, V. Cerdà, A miniaturized analyzer for the catalytic determination of iodide in seawater and pharmaceutical samples, Talanta 108 (2013) 92–102.

[29] D. Nacapricha, P. Sastranurak, T. Mantim, N. Amornthammarong, K. Uraisin, C. Boonpanaid, C. Chuyprasartwattana, P. Wilairat, Cross injection analysis: concept and operation for simultaneous injection of sample and reagents in flow analysis, Talanta 110 (2013) 89–95.

[30] F. Maya, B. Horstkotte, J.M. Estela, V. Cerdà, Lab in a syringe: fully automated dispersive liquid-liquid microextraction with integrated spectrophotometric detection, Anal. Bioanal. Chem. 404 (2012) 909–917.

[31] B. Horstkotte, R. Suárez, P. Solich, V. Cerdà, In-syringe-stirring: a novel approach for magnetic stirring-assisted dispersive liquid-liquid microextraction, Anal. Chim. Acta 788 (2013) 52–60.

[32] J. Ruzicka, E.H. Hansen, Flow Injection Analysis, in: Chemical Analysis: A Series of Monographs on Analytical Chemistry and Its Applications, vol. 62, Wiley, New York, 1988.

[33] A. Cladera, C. Tomás, E. Gómez, J.M. Estela, V. Cerdà, A new instrumental implementation of sequential injection analysis, Anal. Chim. Acta 302 (1995) 297–308.

[34] J. Ruzicka, E.H. Hanssen, Rapid determination of protein in plant material by flow injection spectrophotometry with trinitrobenzenesulfonic acid, Anal. Chim. Acta 97 (1978) 327–333.

CHAPTER

Online Analytical Determination Modes

2

CHAPTER OUTLINE

Most uses of flow techniques involve the quantitative determination of some target species. This chapter describes various ways of using flow techniques with quantitative purposes, such as calibration curves, based on peak height or peak area, and titrations, based on distance between equivalence points or single-point method. Stopped-flow technique can be used for both, quantitative approach, for example in kinetic methods, or for qualitative determinations inasmuch as it allows spectral and potential scans to be performed. Multiparameter analysis is presented in two forms to be carried out, by multivariate chemometric techniques or by applying sandwich technique. Finally, smart systems are presented as a step forward in automation, commonly used in multiparameter analysis.

Flow Analysis. http://dx.doi.org/10.1016/B978-0-444-59596-6.00002-4

2.1 Determinations based on online calibration curve

2.1.1 Peak height-based online calibration

2.1.1.1 Conventional calibration curves

The usual calibration curve is constructed by plotting peak height against concentration or mass in most analytical methodologies (Figure 2.1) or vs the logarithmic concentration in potentiometric determinations.

Specific software facilitates data acquisition and is able to plot peak areas instead of heights, which in some cases provides better results. The selection of the parameter that gives the better fit in each situation should be determined by testing.

2.1.1.2 Electronic dilution curves

Advances in electronics and software have greatly facilitated the processing of analytical data. Thus, the use of a CCD (charge coupled device) detector capable of storing the whole spectrum at each point of the analytical profile allows the most suitable sensitivity level to be selected at will by the analyst. Moreover, the reproducibility of hydrodynamic variables achieved by a flow-based system affords the use of heights instead of the peak maximum in a procedure known as electronic dilution (Figure 2.2).

In the electronic dilution method, calibration curves are constructed from a single standard injection provided that a previous calibration curve involving the conventional injection of several standards has been done. Peak heights obtained from various concentrations are projected onto the expanded curve for the most concentrated standard in order to identify the time elapsed from injection to the appearance of peaks, the height of which is identical with that of the standard peaks. Subsequent calibration curves can be constructed by injecting the most concentrated standard and measuring heights at the previously established times. The process is illustrated in Figure 2.3.

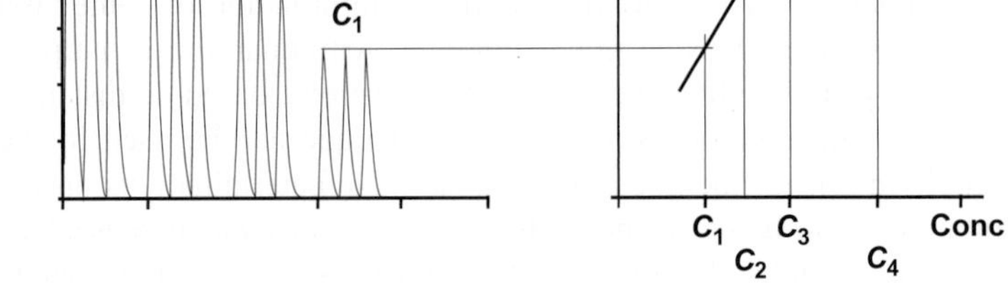

FIGURE 2.1

Peak height-based and peak area–calibration curves. *A*: area; *C*: concentration; *h*: height.

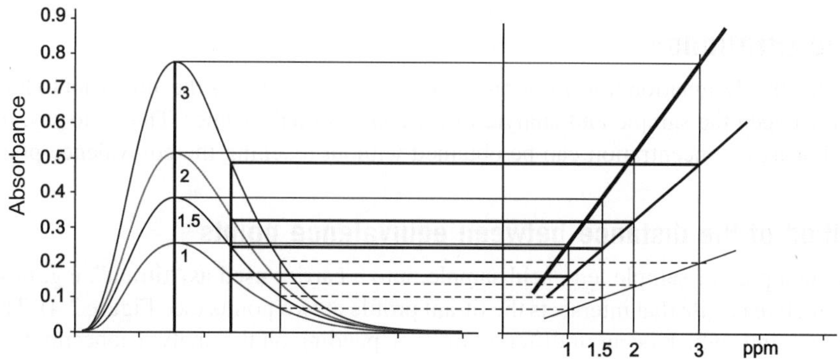

FIGURE 2.2

Construction of a calibration curve using electronic dilution. (For color version of this figure, the reader is referred to the online version of this book.)

2.1.2 Peak width-based online calibration

Peak widths are also used, although much less often than heights, to construct calibration curves. This method can be especially useful when there are double peaks in order to improve resolution [1]. Thus, the width (*W*) of a peak at a given height is measured and plotted vs standard concentration (*C*), usually in logarithmic form:

$$W = a + b \log C \tag{2.1}$$

2.1.3 Peak area-based online calibration

Specific software for data acquisition and processing has facilitated the construction of a calibration curve from the area under a peak. Some software packages, e.g. AutoAnalysis, allow both peak parameters to be measured, so the user is given the opportunity to select the better choice in terms of linearity.

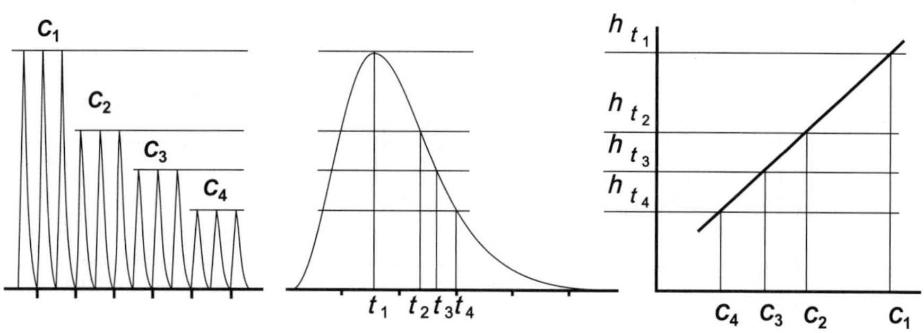

FIGURE 2.3

Procedure used to implement the electronic dilution method. *C*: concentration; *h*: height; *t*: time.

2.2 Online titrations

Please, note that the designation flow titration is misleading inasmuch as the aim is not to find an equivalence point between the sample and analyte in the conventional manner. That is to say, in online titrations, acid or base concentration can be obtained without reaching the equivalence point.

2.2.1 Method of the distance between equivalence points

When inserting a plug of sample, e.g. acid sample, into a carrier used as "titrant", e.g. base solution, produces an analyte profile that intersects the titrant profile at two points (see Figure 2.4). These are the stoichiometric points, which occur at different times depending on the analyte concentration. The distance (Δt) between these two "equivalence points" is related to the analyte concentration (C) by the following expression:

$$\Delta t = a + b \log C \tag{2.2}$$

Parameters a and b should be determined by calibration with standards that depend on variables including the titrant concentration and injected sample volume.

To illustrate the method of the distance between equivalence points, Figure 2.5 depicts a real sequential injection analysis (SIA) titration where an indicator was injected into the carrier in order to measure Δt [2].

2.2.2 Single-point method

If the objective of the flow-based system is to determine the concentration of acid or base that contains a sample, single-point method could be used. It consists in injecting an acid (or basic) sample into a

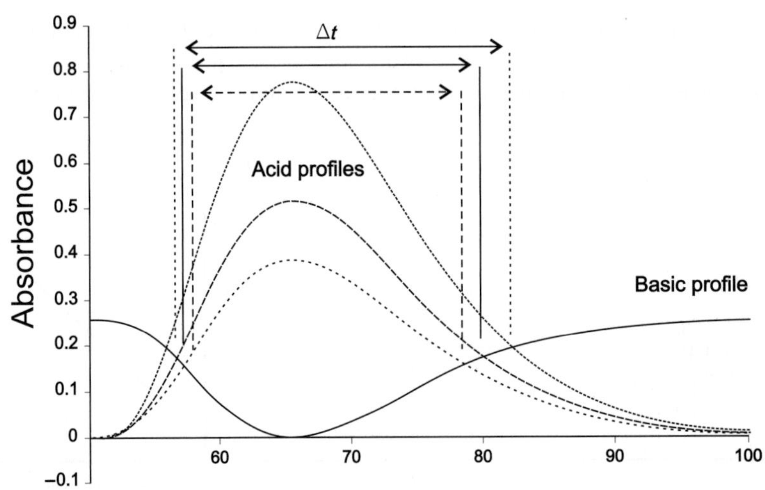

FIGURE 2.4

The distance Δt between the two points where the concentration profile for the sample (an acid) and "titrant" (a base used as carrier) intersect allows the analyte concentration to be determined.

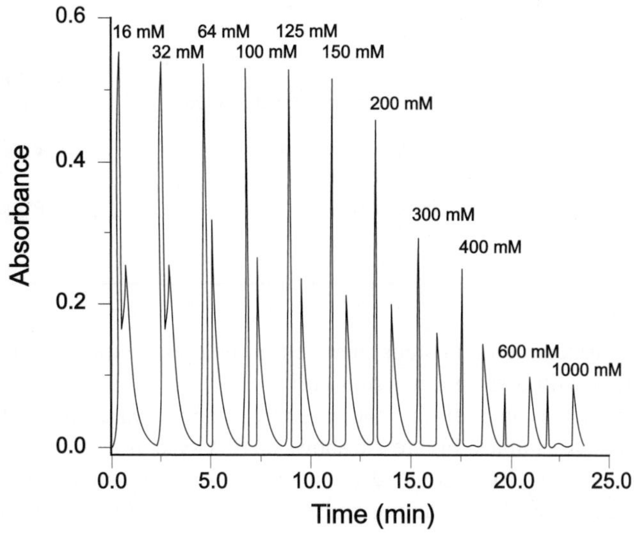

FIGURE 2.5

Real SIA titration curve.

water carrier stream for subsequent merging with a reagent consisting of an acid–base pair (HB/B) form and an indicator (HI/I). This produces a peak the height of which is proportional to the acid (or base) concentration in the sample [3,4].

2.3 Stopped-flow techniques

A stopped-flow technique consists in stopping the flow, at a point in the process, for a specific purpose. Most often, the flow is halted during the emergence of a peak and, especially, at its maximum. This can be used for a variety of purposes, the most salient of which are discussed below.

The inception of computers has facilitated the use of the stopped-flow technique by affording accurate, reproducible halting of the flow in an unattended manner.

One of the earliest uses of the stopped-flow technique was to avoid dispersion. With slow reactions, it is possible to use long enough tubing to ensure adequate reaction development and detection improving sensitivity. This, however, has a double effect as it increases the signal through a prolonged reaction time but also raises dispersion through an increased path length to the detector and ultimately reduces the analyte signal. For this reason, it may be more efficient to increase the reaction time simply by stopping the flow over an adequate period and then drive the sample to the detector along the shortest possible tube length, minimizing dispersion.

In flow injection analysis (FIA), this solution invariably detracts from throughput; while in SIA, the loss can be minimized by stopping the flow at the reaction coil, located behind the switching valve, and starting processing of the next sample. When the second sample reaches the reaction coil, it will push the first, which by then will have reacted to an adequate extent, to the detector.

Table 2.1 Optimization of the Stopped-Flow Technique for Use in SIA

Step	Solution in Reaction Coil	Solution Passing through the Detector
1. Aspiration of sample and reagents into the holding coil	Previous sample	Carrier
2. Delivery of sample and reagents to the reaction coil	Next sample	Previous sample
3. Repetition of step 1		

In Table 2.1 are summarized the main steps involved in the optimization of the stopped-flow technique in an SIA system with a simple reaction coil.

In addition, reaction times can be further increased by using several channels for the reaction coils in order to insert successive samples. Figure 2.6 shows an SIA system with three reaction coils. In this system, the three reaction coils (RCs) are loaded with a reaction plug and the first plug is driven to the detector after loading the other three reaction coils, optimizing the throughput of the system.

2.3.1 Application to kinetic methods

Flow techniques are kinetic-based techniques of the fixed-timed type. In fact, most often the reaction is not allowed to reach equilibrium as it is assumed to develop to a highly reproducible extent by virtue of the high reproducibility of hydrodynamic variables. Whether or not a given flow technique is influenced by reaction kinetics can be readily checked by using it in the stopped-flow mode. If the resulting peak increases with holding time, then, the reaction can be deemed incomplete (Figure 2.7) and a compromise will have to be made between sensitivity and throughput.

The stopped-flow mode facilitates implementation of the initial rate method, which is the most widely used kinetic methodology. In fact, if the flow is stopped while a peak is forming, the resulting

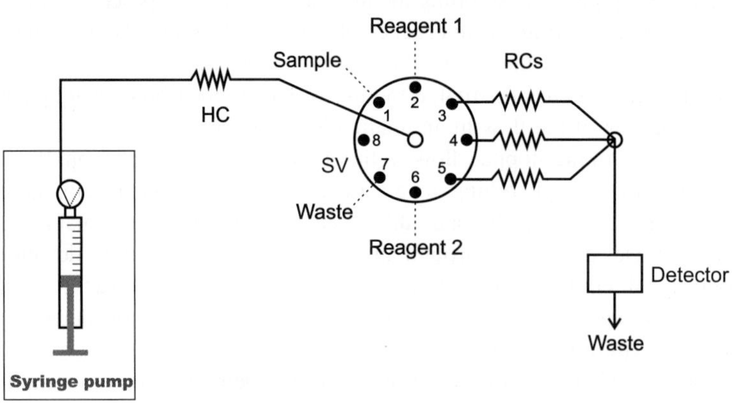

FIGURE 2.6

SIA system exploiting stopped-flow technique. HC: holding coil; RC: reaction coil; SV: selection valve.

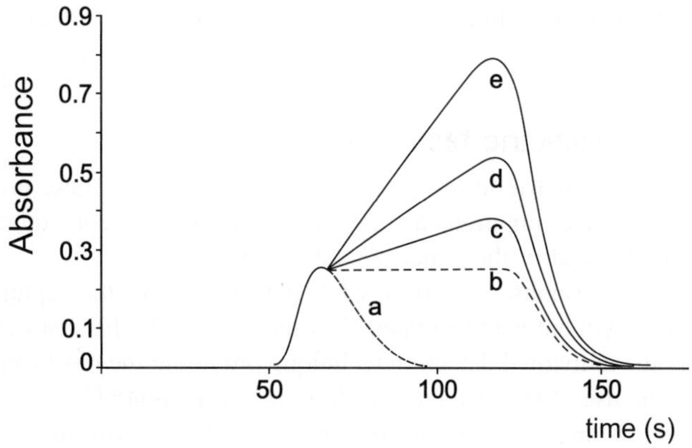

FIGURE 2.7

Implementation of the initial-rate method in a flow manifold. Curves obtained a) without stopping the flow, b) after chemical equilibrium has been reached. c), d) and e) are kinetic curves of variable slope dependent on the catalyst concentration.

signal will vary depending on the particular situation. Thus, if the reaction is already at equilibrium, the signal will be flat as the sample, stopped at the detector, will evolve no further. On the other hand, if the sample is stopped before the reaction has completed, the signal will continue to evolve and if subsequent changes are due to the contribution of the reaction product, then the signal will increase with time in the way it does in the slope method. When the flow is resumed, the signal will return to the baseline.

2.3.2 Application to scanning techniques

One other interesting application of stopped-flow techniques is in scanning processes used to expand available analytical information. Thus, by stopping the flow at the peak maximum, it is possible to obtain UV—Vis spectra for purposes such as spectral identification, purity assessment, or multivariate analysis.

Although the stopped-flow technique is of little use with current CCD and diode array UV—Vis spectrophotometers, which allow the spectrum at each point along a profile to be obtained without halting the flow, it continues to be useful for other scanning techniques, e.g. fluorimetry, polarography, and voltammetric stripping.

2.4 Online multiparameter analysis

From the beginning, flow techniques have been designed with the purpose not only to facilitate automation of analytical processes, but also to determine several parameters at once. The early Technicon® segmented-flow autoanalyzer proved to be highly effective for multiparameter determinations. However, being expensive to achieve as an individual segmented flow analysis (SFA) manifold had to be constructed for each target parameter. As new flow techniques have emerged, the ability to determine several parameters simultaneously without excessive technical complications has continued to be explored.

2.4.1 Multivariate chemometric techniques

Given that multivariate chemometric techniques rely on computations, these provide interesting advantages in multiparameter determinations as they afford the simultaneous quantification of several species without the need to modify the experimental setup.

As noted earlier, the fast response of current CCD detectors allow the capture of the spectrum at each point along a profile, without using stopped-flow techniques. In this sense, CCD-based spectrophotometry is probably the individual detection technique providing the most robust and reproducible measurements. Hence, the use of this detection technique is preferential in multivariate analysis.

Other types of detectors are also used in this context, although with major constraints in some cases. One especially attractive technique here is variable-angle fluorimetry. By stopping the flow at the maximum of the peak, it is possible to choose the pathway in the three-dimensional spectrum for a mixture of fluorescent compounds providing the best possible compromise between the selectivity and sensitivity required to quantify all target species. Figure 2.8 shows the contour map for a three-dimensional spectrum and the variable-angle pathway leading to the best possible determination of fuberidazole and thiabendazole [5].

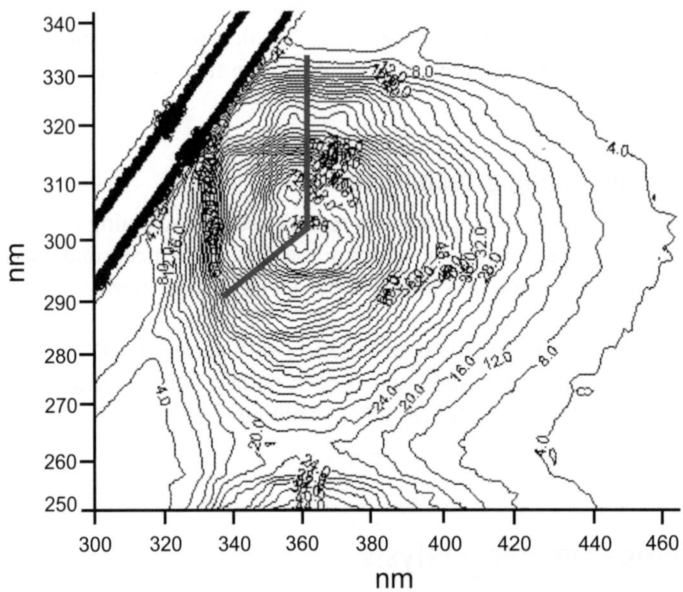

FIGURE 2.8

Contour map for an isoemissive mixture of two pesticides (fuberidazole and thiabendazole) showing the variable-angle pathway used for their determination by SIA. (For color version of this figure, the reader is referred to the online version of this book.)

However, multivariate chemometric techniques are compatible with only a few types of flow-through detectors. Electrochemical techniques such as differential pulse polarography (DPP) and anodic stripping voltammetry (ASV) pose the greatest impediment to application of multivariate methods, and only the *generalized standard addition method* can be used, since electrodes cannot be removed from the solution during the application of the analytical method [6]. In any case, implementing ASV in a flow system has the advantage over the use of its batch counterpart that no stripping in the same preconcentration medium is necessary, so the curve to be measured can be obtained with a more suitable method.

2.4.1.1 Multiple linear regression analysis

All scanning techniques, e.g. UV–Vis spectrophotometry, variable-angle fluorimetry, DPP, and stripping voltammetry, obey the law of additivity of contributions:

$$M_j = \sum_{i=1}^{n} k_i^j C_i \tag{2.3}$$

where M_j is the measurement value obtained at point j in the scan and k_j^i the proportionality constant for the contribution of component i at point j. This equation can be written in matrix form as follows:

$$M = k\,C \tag{2.4}$$

where M, k, and C are the matrices of measurements, proportionality constants, and concentrations of components contributing to the measurements, respectively.

Calibration is done by determining matrix k from standards of known composition, which provide matrix C, and scanning them to obtain matrix M. After the system is calibrated, an unknown sample can be scanned to obtain matrix M and calculate the concentrations of its components from the following equation:

$$C = (k'k)^{-1}k'M \tag{2.5}$$

In the case of spectrophotometric detection, M consists of the absorbances at each point and k of the molar absorptivity of each component at each point in the spectrum. For convenience, Eqn (2.3) can thus be expressed as follows:

$$A_j = \sum_{i=1}^{n} a_j + \varepsilon_i^j C_i \tag{2.6}$$

One frequently overlooked issue when using multivariate analysis methods is the calibration method used to obtain the proportionality constants. Thus, *the single standard method* scans a standard and calculates constants from the following expression:

$$k_j^i = \frac{m_i^s}{C_i^s} \tag{2.7}$$

where k_j^i are the proportionality constants, m_i^s are the measurements made from the scan for the standard of component i, and C_i^s is the concentration of the standard of component i.

The *normalized standard method* uses the same procedure as the single standard method but several concentrations of each standard have to be scanned. This allows average proportionality constants to be calculated for all the scan developed and small oscillations in the constants within the concentration range studied to be averaged as well.

The *multiple standard method* uses mixtures of all components at known concentrations in combination with the following expression:

$$m_j^s = z_j \sum k_i^j C_i^s; \quad \forall s = 1 \ldots n_s \tag{2.8}$$

where z_j is the fitting parameter for each scan j, m_j^s is the mean obtained in scan j for standard s, and n_s is the standard n, of known concentration of each compound.

The *multiple standard addition method*, also known as the generalized standard addition method, involves scanning the unknown sample and then supplying it with known volumes of a mixture of standards in order to obtain a new scan without altering the experimental setup.

With all methods, the measurements obtained in a scan can vary at each point in it. Therefore, it seems logical to increase the weight of a point in proportion to its contribution to the scan, e.g. to favor peak maxima over minima. One way of calculating the weight to be given to a point is by using the following equation:

$$d^j = \sqrt{\frac{\sum_{s=1}^{n_s} \left(m_{s,\,\mathrm{exp}}^j - m_{s,\,\mathrm{calc}}^j \right)^2}{n_s - n}} \tag{2.9}$$

which allows Eqn (2.3) to be rewritten as follow:

$$\frac{m^j}{d^j} = k_0' + \sum_{i=1}^{n} \frac{k_i^j}{d^j} C_i; \quad \forall j = 1 \ldots n_j \tag{2.10}$$

Figure 2.9 shows a three-dimensional SIA recording obtained by using a diode array spectrophotometer to determine metal ions with PAR (4-(2-pyridylazo) resorcinol) [7]. The graph is a plot of absorbance on the y-axis vs time on the x-axis and exposes several interesting phenomena. Thus, it exhibits a signal at the lowest wavelength, immediately above the x-axis, where none of the substances present absorbs any light. Such a signal can be ascribed to spurious peaks resulting from changes in refractive index in the plugs, i.e. schlieren effect. Since changes are virtually independent from the wavelength, the SIA profile exhibits a constant, negligible signal throughout due to the contribution of nonabsorbing species. The advantage of CCD detectors in stopped-flow techniques is that they allow these spurious signals to be subtracted from the total spectrum in order to exclude false contributions from changes in refractive index. This shortcoming can also be easily circumvented by injecting the sample into a stream of liquid with an identical refractive index. In this way, the corrected profile will only contain the signals due to PAR and its complex, e.g. magnesium—PAR complex. If PAR is added in a large excess with respect to magnesium, its concentration will be scarcely altered by the presence of the metal. However, because the molar absorptivity coefficient for the complex is much greater than that for the ligand, the overall spectrum will exhibit an increase due to the contribution of the metal. By subtracting the PAR spectrum from the total spectrum, it is possible to obtain the complex spectrum.

Figure 2.10 shows the spectrum for the ligand, as well as those for its magnesium and calcium complexes obtained by difference. As can be seen, the spectra for the two complexes are strongly overlapped, so they preclude the direct determination of the two metals, which must be addressed by using a multivariate analysis technique instead. For example, in a SIA system exploiting multivariate analysis

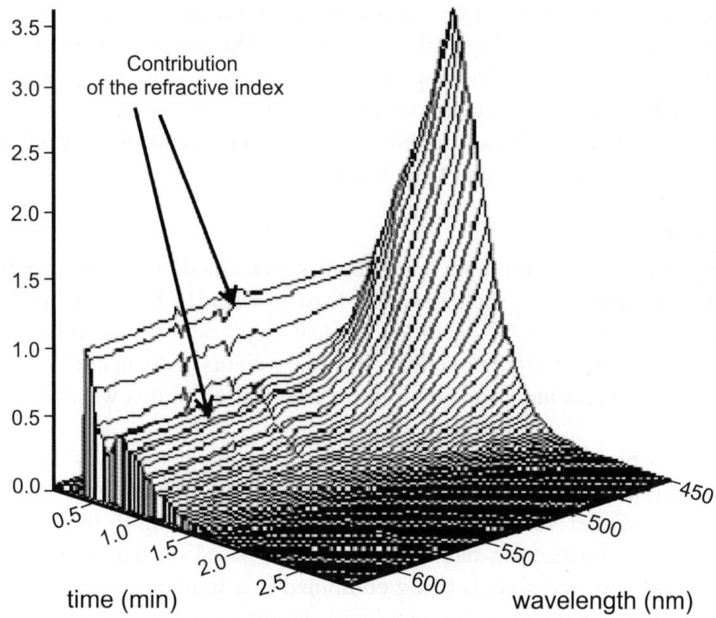

FIGURE 2.9

Three-dimensional SIA recording for the PAR–magnesium complex.

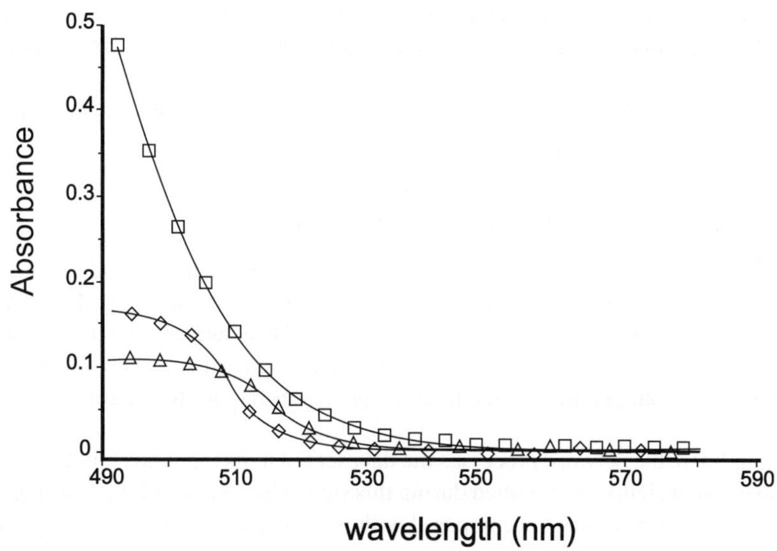

FIGURE 2.10

Spectrum at the maximum of the peak for (□) PAR after subtracting the contribution of the refractive index, and (◇) magnesium and (△) calcium after subtracting the contributions of the refractive index and ligand.

technique statistical satisfactory results of a simultaneous determination of calcium and magnesium in water samples were obtained comparing with those obtained by the reference method [7].

Although other metal ions form PAR complexes the spectra of which overlap with those of calcium and magnesium complexes, they are usually present at much lower concentrations in water and rarely interfere with their determination. If they do, however, it is more practical to include them among the analytes to be determined than to attempt their masking.

2.4.1.2 Neural networks

In addition to the above-described multiple linear regression method, metal ions can be determined by using other multivariate approaches such as partial least squares (PLS) and neural networks. Neural networks produce their own rules of operation by learning from previously processed examples. The learning process relies on a training rule that alters the weights of the neural connections as a function of the response to previous inputs and their desired responses. In this way, neural software learns from experience.

The analogue of a biological neuron in an artificial neural network is a processing element (PE), which encompasses a large number of inputs and outputs. Thus, the output of a PE is connected to the input of another via a connecting weight corresponding to the synapsis force of the neural connections. Each connection has its own weight, based on which inputs of each PE are modified, the resulting weighted signals being combined in a summation and each combined input then being applied a transfer function that can be of the threshold or continuous type. Finally, the output of the transfer function is fed to the output of the PE concerned.

A neural network comprises a large number of PE connected to one another, thus the elements are arranged in groups called "layers". A network consists of a sequence of layers that are mutually connected at random. Data are fed to the input layer and the response of the network delivered via the output layer. A variable number of additional layers called "hidden layers" can be used between the previous two.

When each neuron in a PE is triggered or processed, the following process takes place:

1. Signals are received from another PE;
2. A weighted summation of each signal is calculated; and
3. The transformed summation is fed to other neurons or PEs.

The operation of a neuron involves two principal steps, namely: learning and calling. The learning or training step is the process by which connection weights are modified in response to the data received at the input layer and, optionally, to the output data. Data delivered by the output layer constitute the target response for a given input set. The learning rule used specifies how weights are to be adapted to the training example used. Some learning processes require feeding the network with thousands of examples.

In the "calling" step, the network processes the data set fed to its input layer and delivers a response in the output layer. No weights are modified during this step, which is part of the learning process; thus, a desired response from the network is compared with its actual response in order to generate an error signal. Back-propagation (BP) and general regression (GR) neural networks are especially useful for modeling, precision, and sorting processes. A BP network, which contains an input layer, an output layer, and at least one hidden layer, uses an all-purpose nonlinear regression procedure to minimize errors. The network learns by calculating an error between the desired and actual responses, and then propagating the error back at each node. The process by which connection weights are altered in order to achieve some desired result is called "learning" or "adaptation". The most widely used learning rule is the delta rule, which usually leads to fast convergence. With this rule, the weight for each PE is modified in proportion to the magnitude of the error and the input of the connection.

The optimum number of PEs can usually be estimated by trial and error, using an additional PE in each iteration. Initially, as the number of PEs is increased, the network capabilities improve to a maximum beyond which they start to degrade. Roughly, the optimum number of PEs to be included in the hidden layer can be calculated from the following equation:

$$H = \frac{N}{5\,(i + \sigma)} \tag{2.11}$$

where H, i, and σ are the numbers of PE in the hidden, input, and output layers, respectively, and N is the number of learning examples used.

There are various types of transfer functions, the most widely used of which is the sigmoid function, by which the combined inputs are transformed into a smoothed value of 0–1. Input and output data are always scaled in order to bring them within the range spanned by the transfer function.

Neural computations have been successfully used in various areas. In chemical analysis, for example, they allow UV–Vis, IR, and NIR spectral data for mixture components to be used as inputs. Neural networks have also been used to resolve strongly overlapped signals in differential pulse ASV [8,9].

2.4.2 Sandwich technique in multiparameter analysis

Sandwich technique is a convenient operational mode for determining two parameters using the same manifold. The sample is sandwiched between two reagents each of which is used to determine a different parameter. Usually, the sample volume is large enough to prevent the reagents from interfering with the determination of each other. This allows two parameters to be simultaneously determined with a single injection.

2.4.2.1 SIA sandwich systems

One typical example of SIA sandwich systems is that used in the simultaneous determination of Fe(III) and nitrite in water [2]. The reaction ingredients are aspirated in the following sequence: thiocyanate as chromogenic reagent for iron, water in a large volume, and the Griess reagent, as chromogenic reagent for nitrite ions determination. Then, the content of the holding coil is driven to a photometric detector. A CCD-based detector allows each species to be measured at its optimum wavelength allowing the schlieren effect to be corrected.

Figure 2.11 shows the response obtained in the determination of increasing amounts of nitrite in the presence of a constant amount of Fe(III). As can be seen, the nitrite peak was very sharp by effect of its product being the first to reach the detector; by contrast, the large volume of sample used resulted in increased dispersion in the Fe(III) peak, which delayed overlap with the thiocyanate peak.

Figure 2.12 illustrates the SIA determination of nitrate and nitrite in water [10] using a copperized cadmium column. If the column is inserted between two reaction coils, the system can first aspirate the Griess reagent, followed by a large volume of sample in order to ensure that one end will penetrate the reducing column while the other remains outside. Finally, the Griess reagent is aspirated and

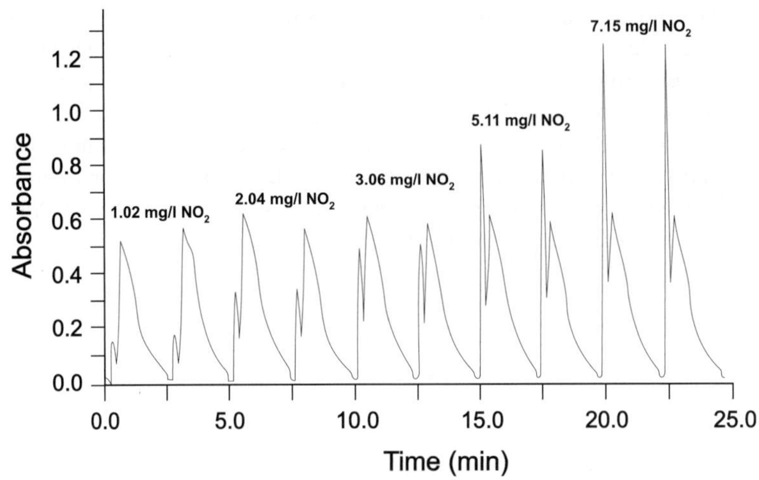

FIGURE 2.11

Simultaneous determination of Fe(III) and nitrite using the sandwich technique with increasing amounts of nitrite (first peak) and a constant amount of Fe(III) (second peak).

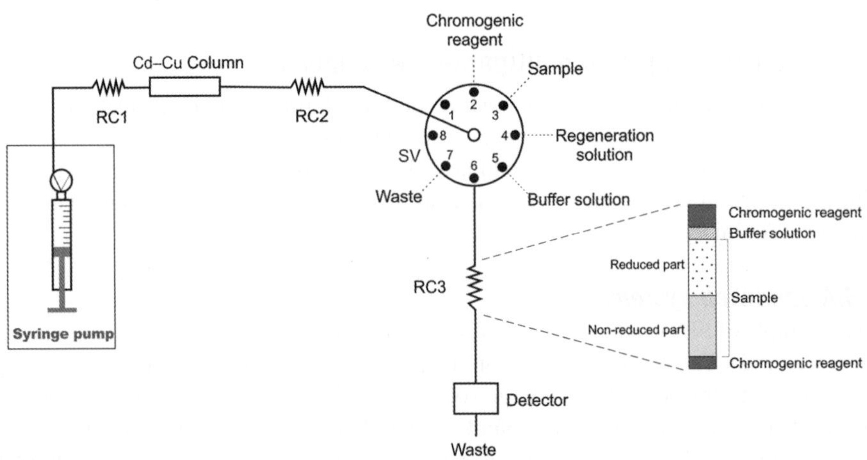

FIGURE 2.12

SIA system for the determination of nitrate and nitrite using the sandwich technique. SV: selection valve; RC: reaction coil.

the content of the holding coil is driven to the detector. In this way, the portion of sample not entering the column gives a peak due to the dye formed with nitrite ions in the sample, followed by another due to the reaction of nitrite ions originally present in the sample and those formed in the reduction of nitrate ions entering the column.

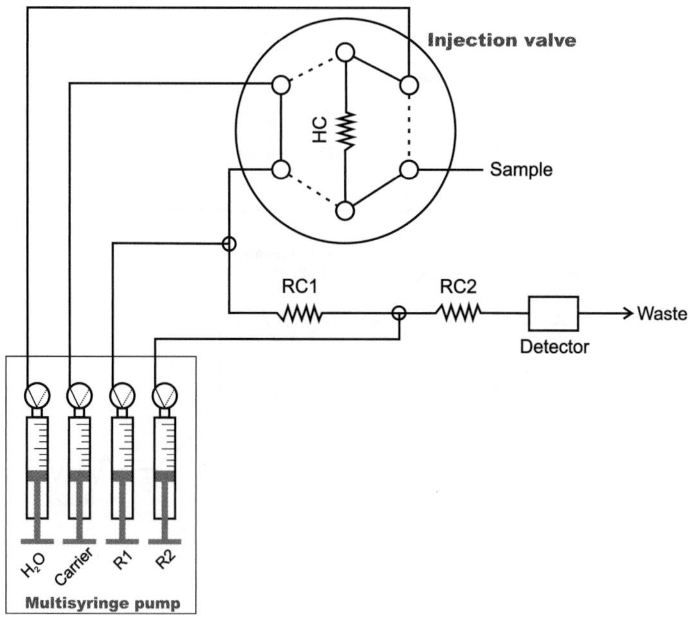

FIGURE 2.13

Implementation of the MSFIA sandwich technique with simultaneous injection of the reagents at the head and tail of the sample plug.

2.4.2.2 MSFIA sandwich systems

MSFIA systems also afford operation in the sandwich mode by injecting one reagent into the sample stream via a burette and then the other reagent when the tailing end of the sample plug reaches its insertion point. By using an appropriate length of tubing between the two insertion points, it is possible to inject both reagents simultaneously and simplify the operations involved in the analytical process (Figure 2.13).

2.4.3 Multiparameter determinations using FIA

By its own operation mode, flow injection analysis is the least suitable flow technique for multiparameter analysis as it frequently requires adapting the manifold for each new parameter to be determined. Nevertheless, Figure 2.14 depicts an unusual FIA multiparameter system that affords the determination of three parameters, viz. nitrate, nitrite, and total nitrogen, in wastewater. The system consists of two peristaltic pumps that establish two distinct zones. The upper zone is used to select the way the sample is inserted and the lower zone for measurements. A sample inserted via solenoid valve 1 (V1) is passed through a C18 resin cartridge to remove potentially present colored organic interferents. When the sample is injected while V3 is in its water position, the sample is merged with the water and then with chromogenic reagent (reagent 3); as a result, nitrite ions form a complex the absorbance of which allows the nitrite content in the sample to be determined.

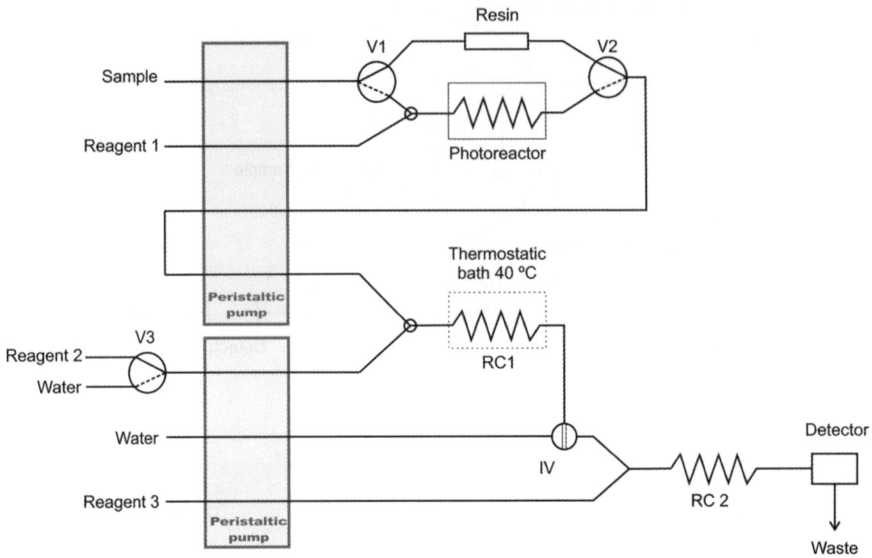

FIGURE 2.14

FIA system for the determination of nitrate, nitrite, and total nitrogen in wastewater. IV: injection valve; RC: reaction coil; V: solenoid valve.

If the process is repeated with V3 in its hydrazine position (reagent 2), nitrate is reduced by the reagent, in a thermostated bath to expedite the process. Nitrite ions originally present in the sample and those formed by reduction of nitrate are then reacted with chromogenic reagent to determine the total (nitrate + nitrite) content of the sample from the absorbance of the resulting dye. If V1 is then switched to propel toward the photoreactor, the sample is merged with persulfate (reagent 1) and undergoes photooxidation into nitrate under a UV lamp. The debubbler removes oxygen bubbles formed in the decomposition of persulfate and the mineralized sample gradually fills the loop of the injection valve. By injecting the sample thus treated with V3 in its hydrazine (reagent 2) position, it is finally possible to determine the total nitrogen content of the sample.

2.4.4 Multiparameter determinations using SIA

By virtue of its operational foundation, SIA is the most suitable flow technique for multiparameter analysis. In fact, a switching valve furnished with enough side ports can be used to aspirate the sample and as many reagents as required to sequentially determine several parameters.

Figure 2.15 shows an SIA system for the sequential determination of two parameters. Initially, the sample and the reagents needed to determine the first parameter are aspirated through the appropriate ports; then, the process is repeated with the ingredients involved in the determination of the second parameter. The whole process is computer-controlled, so the system can be programmed to conduct each determination at a different frequency. No other flow technique can match SIA in multiparametric determination capabilities. Its operation mode and the availability of switching valves, with many side

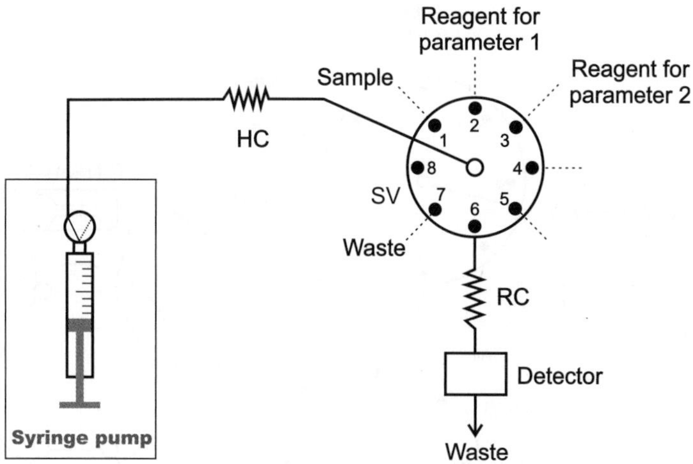

FIGURE 2.15

Sequential determination of two parameters by SIA.

ports, allow SIA systems to be used to determine a large number of analytical parameters in a single sample.

Figure 2.16 shows an SIA system for monitoring wastewater [11] that affords the sequential determination of up to 12 parameters with a view to characterizing the incoming and outgoing flows of a water purifier. Initially, the sample is aspirated via port 1 (valve A) in the upper selection valve and driven to a diode array detector (Pastel UV, Secomam, Alès, France) connected to port 7 of the lower valve (valve B) which, following application of a multivariate analysis method, provides the values of seven major parameters including BOD, detergents, nitrate, and total suspended solids (TSS) without the need for a chemical reagent. Subsequently, previously optimized individual methods are implemented by using appropriate reagents delivered via the selection valves in order to determine other major parameters such as the ammonium content using a gas diffusion cell, nitrite (using the modified Griess reagent), total nitrogen (by photooxidation with persulfate), orthophosphate (with the molybdenum blue reaction), and total phosphorus (by photooxidation with persulfate and formation of molybdenum blue). The whole process takes about 15 min per sample.

2.4.5 Multiparameter determinations using MCFIA

The multicommutated flow technique lends itself more readily than FIA to multiparameter analysis, as it allows streams to be redirected in an automatic manner in order to perform as many sample pretreatments as required. The process of Figure 2.14 was also conducted in a manual MCFIA system that can be readily automated by replacing all manual valves with solenoid valves. This affords time-based injection, with which the amount of sample injected does not depend on the size of the loop but on the time the valve is switched On. In this way, the sample volume can be easily changed as needed without the need to alter the experimental setup. Figure 2.17 shows an MCFIA system for the determination of nitrate and nitrite using a copperized cadmium-reducing column.

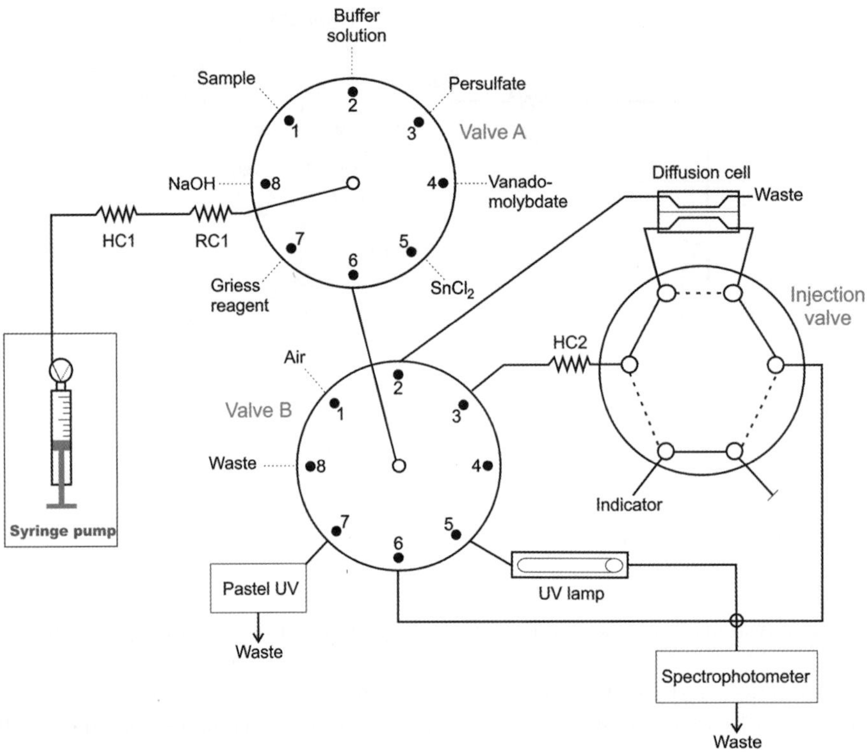

FIGURE 2.16

SIA multiparameter determination system for wastewater monitoring.

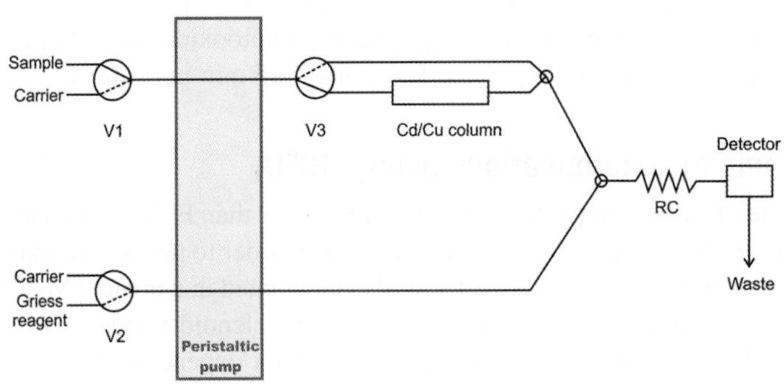

FIGURE 2.17

MCFIA system for the determination of nitrate and nitrite using a copperized cadmium-reducing column.

2.5 Smart systems

The progress of instrumentation along with the development of specific software allows the full automation of chemical analysis. An advance in this field is the development of smart systems [12], which are able by themselves to choose the best strategy to quantify the analyte through the yield of a binary answer, such as yes/no, absent/present, and lower/higher than a preset threshold value. These systems are defined within the field of the artificial intelligence like computer programs into which the user enters expertise specific to the application, which is taken into consideration for implementing eventual modifications in the course of further analytical steps. In the last years, smart systems based on flow analyzers, such as FIA, SIA, MSFIA, or MPFS, have been developed for the determination of environmental parameters [13–17], control of industrial processes [18], and quality control of food [19,20].

Figure 2.18 shows an MPFS system capable of performing the speciation of Fe(II) and Fe(III) with and without preconcentration depending on the results successively obtained [21]. The system uses five micro-solenoid pumps (MPs) and a solenoid valve (V). Micropumps are used to insert the sample, carrier, eluent (HCl), oxidant (hydrogen peroxide), and chromogenic reagent (ammonium thiocyanate).

The process starts with the determination of Fe(III) without preconcentration, propelling a preset amount of sample that is then driven by the carrier to the detector while V is On, to prevent the sample from passing through the chelating filter. Sample plug is mixed with thiocyanate in order to obtain a red complex which is measured by the CCD detector.

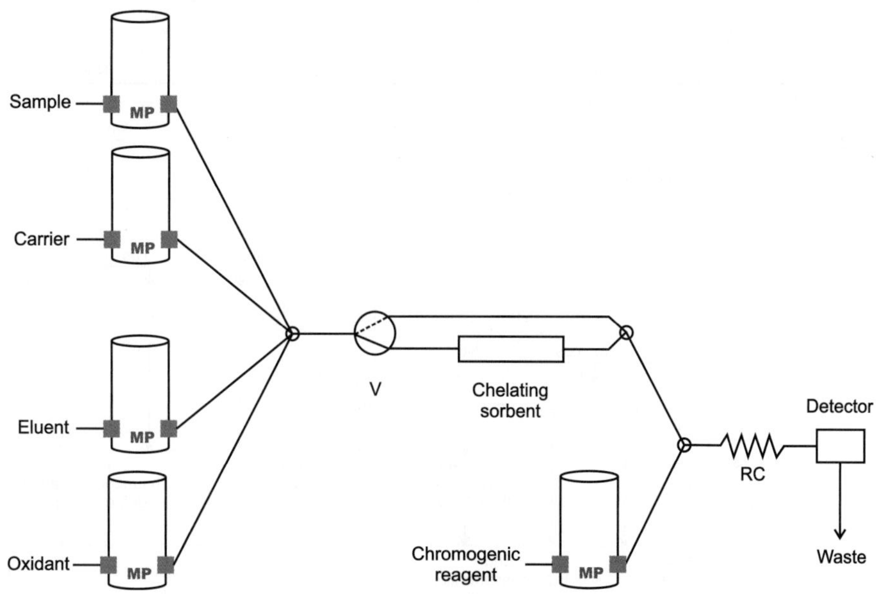

FIGURE 2.18

Smart MPFS systems for the speciation of Fe(II) and Fe(III) with and without preconcentration. MP: solenoid micropump, RC: reaction coil, V: external solenoid valve.

The process is repeated, but inserting hydrogen peroxide in order to convert Fe(II) into Fe(III) and determine total iron.

If the software detects that the resulting peaks are very low, it commands the system to repeat both processes, although with V in Off position in order to preconcentrate Fe(III) in the chelating filter, which contains iminodiacetic groups. The Fe(III) retained by the chelating filter is subsequently eluted with HCl propelled and supplied with thiocyanate.

To illustrate the sequence of procedures and conditionals involved in a smart system, a flow diagram corresponding to a method for uranium and thorium determination is shown in Figure 2.19. The hyphenation of lab-on-valve (LOV) and multisyringe flow injection analysis (MSFIA), coupled to a long path length liquid waveguide capillary cell, allowed the spectrophotometric determination of thorium and uranium in different types of environmental sample matrices achieving high selectivity and sensitivity levels [17]. Thus, the smart LOV−MSFIA system designed is able to make decisions for itself, selecting the adequate sample volume to analyze, allowing the dilution or the

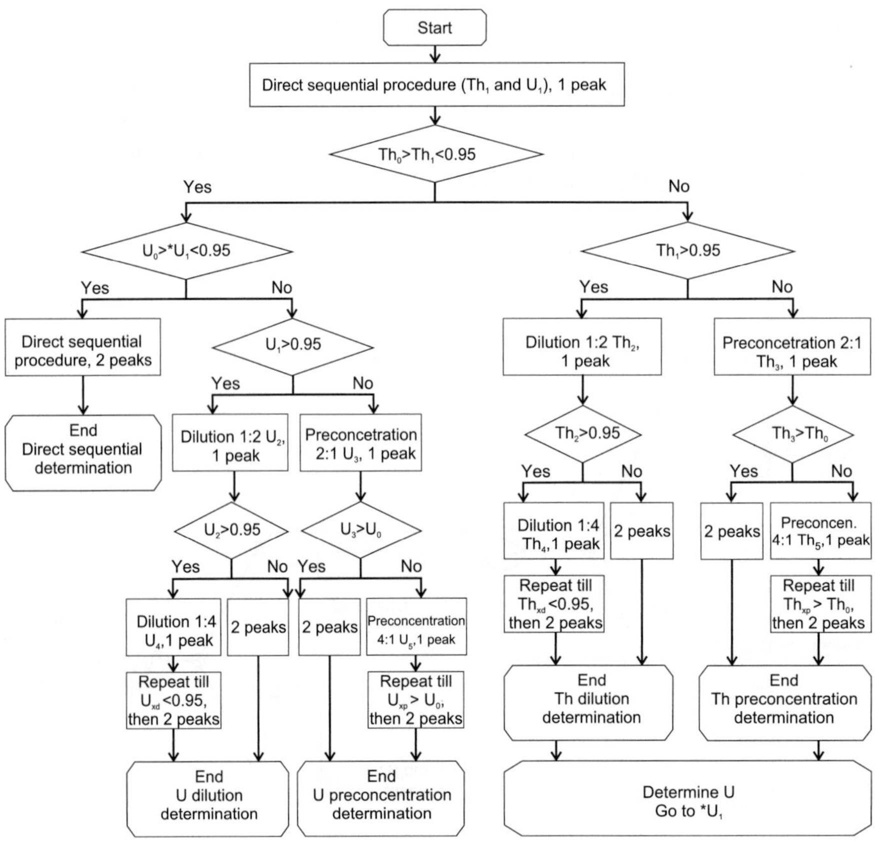

FIGURE 2.19

Software flow diagram for the smart determination of thorium and uranium.

preconcentration protocols. In order to run this method, a series of variables that correspond to the peak height absorbance measurements (Th_x and/or U_x) are introduced as feedback tools. To design a smart system with the software AutoAnalysis, a method based on conditionals is used, whose values can be modified by the users according to analytical requirements, e.g. preconcentrating 4:1 instead of preconcentrating 2:1.

As an example, one peak for each analyte through simultaneous procedure is obtained (Figure 2.19, Th_1 and U_1). If the absorbances of both analytes are between 0.95 absorbance units (AU) and its blank signals, the sample is analyzed through the direct procedures. If the sample has a high thorium concentration, the smart system analyzes it through the dilution procedure (e.g. first dilution 1:2). One peak height absorbance is recorded again (Figure 2.19, Th_2). If the peak height absorbance is not >0.95 AU, more peaks will be obtained with this dilution and the method will be finished for this sample. If the peak height absorbance is >0.95 AU, the sample has a high thorium concentration yet, and another procedure for thorium determination with dilution (e.g. second dilution 1:4) will be started. If the peak height absorbance is <0.95 AU, more peaks will be obtained with this dilution and the method will be finished for this sample. If the peak height absorbance (Figure 2.19, Th_1) is not between its blank signal and 0.95 AU and is not >0.95 AU, then Th_1 < blank signal AU, and the procedure involving preconcentration of analyte will be started. And so on, till the concentration could be quantified. Once the thorium determination is ready, the same methodology is applied for uranium determination.

References

[1] J. Ruzicka, E.H. Hansen, Flow Injection Analysis, second ed., J. Wiley, 1988.

[2] J.M. Estela, F. Mas, A. Cladera, V. Cerdà, Study on the implementation of flow titrations in sequential injection analysis, Lab. Rob. Autom. 11 (1999) 207–216.

[3] F. Albertús, I. Cortés, M. Alvarez, V. Cerdà, Non-linear calibration in single point flow titration of protolytes. A comparison of model options by using delta Test, Anal. Chim. Acta 414 (2000) 221–237.

[4] F. Albertús, B. Horstkotte, A. Cladera, V. Cerdà, A robust multisyringe system for process flow analysis. Part I. On-line dilution and single point titration of protolytes, Analyst 124 (1999) 1373–1381.

[5] G. De Armas, E. Becerra, A. Cladera, J.M. Estela, V. Cerdà, Sequential injection analysis for the determination of fuberidazole and thiabendazole by variable-angle scanning fluorescence spectrometry, Anal. Chim. Acta 427 (2001) 83–92.

[6] G. Turnes, A. Cladera, E. Gómez, J.M. Estela, V. Cerdà, Resolution of highly overlapped differential pulse polarographic (DPP) and anodic stripping voltammetric (ASV) peaks by multicomponent analysis, J. Electroanal. Chem. Interfacial Electrochem. 338 (1992) 49–60.

[7] E. Gómez, J.M. Estela, V. Cerdà, M. Blanco, Simultaneous spectrophotometric determination of metal ions with 4-(pyridyl-2-azo)resorcinol (PAR), Fresenius J. Anal. Chem. 342 (1992) 318–321.

[8] A. Cladera, J. Alpízar, J.M. Estela, V. Cerdà, M. Catasús, E. Lastres, L. García, Resolution of highly overlapping differential pulse anodic stripping voltammetric signals using multicomponent analysis and neural networks, Anal. Chim. Acta 350 (1997) 163–169.

[9] E. Lastres, G. de Armas, M. Catasús, J. Alpízar, L. García, V. Cerdà, Use of neural networks in solving interferences caused by formation of intermetallic compounds in anodic stripping voltammetry, Electroanalysis 9 (1997) 251–254.

[10] A. Cerdà, M.T. Oms, R. Forteza, V. Cerdà, Sequential injection sandwich technique for the simultaneous determination of nitrate and nitrite, Anal. Chim. Acta 371 (1998) 63–71.

[11] O. Thomas, F. Theraulaz, V. Cerdà, D. Constant, Ph. Quevauviller, Wastewater quality monitoring, Trends Anal. Chem. 16 (1997) 419–424.

[12] A. Bonastre, R. Ors, M. Peris, Distributed expert systems as a new tool in analytical chemistry, Trends Anal. Chem. 20 (2001) 263–271.

[13] A. Rius, M.P. Callao, F.X. Rius, Self-configuration of sequential injection analytical systems, Anal. Chim. Acta 316 (1995) 27–37.

[14] C.W.K. Chow, D.E. Davey, D.E. Mulcahy, An intelligent sensor system for the determination of ammonia using flow injection analysis, Lab. Autom. Inf. Manage. 33 (1997) 17–27.

[15] V. Grassi, A.C.B. Dias, E.A.G. Zagatto, Flow systems exploiting in-line prior assays, Talanta 64 (2004) 1114–1118.

[16] L. Ferrer, J.M. Estela, V. Cerdà, A smart multisyringe flow injection system for analysis of sample batches with high variability in sulfide concentration, Anal. Chim. Acta 573–574 (2006) 391–398.

[17] J. Avivar, L. Ferrer, M. Casas, V. Cerdà, Smart thorium and uranium determination exploiting renewable solid-phase extraction applied to environmental samples in a wide concentration range, Anal. Bioanal. Chem. 400 (2011) 3585–3594.

[18] A. Bonastre, R. Ors, M. Peris, Monitoring of a wort fermentation process by means of a distributed expert system, Chemom. Intell. Lab. Syst. 50 (2000) 235–242.

[19] M. Peris, Present and future of expert systems in food analysis, Anal. Chim. Acta 454 (2002) 1–11.

[20] A. Bonastre, R. Ors, M. Peris, Advanced automation of a flow injection analysis system for quality control of olive oil through the use of a distributed expert system, Anal. Chim. Acta 506 (2004) 189–195.

[21] C. Pons, R. Forteza, V. Cerdà, Multi-pumping flow system for the determination, solid-phase extraction and speciation analysis of iron. Anal. Chim. Acta 550 (2005) 33–39.

CHAPTER

3 Online Separation and Preconcentration Methods

CHAPTER OUTLINE

Flow Analysis. http://dx.doi.org/10.1016/B978-0-444-59596-6.00003-6

Method's selectivity and sensitivity play a major role in analytical determinations. The presence of interfering materials and the low concentration of some analytes of interest in the samples may preclude their determination by direct sample injection in conventional detectors. Sample pretreatment accounts for over 60–80% of the total analysis time and normally is the main contributor to analytical uncertainty. Flow-based techniques play a major role in automating, simplifying optimization and miniaturization of solution handling in sample pretreatment. Thus, automation of sample preparation is of great relevance in order to maximize selectivity, sensitivity, reproducibility, sample throughput and to reduce costs, time, and analyst risks due to chemicals exposure. Despite flow analysis equipment is fairly inexpensive it cannot by itself separate analytes in a mixture. Therefore, online coupling of separation techniques to different detectors represents the automation milestone. The advantages of combining flow techniques with separation techniques are noteworthy. Even when using selective detection techniques such as inductively coupled plasma mass spectrometry (ICP-MS), the analytical performance can be enhanced by online preconcentration and sometimes the separation from the matrix is mandatory for the good performance of the instrument.

On the one hand, selectivity can be improved by using a classical procedure in a homogeneous medium (e.g. by adding a masking agent) or, alternatively, a heterogeneous medium such as those employed in liquid–liquid extraction (LLE), solid-phase extraction (SPE), liquid chromatography (LC), gas chromatography (GC), gas diffusion (GD), dialysis, or capillary electrophoresis (CE).

On the other hand, improving sensitivity frequently requires preconcentration with, for example, SPE materials such as a column packed with an ion exchanger, chelating agent or C18 material which additionally allow the simultaneous removal of potential interferents.

This chapter discusses the most salient separation and preconcentration techniques available for online implementation.

3.1 Online liquid-liquid extraction

LLE is a classical and widely used technique for sample matrix separation and preconcentration prior detection, having been applied to various analytical fields. Manual LLE requires large amounts of organic solvents and time-consuming multistage manipulations. Therefore, LLE was among the earliest techniques implemented in flow assemblies in order to overcome these inherent drawbacks, i.e. to reduce organic solvent consumption and to speed up extractions. As shown below, the way it is implemented varies from one flow technique to another. Flow-based LLE has been applied to various areas, such as environmental, pharmaceutical, clinical and food analysis, among others. It has been mostly coupled to optical detectors, due to the influence of the organic phase is minimized in such systems.

3.1.1 LLE in FIA technique (FIA–LLE)

Figure 3.1 depicts a typical assembly for LLE in FIA. First flow based systems for the online LLE were proposed in 1978 exploiting FIA [1,2]. The usual procedure is as follows, first an organic extractant is merged with an aqueous carrier containing the sample and, after a time enough up to complete extraction, the two phases are separated and the organic one is driven to the detection cell for measurement. Classical FIA–LLE manifold is characterized by three main components: phase segmenters for the

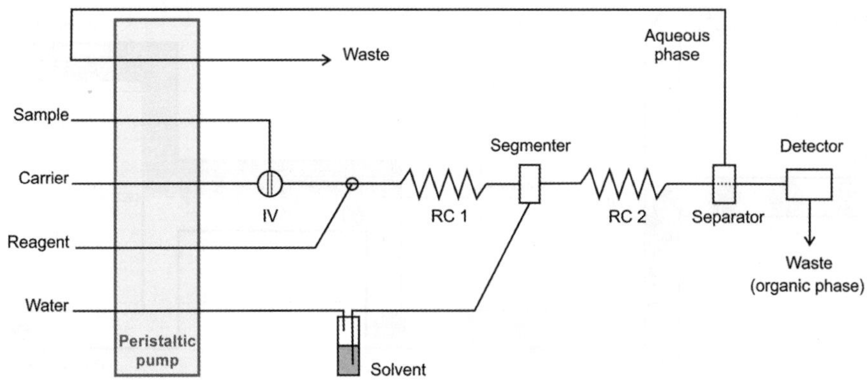

FIGURE 3.1

FIA system for implementation of liquid–liquid extraction. IV: injection valve; RC: reaction coil.

mixing of the organic and aqueous streams, extraction coils where the mass transfer takes place and phase separators to split into individual streams the aqueous and the organic phases (see Figure 3.1).

Segmenters can be divided in two different groups: continuous flow segmenters and mechanical segmenters. As an example in Figure 3.2 a T-shaped segmenter is shown, in which the organic phase is inserted into the carrier stream. The distance between the outlet tube and the solvent penetration point at the T-piece dictates the size of the aqueous and organic segments formed. If the outlet tube is made of glass, the aqueous drops and segments are convex and concave, respectively (see Figure 3.2); on the other hand, if the tube is made of a hydrophobic material such as Teflon, the two shapes are reversed. The key to successful extraction in FIA lies in ensuring reproducibility in the organic segments dispersed in the carrier.

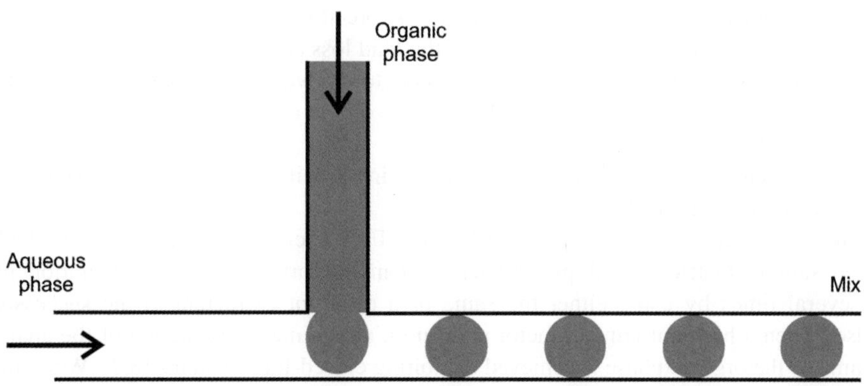

FIGURE 3.2

Segmenter for use in liquid–liquid extraction.

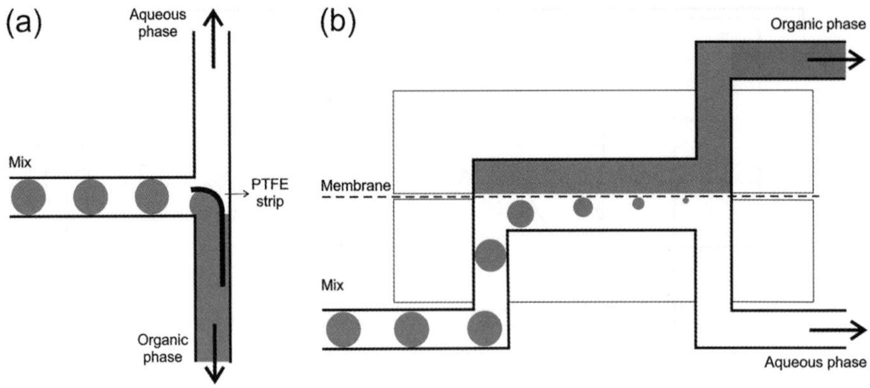

FIGURE 3.3

Typical phase separators. (a) T-shape separator. (b) Membrane-based separator.

The phase separator is probably the most significant component of an LLE flow system. Once extraction has been completed, the aqueous and organic phase can be separated in various ways. One involves passing the mixture through a T-piece (Figure 3.3(a)). If the organic phase is denser than water, then the former leaves the tee through the bottom. This can be facilitated by inserting a piece of Teflon between the side and top branches (see Figure 3.3) to have the nonpolar organic solvent adhered to the Teflon walls while the water leaves the T-piece via the top channel. It is advisable to aspirate the water through a piece of tubing leading to the peristaltic pump. The tube length used should ensure an adequate aspiration flow rate (viz one coinciding with the combined flow rates of the carrier and reagent) so that the excess rate helps drive the organic solvent to the detector.

Another way of separating the organic and aqueous phase is by using a hydrophobic porous membrane sandwiched between two blocks having a carved circulation channel on one side each. Gas diffusion or dialysis cells can be used for this purpose provided any plastic materials potentially attacked by the solvent are avoided (Figure 3.3(b)). The mixture is forced through one channel whose end is connected to a long and thin piece of tubing to obtain the load loss and pressure required to force the pass of part of the liquid through the membrane pores. Due to its hydrophobic nature, the membrane will only allow the organic solvent to pass through and to be driven to the detector, the liquid not crossing the membrane is sent to waste.

There are also alternative FIA–LLE systems, e.g. using gravity phase separators [3] or even neither segmenter nor phase separator [4].

There are other operational modes for LLE using FIA, such as for example when dealing with complicated sample matrices, multiple extractions can be carried out. The separation process is repeated several times by using either the same or a different extractant in the successive stages [5–7]. Also, when a high enrichment factor is desired, a continuous extraction of the analyte into a small volume of the organic phase is achieved exploiting closed-loop systems [8,9]. When interfacing LLE to some detectors, the use of organic solvents is not recommended or even prohibitive e.g. inductively coupled plasma (ICP) or electrothermal atomic absorption spectrometry (ETAAS). In order to overcome this, back-extraction technique can be exploited. Back extraction systems are based on multistage extraction where the analyte is first extracted into an organic solvent and then back-extracted into an aqueous solution, which is then introduced in the detector [10].

These FIA–LLE systems have the disadvantage that the flexible peristaltic pump tubing is vulnerable to solvents and breaks easily lengthwise. This can be avoided by using the displacement technique, which involves passing water through the pump channel corresponding to the solvent and entering a tightly closed vessel containing the solvent, the water displaces an equivalent amount for insertion into the manifold (see Figure 3.1). If the solvent is denser than water, then the latter is introduced at the top of the vessel; otherwise, it is introduced at the bottom.

3.1.2 LLE in sequential injection analysis (SIA) and multisyringe flow injection analysis (MSFIA) techniques

These two flow techniques differ markedly from FIA in the way they implement LLE. Thus, Sequential injection analysis (SIA) and Multisyringe flow injection analysis (MSFIA) are subject to none of the limitations of FIA regarding to the use of organic solvents as they employ piston pumps avoiding peristaltic pump tubes. Moreover, SIA and MSFIA provide even greater advantages than FIA over manual methods since reagents are only injected to the system when required, affording substantial savings in samples, reagents and solvents in extraction processes and minimizing the environmental impact per analysis. Furthermore, both SIA and MSFIA can exploit the fact that the organic solvent adheres to the walls of hydrophobic tubing, forming films of organic solvent on them to carry out the LLE. The solvent, which can be a mixture, should be of an appropriate viscosity so that the film is neither too thick (to avoid interfering with the back-extraction (stripping)) nor too thin (so it will not break easily).

The way SIA and MSFIA operate provides more versatility to the implementation of LLE, together with simpler manifold designs and higher robustness. There are reports in bibliography in which conical separating chambers attached at one port of the valve are employed to carry out the LLE procedure [11]. In addition, the possibility of using stop flow and flow reversal is advantageous since better extraction efficiencies can be achieved. Also the combination of the sandwich technique and flow inversion allows the reduction of organic phase consumption and sensitivity improvements [4,12]. SIA and MSFIA systems are therefore, most suitable platforms for implementing online LLE.

As said before some detectors cannot handle organic solvents what make necessary back extraction procedures when exploiting LLE. There are in bibliography some SIA–LLE back extraction systems for example for cadmium determination by ETAAS [13], copper and lead determination by ICP-MS [14] exploiting dual-conical gravitational phase separators. As an example we will briefly describe an SIA system for the determination of phenols in water [15] using acidity adjustment to effect extraction-back-extraction (Figure 3.4). A preset volume of solvent of appropriate viscosity (i.e. 1 cp) is initially dispensed into the reaction coil, which is used as an extraction coil in the process. As the carrier (water) flows, a film of solvent is formed on the walls of the coil. Then, the system aspirates acetonitrile, an air bubble, sodium hydroxide and an appropriate volume of the sample to be extracted—which must be previously adjusted to the required pH—in this sequence. As these stacked liquids are propelled to the extraction coil, they enter it in the opposite sequence, i.e. last in first out. Because phenols in the sample are undissociated at the pH of the medium, they are extracted by the organic solvent film together with other low-polar substances present in the sample. When the next reagent in the sequence (sodium hydroxide) arrives, the phenols are converted into phenolates and back-extracted for delivery to the detector (a diode array instrument which provides the whole spectrum for the sample at the peak maximum). Any

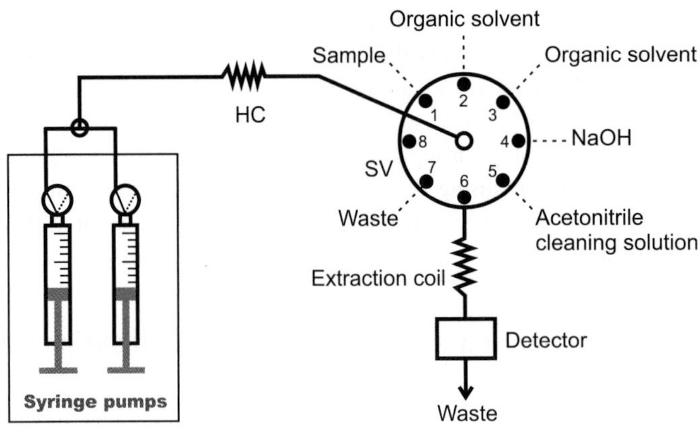

FIGURE 3.4

SIA system for the determination of phenols by liquid–liquid extraction. HC: holding coil. SV: selection valve.

polar compounds undergoing no change when the pH is adjusted are back-extracted to a negligible extent, so they remain in the organic phase. The air bubble inserted between sodium hydroxide and acetonitrile is intended to reduce dispersion in the plug of back-extracted phenols, and the acetonitrile to remove the solvent film formed and restore the initial conditions. In fact, the solvent film is not still during extraction and back-extraction; rather, it moves as the aqueous phase flows, but at a lower rate. This requires using an appropriate length of extraction coil in order to avoid losses of organic phase through the detector before the analytes are determined in suitable forms.

By using standards to calibrate the system and spectra obtained at the maxima of the SIA peaks, one can simultaneously determine several phenols without the need to separate them simply by using a multivariate chemometric technique, e.g. multiple linear regresion (MLR), as explained in a previous section (see Chapter 2).

3.1.3 Dispersive liquid–liquid microextraction and lab in a syringe

The dispersive liquid–liquid microextraction technique (DLLME) [16] has attracted much attention due to its simplicity and to the high enrichment factors that can be achieved. DLLME is a fast micro-extraction technique based on the use of a ternary mixture, composed by an aqueous phase, an organic phase (extractant) and an additional organic solvent also named as disperser solvent, which is miscible in both phases. The disperser is initially mixed with the extractant and then rapidly injected into the sample. By the fast dissolution of the disperser into the aqueous phase, the extractant is disrupted into small droplets enhancing the effective surface area of extraction. The separated extractant droplets are then sedimented at the bottom of the vial or float upon the aqueous sample-disperser phase, depending on the density of the extractant used.

Required extraction times are usually short in DLLME, since the extraction equilibrium is quickly reached due to the large transfer area for the extraction procedure. The efficient recovery of the sedimented or floating extractant and its further injection into analytical instrumentation for the quantification of the target compounds is the most troublesome part of the procedure, mainly

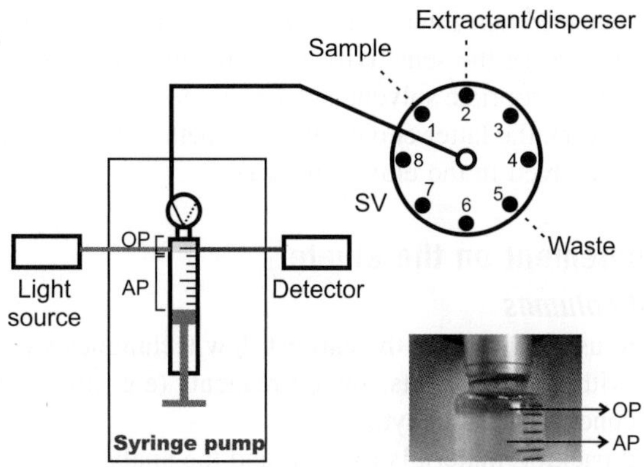

FIGURE 3.5

Lab in a syringe system. AP: aqueous phase; OP: organic phase; SV: selection valve. (For color version of this figure, the reader is referred to the online version of this book.)

when the extractant is floating on top of the sample, being automation one of the challenges of the DLLME technique.

Some advances have been accomplished exploiting SIA and an additional peristaltic pump, in order to mix by confluence the sample and the disperser/extractant mixture. However, additional steps are required such as the need of a solid support for the retention/collection of the dispersed organic microdroplets, and the re-elution of the retained microdroplets from the polytetrafluoroethylene (PTFE) support prior to detection. Other SIA systems have been proposed by using a conical tube and adding an auxiliary solvent to adjust the density of the extraction solvent when this is lighter than water density [17–19]. Other authors use micro columns packed with hydrophobic sorbent materials after extraction as phase separators requiring a second extraction [20–22].

The complete automation of DLLME has been achieved using the MSFIA technique [23], carrying out the DLLME inside the syringe (see Figure 3.5) and it has also been successfully combined with a hyphenated LC procedure in the same syringe pump, which is known as multisyringe chromatography (MSC), described below.

Later DLLME with spectrophotometric detection was integrated and fully automated inside a glass syringe [24], which acted as the container for the sample treatment and as the detection cell, enabling the accomplishment of the whole procedure (microextraction + detection) inside the syringe. This technique has been called lab in a syringe. This system is shown in Figure 3.5.

3.2 Online solid phase extraction

SPE is one of the most important preconcentration/separation procedures to trace heavy-metal ions and organic pollutants, due to its simplicity and limited usage of the organic solvents. Among many others, ion exchange and selective resins are very popular methods due to their applicability to both preconcentration and separation. The implementation of highly selective columns to flow-based methodologies allowed the automation of many analytical methods.

SPE has been used either to retain interferents or the target analyte, which can thus be simultaneously preconcentrated to increase the sensitivity. Once the analyte is retained by the solid support, it can be either eluted with an appropriate solvent and determined in the eluate, or measured directly on the support (with an optosensor), the latter choice provides better detection limits by the avoidance of the dispersion phenomena involved in the elution process.

3.2.1 SPE with measurement on the eluate

3.2.1.1 Use of packed columns

Packed columns have been used ever since the earliest flow techniques were developed for purposes such as treating samples with solid reagents, saving reagents (e.g. immobilized enzymes) or, especially, separating and preconcentrating analytes.

Table 3.1 lists selected packing materials for SPE and the analytes with which they can be used. One of the greatest problems posed by these packing materials is the overpressure caused in the extraction system. The effect, however, can be reduced by placing a frit at the end of the column; the frit pores should be neither too large (to prevent packing particles from passing through or clogging them) nor too small (to avoid excessive overpressure). Preparing and handling efficient packed columns requires some skills and practice. Some commercially available columns come with a frit or are prepacked with a specific phase [25,26].

Respect to particle size, our experience in performing online SPE allows us to recommend the use of beads between 50 and 100 μm or 100−150 μm in order to avoid overpressure and achieve the maximum efficiency of extraction.

3.2.1.2 Use of modified filters

The principal shortcoming of packed columns is that the packing material frequently shrinks and this results in undesirable overpressure. Also, packing columns in a reproducible manner requires some practice. One way of circumventing these shortcomings is by using modified filtering discs [27] containing the extracting phase (e.g. Empore discs). Figure 3.6 shows an example of disk holder, the filter

Table 3.1 Packing Materials for Online Solid-Phase Extraction

Type	Principle	Target Analytes	Eluent
Polymeric	Partitioning	–	80% methanol
C18	Partitioning	Low-polar	80% methanol
Cationic	Ion exchange	Cations	–
Anionic	Ion exchange	Anions	–
Chelating	Chelation	Metals	Acid
Sr resin	Selective resin	Sr, Pb	Nitric acid, ammonium oxalate
TRU resin	Selective resin	Fe, actinides	Nitric acid
UTEVA® resin	Selective resin	U, actinides (IV)	Nitric acid, oxalic acid
TEVA® resin	Selective resin	Tc, actinides (IV), Am, lanthanides	Nitric acid, hydrochloric acid

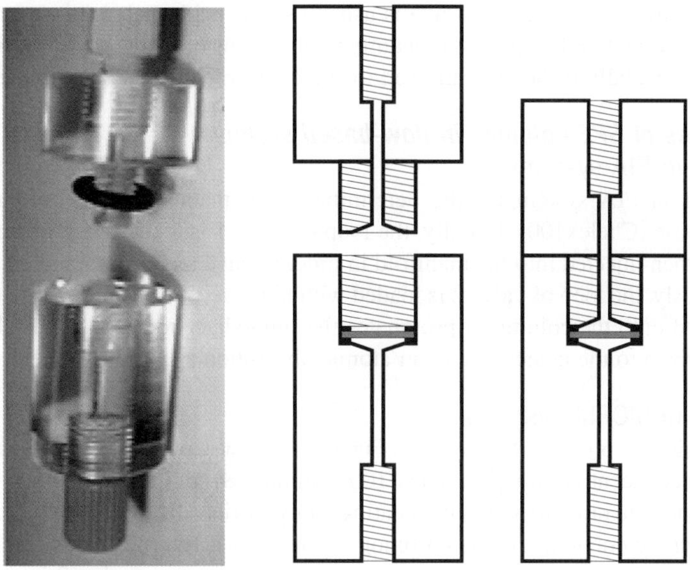

FIGURE 3.6

Disk holder designed by Sciware Systems S.L. Extraction filters are sandwiched between two O-rings. (For color version of this figure, the reader is referred to the online version of this book.)

being located between two O-rings. These filters avoid overpressure, can be reused many times, are easily replaced and perform highly reproducibly.

3M Empore extraction discs come with a wide variety of stationary phases spanning most imaginable applications.

Reverse phase extraction

- C18, or C8 Empore discs
- SDB-XC Empore discs
- Oil and fat loaded Empore discs

Mixed-phase extraction

- SDB-RPS Empore discs

Ion-exchange-based extraction

- SR anion-exchanger discs

Chelation-based extraction

- Chelex 100 chelating discs

Discs to be used to determine metals in trace amounts (ppb) must be first conditioned by wetting in water; washing with nitric acid 3 mol/l or HCl 3 mol/l twice and then with water again allowing the disc to dry after each operation. Then it is recommended to wash them with 0.1 mol/l ammonium acetate buffer at pH 5.3 and then several times with distilled water in order to convert them to its most active form (ammonium form). Once the disk is ready, the sample is passed through and the analytes of interest are retained. Care should be exercised to remove as much residual water from the discs as possible before elution.

Eluting metals from a chelating disc entails using a strong acid, such as nitric acid 3 mol/l or HCl 3 mol/l. Most metals can thus be recovered quantitatively. A few of them (e.g. chromium) however, are difficult to elute, especially if they are allowed to stand adsorbed on the membrane for hours.

3.2.1.3 Examples of SPE columns in flow-based systems

3.2.1.3.1 SPE in FIA systems

Figure 3.7 shows an FIA system for the determination of metals using a column packed with a chelating solid phase (Chelex100). Initially, the loop of valve 1 is loaded with an appropriate volume of sample that is then injected into the manifold in order to have metals retained on the chelating column. Simultaneously, the loop of valve 2 is loaded with nitric acid of an appropriate concentration. As the acid is injected onto the column, it protonates the iminodiacetic groups and releases the metals, which are then driven to the detector (e.g. an atomic absorption spectrophotometer).

3.2.1.3.2 SPE in MCFIA systems

For comparison with the previous FIA system, Figure 3.8 shows an MCFIA system also used to determine Fe(III) by preconcentration on a Chelex 100 column [28]. In the standby position, the solenoid valves are in Off mode in order to avoid overheating. Only carrier (distilled water) is thus aspirated, but not passed through the column, and the chromogenic reagent is returned to its reservoir.

On the one hand, in the determination without preconcentration, valve 2 is switched On while valve 3 is switched Off to inject an appropriate volume of sample (time-based injection) without passing through the SPE column, and at the same time valve 4 is actuated to inject thiocyanate. The red complex thus formed is measured with the spectrophotometer. On the other hand, in order to increase sensitivity the determination can be carried out with preconcentration. Valve 2 and valve 3 are switched On to inject an appropriate volume of sample (time-based injection) passing through the column. Then valve 1 and valve 2 are switched Off while valve 3 is kept in On position in order to

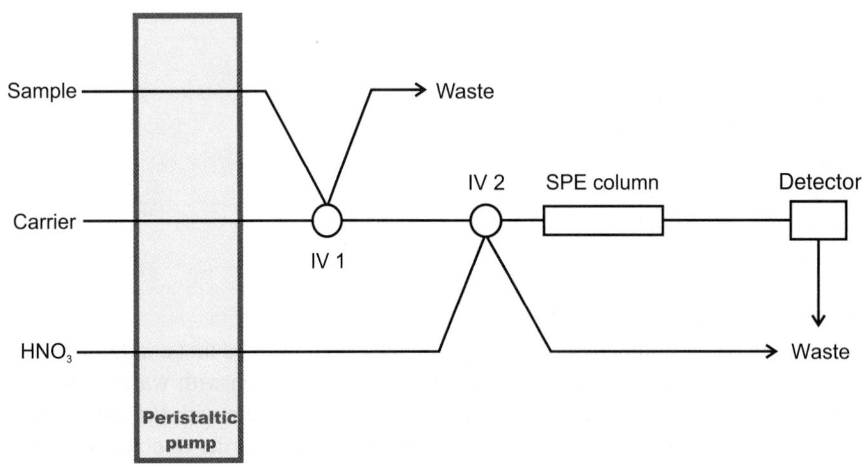

FIGURE 3.7

FIA system for the extraction of metals by use of a column packed with Chelex 100 resin. IV: injection valve.

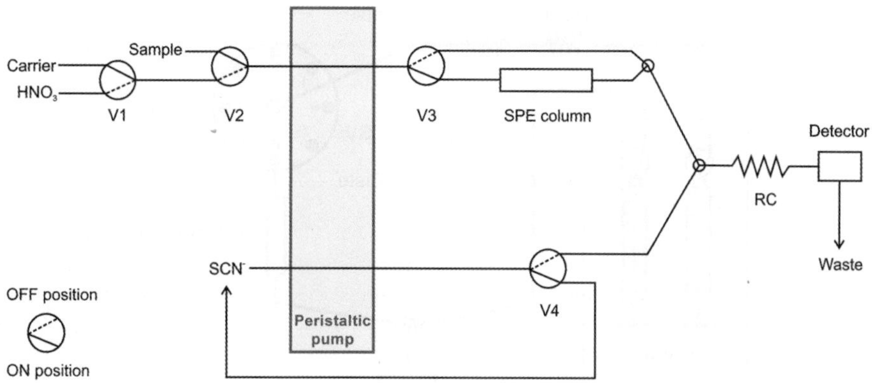

FIGURE 3.8

MCFIA system for the determination of Fe(III) with and without preconcentration. RC: reaction coil; V: solenoid valve.

elute the Fe(III) with nitric acid of an appropriate concentration. As the acid is injected onto the column, it protonates the iminodiacetic groups and releases the metals. At the same time valve 4 is actuated to inject thiocyanate. The red complex thus formed is measured in the detector. These two operational modes allow the determination of metals in a wide range since they can be directly measured in a discrete volume of sample or preconcentrated up to two or three orders of magnitude respect to their original concentration in the sample. Also the fact of using SPE allows the cleanup of the sample prior detection.

3.2.1.3.3 SPE in SIA, MSFIA and LOV systems

Figure 3.9 shows an SIA system for the determination of iron traces in water [29]. The procedure involves aspirating a large volume of sample that is passed through a Chelex column, after which 2 mol/l nitric acid is aspirated in order to protonate the iminodiacetic groups in the packing material and release previously retained Fe(III). At that point, the other burette is actuated in order to inject an appropriate volume of thiocyanate and form a red complex on merging with Fe(III) at a T-piece, the absorbance of the complex being measured at the detector.

Alternatively, Fe(III) in the sample can be directly sent to the nebulizer of an atomic absorption spectrophotometer for measurement [30]. In this case, the second burette is used to keep the nebulizer flow-rate constant and only comes into play when the other burette is aspirating.

Although this assembly was depicted for SIA, it constitutes the first example of an MSFIA system as it uses two syringes simultaneously. The difference is that, while in this SIA system the two syringes are actuated by two independent burettes (i.e. two motors), a single burette actuates both syringes in MSFIA. MSFIA is an ideal multichannel technique for challenging analytical procedures, which require high and precise flow rates, and high-pressure stability such as those with sorbent columns implementation [31], enabling at the same time, the handling of aggressive and volatile solutions.

In a true SIA system (i.e. using a single syringe), preconcentrating a large volume of sample to be previously reacted with a reagent to form the target species can be difficult. One effective solution is

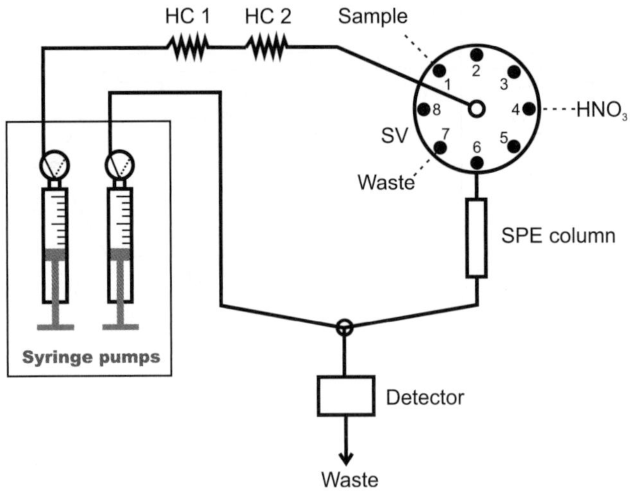

FIGURE 3.9

SIA system for the determination of Fe(III) with preconcentration on a Chelex 100 column. HC: holding coil; SV: selection valve.

to alternately insert small portions of sample and reagent. For example, the procedure for the determination of nitrites using the modified Griess method [32] consists in driving small slices of sample and reagent, which had been previously alternatively aspirated into the holding coil, to a C18 column where the formed complex is retained. Then, 80% methanol is aspirated to flush the complex to the detector in order to measure its absorbance. This method can be more efficient in MCFIA as switching the selection valve to another position takes a relatively long time relative to a multicommutated solenoid valve.

Lab on valve (LOV) is the best option to choose when exploiting SPE since its flexible fluid manipulation is also suitable for delivering beads in flow-based manifolds, being able to automate the column replacement. That is to say, the automated transport of solid materials in such a system allows their renewal at will and thus provides measurement, packing and perfusion of beads, samples and reagents with a high degree of repeatability. To contain the sorbent within the cavity of the LOV module and prevent them from escaping, the outlet of the column can be furnished with a glass fiber prefilter (Millipore) retaining the beads while allowing the solution to flow freely (Figure 3.10) [33].

3.2.2 SPE with direct on-column measurement

Instead of eluting the product retained on a preconcentration column or disk and measure it in the eluate, it can be measured directly on the column or disk in order to obtain better sensitivity and detection limits through decreased dispersion of the target species. This on-column measurement procedure is known as optosensing.

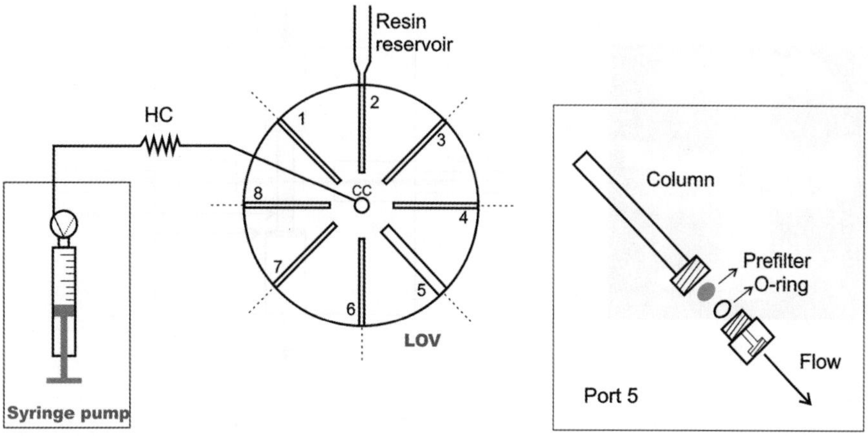

FIGURE 3.10

LOV system and LOV port which serves as column equipped with an O-ring (toric joint) and a prefilter to retain the beads. CC: central conduit; HC: holding coil.

3.2.2.1 Optosensing of packed columns

On-column optosensing measurements can be performed in two ways, namely: by transmittance and by reflectance, i.e. by measuring the light passing through or reflected, respectively, by the solid phase. The former mode is more commonly used than the latter even though it is subject to gross light losses through dispersion across the solid phase. This can be reduced by using very narrow light paths.

Figure 3.11 shows an FIA system for the determination of nitrites by use of a transmittance optosensor developed by Sciware Systems (Figure 3.12) [34]. The sample is merged with modified Griess

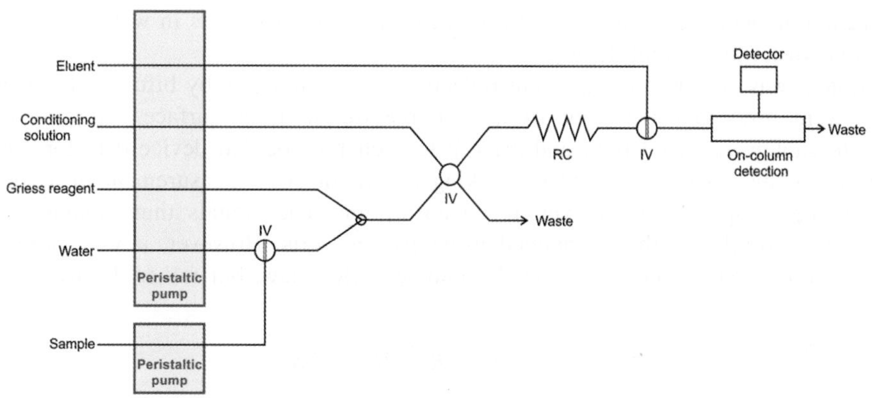

FIGURE 3.11

FIA system for nitrite determination exploiting optosensing detection. IV: injection valve; RC: reaction coil.

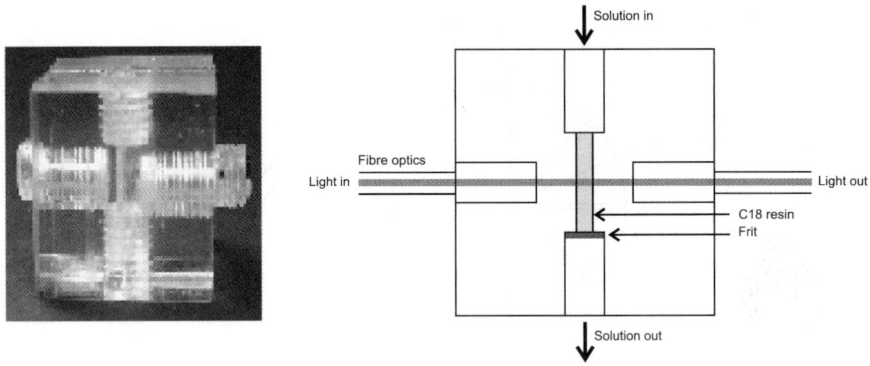

FIGURE 3.12

Optosensor used in the determination of nitrite (see Figure 3.11). (For color version of this figure, the reader is referred to the online version of this book.)

reagent to form a complex that is retained by C18 material packed in a special flow-cell having a prismatic channel 2 mm wide and 1 mm thick (see Figure 3.12).

As the complex reaches the column, it is retained at the top of the packing and its absorbance monitored with a fiber-optic spectrophotometer. The output is a curve where the peak height is proportional to the nitrite concentration in the sample. This procedure affords detection limits in the sub-ppm range.

3.2.2.2 Optosensing of filters

As noted earlier, properly column packing in order to avoid overpressure or clogging requires some practice. This problem can be overcome by using filtering membranes containing appropriate extractants; in addition, these ensure a high reproducibility when they need to be changed after they wear out. Figure 3.13 shows a flow-cell specially designed by Sciware Systems for this type of membrane which has been used in the determination of nitrites in water [35] and sulfide in water and wastewater samples [36].

This system was designed to carry out reflectance measurements by bifurcated optical fibers: six fibers were used to drive light from a source to the filter surface and a seventh was employed to capture reflected light and drive it to a charge coupled device detector. The detector allowed readings to be corrected by making a simultaneous measurement in a spectral region where the complex formed exhibits no absorption. The signals thus obtained and their processing were similar to those obtained using packed cells. However, it was more effective to calculate concentrations not by using the Lambert–Beer law, but rather the Kubelka–Munk formula:

$$F(R) = (1 - R)^2/2R = \varepsilon CS$$

Where R is the ratio of the light intensity reflected in the presence of analyte (i.e. the sample signal minus the blank signal) to that in its absence, ε the molar absorptivity and S a dispersion-dependent parameter.

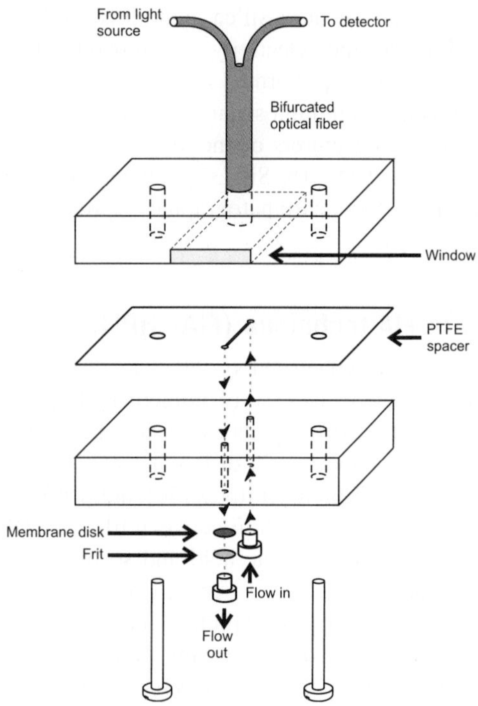

FIGURE 3.13

Flow-cell for optosensing detection incorporating filtering membranes for extraction purposes. (For color version of this figure, the reader is referred to the online version of this book.)

3.3 Online liquid chromatography

The high selectivity of liquid chromatographic separations can be easily incorporated to flow-based systems. Both, high-performance liquid chromatography (HPLC) and monolithic columns raise flow-based systems selectivity to required levels. Moreover, other separation techniques may further be incorporated to these systems for sample cleanup before introducing the sample into the HPLC column or monolithic column. Therefore, automation of sample processing for further or previous connection with HPLC/monolithic column is of utmost interest in order to reduce analysis cost and enhance determination rate.

The advent of monolithic columns has opened new possibilities in flow analysis. Monolithic columns consist of a single, rigid or semirigid, porous piece (monolith). Like other continuous media, monolithic columns fast analysis by bypassing the limitations imposed by pressure via through-pores, which allow higher flow rates than particulate columns at reasonable column backpressures due to the presence of mesopores. Analyte capacity is usually provided within the monolithic structure by smaller micropores. Two types of monolithic columns have been developed for chromatography: organic polymers based on polymethacrylates, polystyrenes or polyacrylamides, and inorganic polymers based on silicates. Silica monoliths can be operated at higher flow rates regardless of back pressure. Moreover the selectivity is comparable to microparticulate columns, allowing chromatographic analyses to be performed in a shorter time. Because of their better mass transfer properties, these columns maintain high separation efficiency, even at high linear flow rates.

Two main manufacturers of monolithic columns are commercially available, i.e. Phenomenex® Onyx™ (C18, C8, HD-C18, Si) [37] and Merck® Chromolith™ [38]. Some examples of the implementation of different chromatographic approaches combined with flow techniques are described below.

3.3.1 LC in FIA technique (FIA–HPLC)

Many FIA–HPLC hyphenated systems have been developed since FIA is the most widely accepted flow technique, due to its ease of implementation and that in the beginnings of automation, the no need of a computer expedited its expansion. FIA–HPLC systems have been proposed for determination of different compounds in different matrices, such as for example chloramphenicol in environmental samples [39], food additives [40] and parabens [41].

To illustrate the hyphenation between FIA and HPLC, a system for food additives (acesulfame-k, saccharin, caffeine, benzoic acid and sorbic acid) determination [40] is shown in Figure 3.14. The system includes a dialysis cell in order to eliminate the sample matrix. Thus, a peristaltic pump dispenses the sample, which is loaded into a holding coil of an injection valve (IV1) and later injected into the donor stream. Both, donor and acceptor streams are also propelled by the peristaltic pump. Once the dialysis has been performed, analytes are loaded into a holding coil of a second injection valve (IV2). At this point, the sample passes to the HPLC system. The sample is injected into a mobile phase stream dispensed by the HPLC pump. So, the pretreated sample passes through the precolumn and column, and the analytes are separated and detected. Thus, the treatment, separation and detection steps are carried out in an efficient and high automated way.

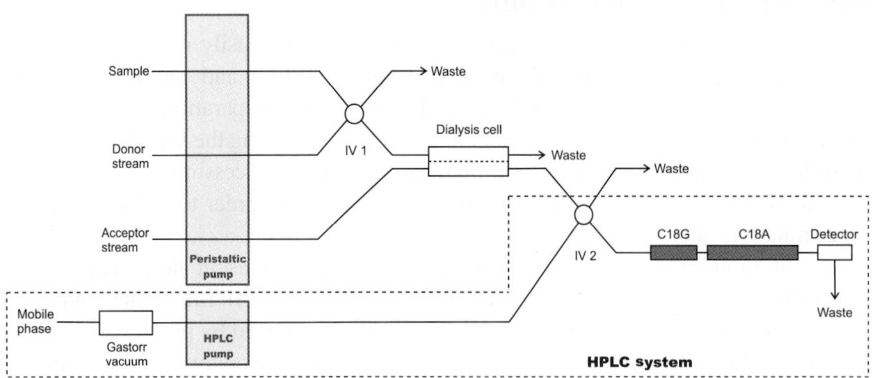

FIGURE 3.14

FIA–HPLC system using a dialysis cell. IV: injection valve.

3.3.2 LC in SIA and LOV techniques (SIA–HPLC, LOV–HPLC)

Both techniques, SIA and LOV, in a sequential injection approach, offer the means of performing automated handling of sample pretreatment, e.g. sample preconcentration and matrix removal, exploiting SPE achieving high enrichment factors (between 20 and 125).

As an example of an LOV–HPLC method, a system for determining carbamate insecticides (isoprocarb, methomyl, carbaryl, carbofuran, methiocarb, promecarb, and propoxur) in food and environmental samples [42] is shown in Figure 3.15. Here, the LOV is responsible for handling the beads, i.e. filling, conditioning and renewal of the column when required, for sample loading and elution of the analytes of interest. Then, the eluate is loaded into a holding coil of the injection valve (IV), and injected into a mobile phase stream. At this point, the eluate is conducted by the HPLC pump through the precolumn and column of the HPLC where the separation is performed and the analytes are detected. So, when coupling LOV to HPLC, the pretreatment, separation and detection steps can be fully automated.

To cite a few examples, HPLC was coupled with SIA for the simultaneous determination of several heavy metals by means of nitro-PAPS (polyfluoroalkyl phosphate esters) complexes [43]. An SIA–HPLC–atomic fluorescence spectrometry (AFS) system was proposed for As speciation in seafood extracts, implementing standard addition method for simultaneous quantification of four As species [44]. An SIA–HPLC with electrochemical detection was proposed using a homemade microcolumn SPE coupled to SIA in order to automate the sample cleanup, extraction and detection of sulfonamides [45].

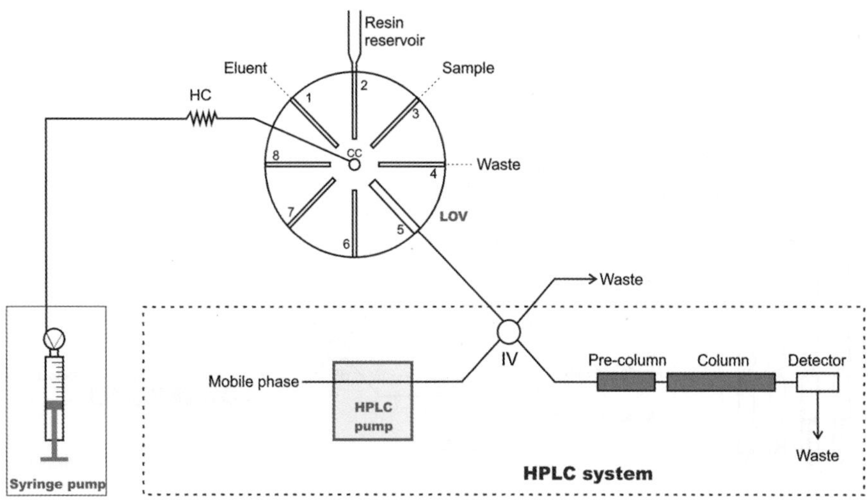

FIGURE 3.15

LOV–HPLC system for determining carbamate insecticides. CC: central conduit; HC: holding coil; IV: injection valve.

3.3.3 LC in MSFIA technique (MSFIA–HPLC)

The MSFIA technique allows working in multicommutated schemes, e.g. all the steps for the matrix removal with SPE, and the conduction of the eluate to the injection valve of the HPLC system can be carried out by the multisyringe burette.

Some MSFIA–HPLC systems have been developed, e.g. an MSFIA–HPLC system exploiting SPE was proposed for screening of phenolic pollutants in waters at ng/ml levels [46]. Nevertheless, a step forward in automation is achieved when coupling MSFIA with LOV, since this allows to add to the benefits of MSFIA, the fully automation of the SPE. Figure 3.16 shows an MSFIA–LOV–HPLC system for pharmaceutical residues determination (ketoprofen, naproxen, bezafibrate, diclofenac, and ibuprofen) [47]. In this system, one of the syringes is used to propel the liquid in the LOV, another for sample loading, and the third syringe is used to conduct the eluate without passing excess of liquid through the SPE column. Like it has been described above, once the eluate is in the holding coil of the injection valve, it is injected in a mobile phase stream. In this case, there are two mobile phases, MeOH/water 20/80 v/v and 95/5 v/v, that are selected to pass through the precolumn and column of the HPLC.

However, since HPLC equipment is expensive, a search for more affordable alternatives was fostered, such as sequential injection chromatography (SIC) and multisyringe chromatography (MSC), both performed with a monolithic column, which have proved to be effective alternatives to HPLC.

3.3.4 Sequential injection chromatography (SIC) and multisyringe chromatography (MSC)

These techniques are result of the online hyphenation of SIA and MSFIA to monolithic columns, respectively. Both have proved to be excellent tools which exploit the capability of monolithic

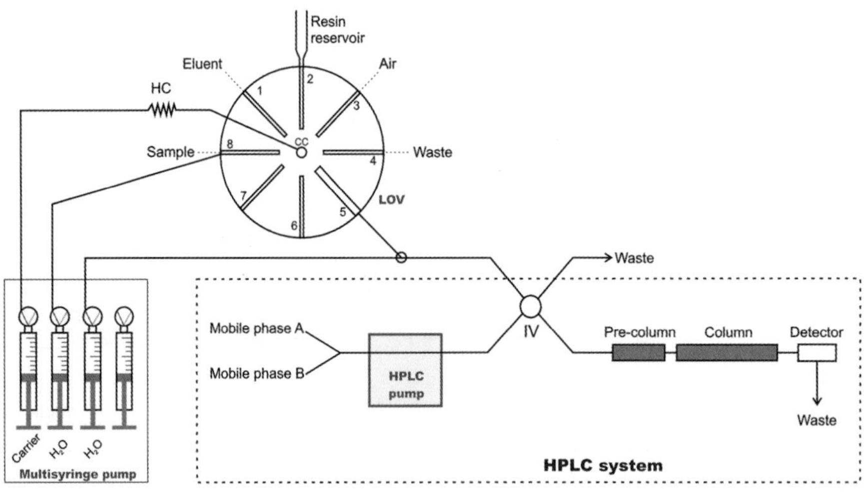

FIGURE 3.16

MSFIA–LOV–HPLC system. CC: central conduit; HC: holding coil; IV: injection valve.

columns of relatively large pore sizes to effect separations without the need for high-pressure pumps [48,49]. The research in the field of LC columns has tremendously accelerated during last years. An important direction of this research is the development of monolithic columns with high porosity sorbent permitting high flow rates of mobile phase at low back pressures without losing efficiency. SIC and MSC offer some advantages over HPLC such as the possibility of two ways and stopped eluent-flow direction, the low eluent consumption (due to the batch eluent-delivery mode), the short time of analysis, their low cost and the possibility of the analyzer portability due to their dimensions allowing on field measurements.

In an SIC system, the monolithic column is coupled on a peripheral port of the selection valve, which may support the pressure when liquid pass through the column. However if there is overpressure, the valve placed on the head of the syringe can be replaced by a two-way connector and an additional solenoid valve high-pressure resistance. Many SIC systems have been mainly applied to pharmaceutical analysis, e.g. salicylic acid and its ester methylsalicylate determination [50] (Figure 3.17), ambroxol hydrochloride, methylparaben and benzoic acid [51] in pharmaceutical preparations. In another application, SIC was used to study the binding between drugs and proteins, in particular between the antibiotic ciprofloxacin as model drug compound and bovine serum albumin, which was strongly retained on the monolith strong anion-exchanger [52]. SIC has also been applied to pesticides determination [53], using a miniaturized 10 mm monolithic column and spectrophotometric detection.

In the case of MSC, when implementing monolithic columns on multisyringe burette it has to be taken into account that standards solenoid valves cannot withstand pressures higher than 2 bars, and a slight modification of the system is usually carried out. The standard solenoid

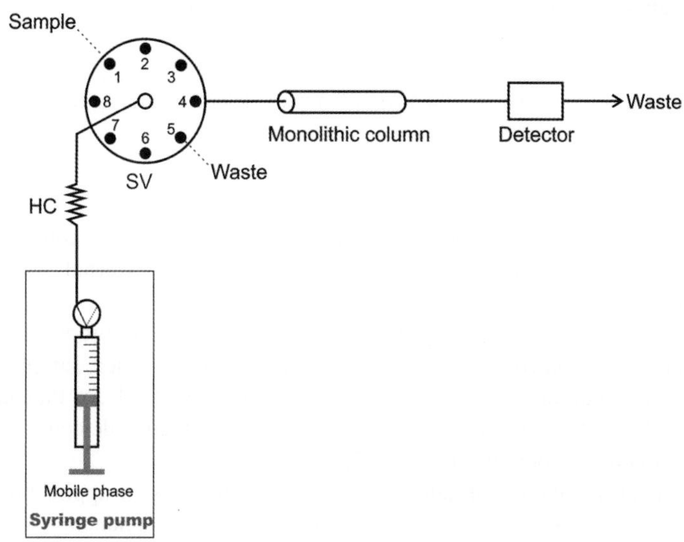

FIGURE 3.17

SIC system for salicylic acid determination. HC: holding coil; SV: selection valve.

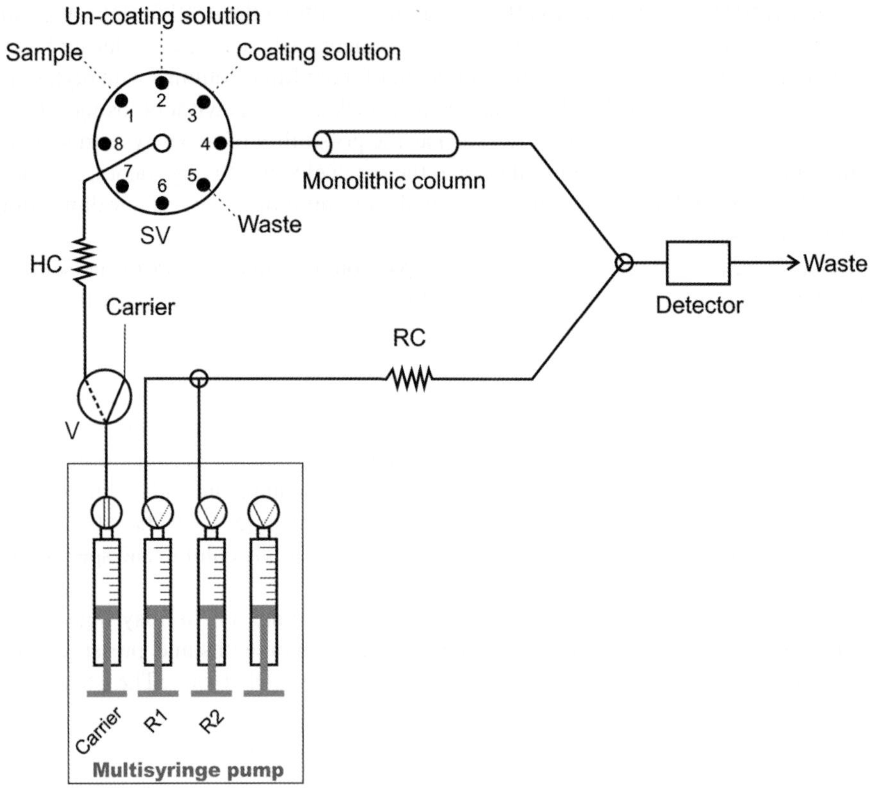

FIGURE 3.18

MSC system for oxalate determination with chemiluminescence detection. HC: holding coil; R: reagent; RC: reaction coil; SV: selection valve; V: solenoid valve.

valve equipped at the head of the syringe is replaced by a two-way connector (Figure 3.18) made of polyoxymethylene which connects the syringe with an external solenoid valve which withstands higher pressure.

MSC has been widely used in the last years, e.g. MSC systems were proposed for the online SPE and determination of hydrochlorothiazide and losartan potassium in water samples [54], for oxalate determination with chemiluminescence detection in beer and urine [55], for thiazide diuretics determination exploiting SPE [56] and also coupled to spectrophotometric detectors for sulfonated azo dyes determination in environmental samples [57].

The combination of sample pretreatments in flow systems expands the applicability of low pressure LC due to the isolation/preconcentration of the target compounds, enhancing selectivity and sensitivity. For example an MSC system was applied to the simultaneous analysis of three herbicides (dicamba, 2,4-D and atrazine) by online SPE coupled to MSC using UV detection [58]. A C18 membrane extraction disk and a C18 monolithic column were used for preconcentration and chromatographic separation, respectively.

One additional advantage of MSC is the ability to perform multi-isocratic chromatographic development by the use of different mobile phases avoiding the need for gradients. However, in SIC and MSC the choice of flow rates for the mobile phase is limited by the highest pressures that the valves can withstand and in addition, HPLC provides higher resolution of the peaks and higher robustness being able to analyze samples of higher complexity [59]. Furthermore, SIC and MSC would require a second piston pump, autoburette or syringe to carry out chromatographic separations in the gradient mode.

3.4 Online gas chromatography

Gas chromatography (GC) is an outstanding tool for the analysis of volatile, semivolatile and nonpolar compounds. Automation of sample treatment prior GC analysis is the bottleneck in GC analysis. It is of utmost interest to avoid laborious and time-consuming operations as well as sample contamination.

SPE is one of the most common pretreatment techniques used previous GC. When coupling flow based systems online with GC some technical details have to be taken into account, e.g. the volume of the extract must be kept small, typically it should be less than 1−2 ml or the extract must be concentrated prior injection into the gas chromatograph. The SPE−GC interfaces are similar to those used in large volume injection and online LC−GC, including namely on-column, loop-type interfaces and programmable temperature vaporizer (PTV). Also it is mandatory to avoid water introduction into the GC. Therefore, after retention and before elution of the analytes of interest, the SPE is usually dried with a gas flow, or water is removed from the extract bypassing it through a drying column packed with copper sulfate or silica. On the one hand, removal of water with an online drying cartridge is faster, but the adsorption of polar analytes onto the packing of the column may be problematic. On the other hand, drying using gas flow is time-consuming, especially when dealing with C18-bonded silicas. Thus polymeric SPE columns are the better option [60]. The eluent has to be suitable with GC analysis, i.e. it has to be sufficiently volatile and nonpolar, e.g. ethyl acetate or *n*-propanol [61−63]. Furthermore, it has been reported that the back pressure increases with the progressive tighter packing of the SPE beads when reusing the material which also causes deterioration of peak shape of target analytes [64]. An example of this technique exploiting SPE and coupled with a flow technique is the MSFIA−LOV−GC system applied to polychlorinated biphenyls (PCBs) determination in solid waste leachates [65] (Figure 3.19), which avoids backpressure increase by providing a renewable sorbent column in each analysis cycle. This LOV−GC hyphenation permits the automation of the sample pretreatment prior GC overcoming the difficulties inherent to GC online coupling with pretreatment techniques in an inexpensive way.

There are also reports exploiting FIA to automate the sample pretreatment previous to GC. For example, in bibliography there are some online LLE−GC FIA systems for the determination of chlorinated pesticides in groundwater [66], halocarbons in seawater [67] or organic trace compounds in water [68].

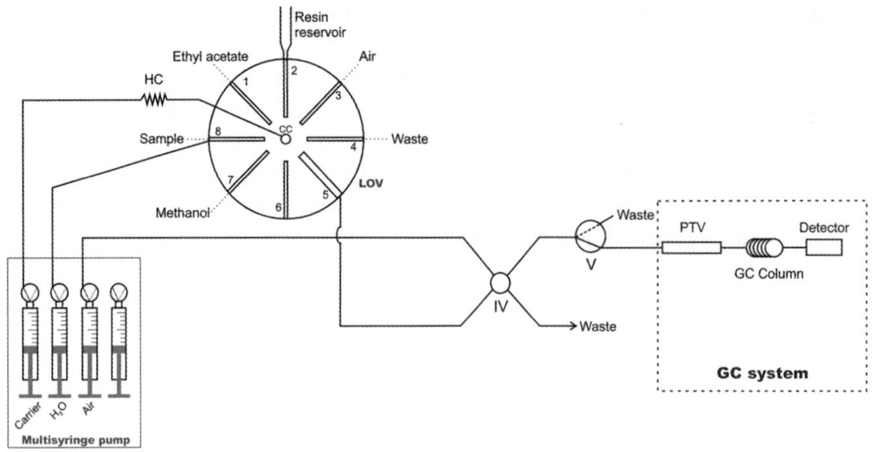

FIGURE 3.19

MSFIA–LOV–GC system for PCBs determination. CC: central conduit; HC: holding coil; IV: injection valve; PTV: Programmable Temperature Vaporizer; V: external solenoid valve.

3.5 Online gas diffusion

Gaseous diffusion is a separation technique which improves efficiency by carrying it out in an online approach. So that, it may use two streams, one donor and one acceptor, separated by a gas permeable membrane. The key of this technique is that only gaseous substances can pass through the membrane from the stream of highest concentration (donor) to the stream of lowest concentration (acceptor). Thus, gaseous diffusion technique for the separation and preconcentration of volatile analytes can be easily incorporated in a flow-based system.

Some membranes allow the selective passage of specific components based on molecular weight (e.g. dialysis membranes, cellulose acetate) or hydrophobicity (e.g. PTFE or PVDF membranes for gas diffusion). Hydrophobic membranes not only repel water, but also solvated ions, and allow passage of dissolved gases such as CO_2, SO_2, H_2S, NH_3 or volatile amines.

In order to maximize the efficiency and durability the selection of the membrane is very important. The permeability of membranes to some gaseous components is very similar (e.g. 15% for Durapore® hydrophobic membranes and 13% for Celgar 2500 membranes and Teflon® plumbing tape). The useful lifetime of Teflon® tape as a membrane material can be as short as a day or even less under stopped-flow conditions. By contrast, Durapore® membranes (PVDF, 0.22 μm pore size) and Celgar 2500 (polypropylene, 0.04 mm pore size), from Celanese Corp., can last over a month. This is because the hydrodynamic pressure generated by the flowing stream under stopped-flow conditions is not offset on the other side of the membrane, where the acceptor stream is static; this frequently disrupts the membrane and eventually causes its breaking.

The composition of the carriers of the donor and acceptor solutions usually depends on whether the analyte is acid or basic. Thus, an acid species such as ammonium ion or a volatile amine requires an NaOH carrier to capture its proton and convert it into its conjugate base; on the other hand, basic compounds (sulfides, carbonates, sulfites) are injected into an acid carrier (e.g. HCl) for conversion into their volatile conjugate acids (H_2S, H_2CO_3, H_2SO_3), which pass through the diffusion membrane either as such or as their gaseous decomposition products (CO_2, SO_2).

The acceptor solution can be pure water, but is more often a solution of opposite character to the other carrier. Thus, if the donor solution is acidic, the acceptor solution is alkaline and vice versa. An appropriate chromogenic reagent (e.g. an indicator) can also be used as carrier to obtain an easily detected complex.

The greatest advantage of gas diffusion cells is their high selectivity due to the small number of gaseous species that can be formed, which allows their use with scarcely selective, but straightforward, inexpensive detectors.

3.5.1 Gas diffusion cells

A gas diffusion cell consists of two symmetric blocks (Figure 3.20), commonly constructed with methacrylate or PVDF, each having a carved channel; the two channels are the mirror image of each other and fit when the blocks are joined together. The channels can vary widely in shape (e.g. linear, spiral, circular or winding), length and depth. The simplest model is the linear channel, which allows two identical blocks to be constructed and the length of which can be adapted to the particular sensitivity level required. The membrane, consisting of semipermeable material, is placed in between the two channels.

The sample is injected into an appropriate carrier and passed through one of the channels to be part of the donor solution. A relative low proportion (15%) of gaseous components passes through the membrane in a reproducible manner and is dissolved in the acceptor solution, which is held in the channel of the other cell block.

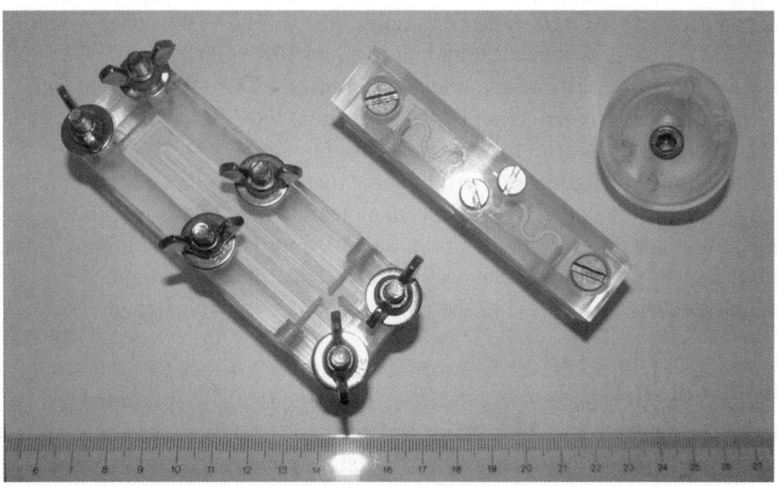

FIGURE 3.20

Rectangular U-shape and sinusoidal-shape and cylindrical gas diffusion cells.

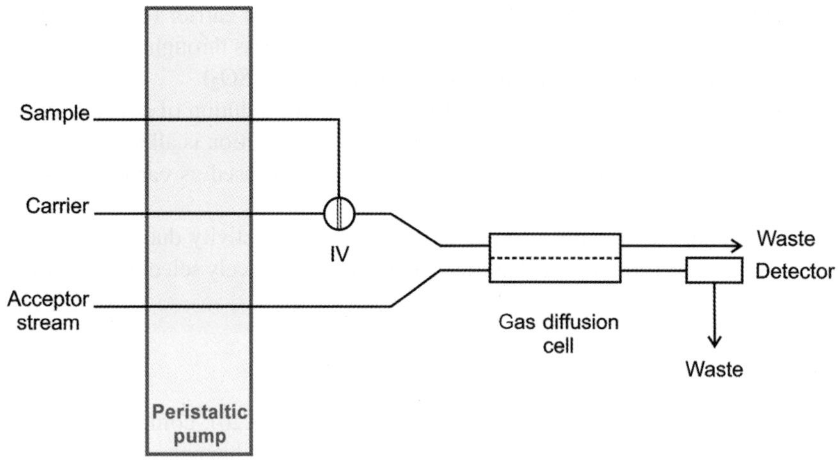

FIGURE 3.21

FIA system equipped with a gas diffusion cell. IV: injection valve.

3.5.2 Gas diffusion in FIA

FIA systems exploiting gas diffusion are highly flexible (Figure 3.21). In FIA systems including a gas diffusion cell, the donor and acceptor solutions are circulated simultaneously on opposite sides of the membrane, so that their respective pressures on the membrane cancel each other. This operation mode prevents the membrane deformation and its consequent decrease of efficiency and durability.

Coupling of FIA with GD is the most frequent. For example for ammonium determination several FIA–GD systems have been developed. Different detection methods coupled with GD–FIA system have been compared elsewhere [69]. Other examples are: an FIA–GD system exploiting conductometric detection for Kjeldahl-produced ammonia [70], an FIA–GD system [71] applied to clinical blood samples using a bulk acoustic wave impedance sensor, a GD–FIA system using a capacitively coupled contactless conductivity detection (C4D) [72] and a GD–FIA with conductometric detection to monitor the ammonium content in open ocean seawater samples [73].

In Table 3.2 are shown the figures of merit of some methods developed for ammonium determination exploiting a gas diffusion cell. These results demonstrate the various possibilities of hyphenation of gas diffusion cells with different detectors in an FIA approach, but obtaining different limits of detection (LODs) and linear working ranges. Therefore, the choice of method and detector has to be made according to those best suited to the type of samples to be analyzed.

3.5.3 Gas diffusion in SIA

By contrast with FIA approach in which donor and acceptor solutions are dispensed simultaneously, SIA systems use a single channel, so the acceptor solution remains still on the other side of the membrane while the donor solution is circulated. This produces an overpressure on the side through which the donor solution is passed that eventually causes the acceptor channel to be evacuated by effect of the flexibility of the membrane unless some effective precaution is adopted. One way of avoiding this

Table 3.2 Comparison of the Analytical Figures of Merit of Various Methods for the Determination of Ammonium Ion Exploiting a Gas Diffusion cell

Method		Linear Range (mg/l)	LOD (3σ) (mg/l)	RSD (%) (n = 10)
Berthelot-1		0.1–10	0.05	1.3
Berthelot-2		1–60	1	1.5
Berthelot-3		0.5–20	0.5	2.5
Indicator-1		2–60	1	0.92
Indicator-2		0.5–20	0.3	2
Conductivity	Deionized water	1–60	0.06	4
	Boric acid	1–60	0.03	2

Berthelot-1: usual FIA method. Berthelot-2: FIA method with a gas diffusion cell. Berthelot-3: FIA stopped-flow method with a gas-diffusion cell. Indicator-1: spectrophotometric method (bromothymolblue) with GD cell. Indicator-2: as Indicator-1, but in the stopped-flow mode.

problem is by using a wide outlet tube for the donor channel in order to reduce overpressure and a narrow one for the acceptor channel to prevent the liquid from being flushed out.

Figure 3.22 illustrates another solution involving the insertion of an additional injection valve for the determination of ammonium in an SIA approach [74]. In this case, an indicator (bromothymol blue) is used to perform a colorimetric determination. Initially, the indicator in its acid form is aspirated by switching the injection valve and the central port of the selection valve is connected with its fourth side port.

Then, the injection valve is actuated in order to fill the acceptor channel with indicator delivered by the burette. At this point, the injection valve is actuated again to form a closed loop in the acceptor

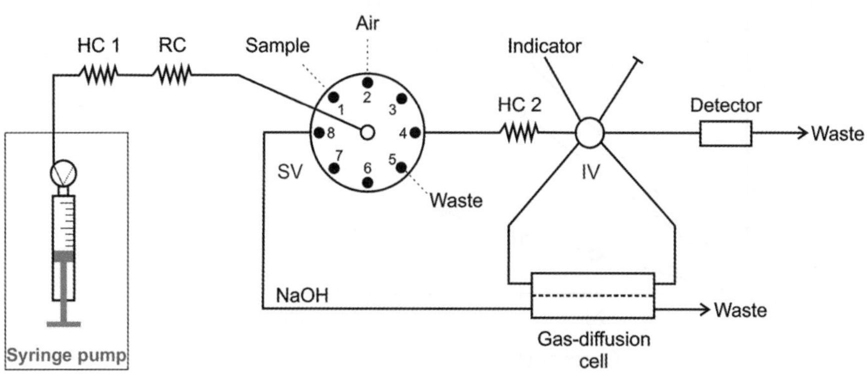

FIGURE 3.22

SIA system for ammonium determination using a gas diffusion cell. HC: holding coil; IV: injection valve; RC: reaction coil; SV: selection valve.

channel and prevent the liquid from leaving it when the sample is propelled through the donor channel. Once the acceptor is ready to be used, the selection valve is actuated to aspirate NaOH and sample; these are driven through the donor channel, where ammonium ion has previously been converted into ammonia during its travel and passed into the acceptor channel in a proportion similar to that obtained by FIA systems. The same SIA–GD system was applied also to ammonium determination but with conductometric detection and boric acid as acceptor solution [75].

Since SIA systems are computer controlled, these can be used for additional operations intended to readily boost their sensitivity. Thus, once the sample has been propelled through the donor channel, the flow direction can be reversed to pass it again through the cell; in this way, the sample is passed not once, but three times across the diffusion membrane and as a result the amount of ammonia crossing it increases. The increased sensitivity is illustrated in Figure 3.23. As can be seen, the slope of the calibration curve depends on the number of times that the sample is brought into contact with the membrane such a number is always odd as each flow reversal involves two passes across the membrane surface. In any case, more than two reversals have little further effect on the sensitivity and the analysis time is too long.

Despite the advantages provided by SIA systems, SIA–GC systems are scarce due to the impossibility of addressing two flow lines simultaneously what results in longer times of analysis.

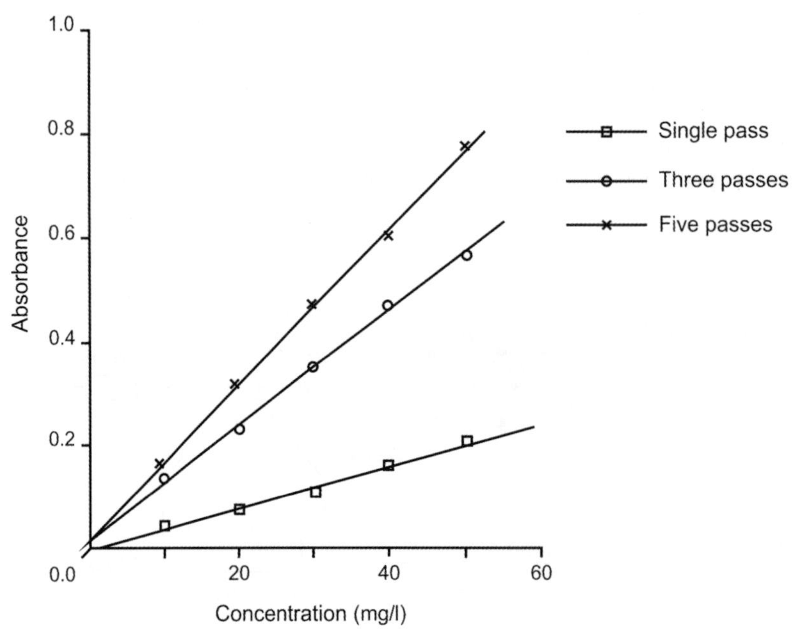

FIGURE 3.23

Calibration curve obtained with an SIA system exploiting direct flow (single pass) and reversal flow (three and five passes through the membrane).

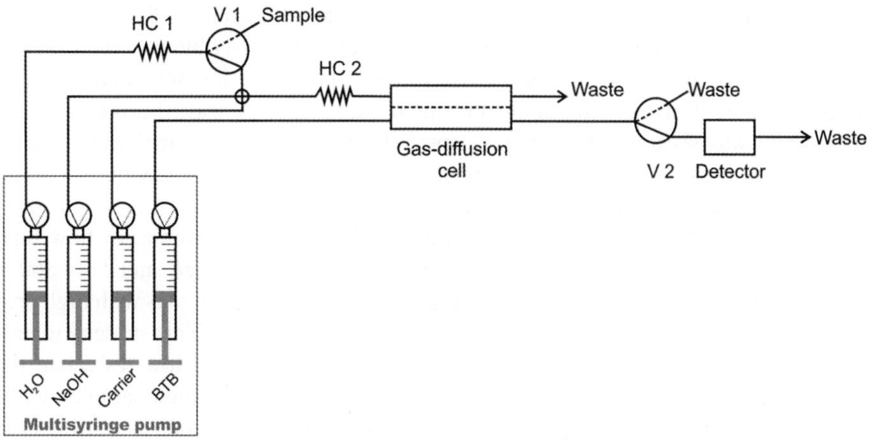

FIGURE 3.24

MSFIA system for the determination of ammonium ion using a gas diffusion cell and spectrophotometric detection. BTB: bromothymol blue; HC: holding coil; V: external solenoid valve.

3.5.4 Gas diffusion in MSFIA

MSFIA–GD systems permit stop flow and flow reversal operation mode as well as implementing two flow lines simultaneously on the GD unit, and also the fact of using piston pumps provides a high precise volume control. Thus MSFIA combines benefits of both FIA and SIA techniques when being coupled to GD. In this section, two methods for determining ammonium ion by MSFIA–GD with spectrophotometric [76] and conductometric [77] determination will be described.

In Figure 3.24 is shown the MSFIA system equipped with a GD cell and exploiting spectrophotometric determination of ammonium via color change of bromothymol blue indicator in the acceptor solution [76]. With samples containing high concentrations of ammonium ion, determinations can be performed at a fairly high throughput: once a large enough volume of sample has been aspirated into the holding coil, the flow direction of the burette is reversed and, while the indicator in its acid form is propelled to the detector via the acceptor channel, the carrier and the sample plus NaOH are alternately injected by actuating the solenoid commutation valves of the syringes in such a way that up to three aliquots are injected without aspiration. Each time an aliquot passes through the cell, the diffusion process occurs and the indicator is converted into its other form to an extent proportional to the ammonium ion concentration in the sample.

If the concentration of ammonium ion is very low, one can always use alternative procedures. Thus, if only a moderately increased sensitivity is required, it will suffice to lower the flow rate of the acceptor stream relative to the donor stream in order to obtain a slight concentration effect in the acceptor. A more marked increase in sensitivity can be obtained by using the solenoid valve at the outlet of the acceptor channel in order to stop the flow while sample plus NaOH are injected through the donor channel.

Another MSFIA–GD system exploiting conductometric detection was also applied to different water samples in a wide concentration range (from 0.075 to 360 mg/l of ammonium) [77], without sample dilution and with a precision better than 2% of relative standard deviation (RSD). The GD cell was made of two identical rectangular poly(methylmethacrylate) (PMMA) blocks, each one containing a U-shape flow channel with 180 μl volume. They were placed overlapping each other, holding in between a 0.22 μm pore size hydrophobic GD membrane, in this way the donor (a mixture of NaOH and sample) was separated from the acceptor (HCl). Six stainless steel screws were used for hermetical sealing of the cell. The cell was connected to the manifold in a way that a counter current flow between donor and acceptor was achieved.

The versatility of the MSFIA system allowed using two different strategies of flow management achieving two wide linear concentration ranges, from 0.075 to 10.0 mg/l and 5.7–360 mg/l with high reproducibility. For samples with high concentration of NH_4^+ (e.g. samples with high organic matter content) continuous flow of acceptor and donor solutions was applied. This strategy together with a high flow rate (2.4 ml/min) and a small sample volume (80 μl) led to a low residence time which implied a decreased sensitivity. This allowed the measurement of undiluted samples with contents of ammonium between 5.7 and 360 mg/l. For low concentration samples such as seawater, the chosen strategy consisted in keeping the acceptor channel stopped while the donor solution was flowing.

This allowed the analyte enrichment on the acceptor side and consequently, an improvement of the sensitivity. Using a flow rate of 0.5 ml/min and a sample volume of 150 μl, a linear working range from 0.1 to 10.0 mg/l was obtained but to the cost of a prolonged time of analysis.

3.6 Online dialysis

Dialysis technique implies the separation of suspended colloidal particles from dissolved ions or molecules of small dimensions by means of their unequal rates of diffusion through the pores of semipermeable membranes, as a function of molecular size.

As in all diffusion processes, the driving force for the transport is the concentration gradient across the membrane, which essentially determines the transfer efficiency. The sensitivity of the determination is additionally influenced by factors such as the temperature and the membrane properties (thickness, porosity, pore size). Thus a right membrane selection guarantees that the low molecular weight fraction of interest is transferred from the sample (donor) to the acceptor solution free of particles or macromolecules potentially interfering with its analytical determination.

Dialysis units can be designed as cells or probes; the latter allow direct sampling by immersion in the medium to be analyzed. A representative portion of the dialyzed compounds is collected by the acceptor solution for injection into the manifold. It is advisable to use an acceptor solution of constant ionic strength in order to facilitate the migration of ions to the other side of the membrane and preserve electroneutrality. In any case, the acceptor solution should be compatible with the analytical reaction to be subsequently used.

Dialysis membranes are classified in terms of pore size, which dictates the size of the molecules to which they are permeable and is expressed in daltons. Standard dialysis membranes typically have a molecular cut-off in the region of 10,000 Da. Many dialysis membranes are made of cellulose acetate or similar materials (e.g. Cuprophane, Cellophane) and range from 1 to 10 μm in pore size. Membrane thickness is also a highly influential variable.

Tests on various types of membranes ranging from some specially designed for dialysis purposes and extensively used in biochemistry to the ordinary ones employed by the food industry have proved to be very effective. Even the Cellophane membranes used to tie marmalade jars are especially effective.

Dialysis membranes must be soaked overnight prior to use. Most membrane manufacturers sell prehydrated packed specimens. Some membranes exhibit an asymmetric response, i.e. their transfer efficiency change depending on the side where the transfer originates.

One less common membrane type is that based on ion exchange. Unlike the previous types, ion-exchange membranes afford preconcentration based on Donnan's theory. Thus, when one is used to separate a solution with a high ionic content from another of low ionic strength, the ions in the more concentrated solution are transferred to the more diluted one. Due to the membrane does not allow ions of opposite charge to pass through, electro-neutrality can only be maintained by passing ions with the same charge from the more diluted solution. If such ions constitute the analyte, they can be preconcentrated provided the volume of the acceptor phase is greater than that of the donor phase. Nafion® ion-exchange membranes are especially commonplace.

3.6.1 Dialysis probes and cells

Figure 3.25 shows a typical dialysis probe consisting of two PVC blocks. The upper block is a cylindrical piece with a linear channel (20 mm long, 3 mm wide and 0.2 mm thick) furnished with inlet and outlet tubes through which the acceptor solution is continuously pumped. The lower block is a PVC disc with a slit the width of which determines the contact area between the membrane and the liquid sample. A cellulose acetate membrane is placed in between the two blocks and supported on a grid in order to minimize displacements potentially caused by pressure differences. The probe components are held together by means of stainless steel bolts.

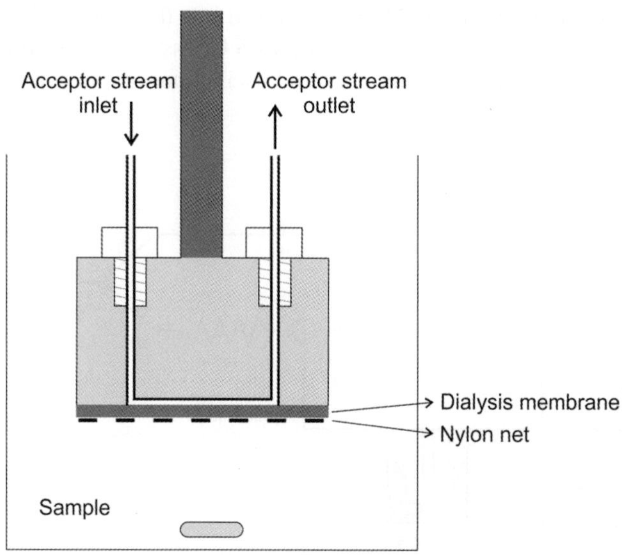

FIGURE 3.25

Dialysis probe.

During the sampling step, ionic analytes are continuously transferred to the acceptor solution through the membrane. Unless the analytes are immediately replenished at the membrane interface, their concentration falls to an adverse level affecting dramatically the dialysis efficiency. By the use of probes, this problem can be avoided by keeping the unknown solution under convective mixing (e.g. with a magnetic stirrer) in order to ensure a constant supply to the membrane surface.

The analyte concentration in the acceptor channel depends not only on the dialysis efficiency, but also on the volume in which the analyte is trapped (the smaller the volume, the greater the concentration).

While dialysis cells are very similar to gas diffusion cells (so that they can even use the same mechanical elements), they employ a rather different type of membrane (hydrophilic and hydrophobic, respectively).

Using a dialysis cell in a manifold for sample pretreatment ensures a very wide determination range at the expense of a low sensitivity by effect of the sample being diluted as a result of the poor transport efficiency across the membrane.

3.6.2 Dialysis probe in FIA

To illustrate the use of dialysis probe, an FIA method for nitrate determination is described. It has been successfully used both in laboratory tests and also for in situ monitoring purposes.

Dialysis probes used in the laboratory are immersed in the sample solution, which is held under magnetic stirring in a beaker (Figure 3.26). A representative portion of solutes of low molecular weight (i.e. dialyzable compounds) are continuously transferred from the bulk sample through the membrane for collection into a circulating KCl acceptor solution that is then used to fill the loop of an injection valve. In this case, for the determination of nitrates, the dialysate is injected into a hydrazinium sulfate carrier and propelled to the reaction coil (RC1), immersed in a thermostatic bath at 40 °C, to reduce nitrate to nitrite. The product formed is then merged with Griess reagent to form the azo dye, which is determined from its absorbance at 540 nm (spectral oscillations due to changes in refractive index

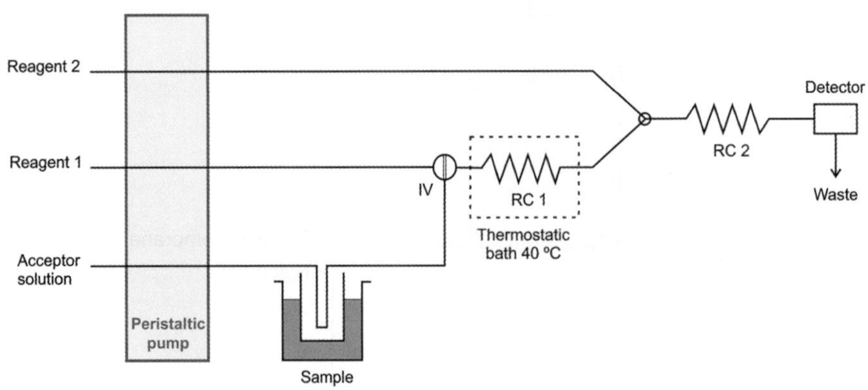

FIGURE 3.26

Dialysis probe used in an FIA system. IV: injection valve; RC: reaction coil.

is corrected by subtracting the absorbance at 420 nm). In this way, the resulting signal is the combination of the contributions of nitrate and nitrite.

On the one hand, the determination of nitrate in routine analyses of wastewater is of paramount importance with a view to assessing nitrification and denitrification in water purifying plants. As a rule, reported methods for the determination of nitrate involve its conversion into nitrite for its subsequent colorimetric detection based on the Griess reaction. One interesting, straightforward alternative is to directly measure the absorbance of nitrate at 210 nm, where the ion exhibits an absorption maximum. This constitutes an official method of analysis applicable to drinking water which, however, poses serious problems to be used with wastewater. Such problems have largely been ascribed to the presence of polymeric substances (e.g. humic acids) absorbing at the analyte wavelength or suspended particles hindering correct detection. This drawback can be overcome by using a dialysis technique previous the detection.

3.6.3 Dialysis cell in SIA

As an example, an SIA system exploiting dialysis technique is presented. The sample is dialyzed online to remove polymeric substances in order to avoid their interference with subsequent UV spectrophotometric measurements. The equipment for this system is similar to the one used for SIA−gas diffusion method, just using a dialysis membrane instead of a diffusion one (see Figure 3.22).

The dialysis cell used is composed of two PTFE blocks with symmetric winding channels (surface area 1 cm^2, volume 25 μl) and holding an intervening cellulose acetate membrane in place. Firstly, the sample is loaded into a holding coil. Then, the acceptor channel of the dialysis cell is loaded with NaOH solution (while the injection valve is kept in its loading position and solution is dispensed to the cell). The sample is dispensed to the donor channel of the dialysis unit. On reaching the cell, nitrate ions diffuse across the membrane and are collected by the acceptor solution. The solution containing diffused nitrate is transferred to the detector and the content of the acceptor channel is replaced with fresh solution, the system thus being made ready for a new measurement cycle.

Stopping the flow was found to be counterproductive as it resulted in poor mixing of the liquid layers in contact with the membrane by effect of the absence of turbulence and hence in deficient transfer.

The influence of the nature of the acceptor solution was examined by using distilled water, NaCl, LiCl, $CaCl_2$ and $AlCl_3$. The dialysis efficiency was found to increase with decreasing size of the hydrated ion and increasing charge of the metal ion. However, $CaCl_2$ and $AlCl_3$ resulted in poorer reproducibility than the other tested acceptor solutions.

3.7 Online capillary electrophoresis

CE technique separates species based on their size to charge ratio in the interior of a small capillary filled with an electrolyte. The separation of compounds by CE is dependent on the differential migration of analytes in an applied electric field. CE has gained importance during the last years as it is an efficient separation methodology for a wide variety of analytes in diverse matrices. Moreover CE achieves low sample and electrolyte consumption, and experimental simplicity. Its sensitivity and precision can be improved by hyphenation with flow-based techniques [78] and temperature control. Coupling of pressure-driven flow and electrophoretic separation methods presents some technical challenges. The design of the interface which connects both analytical techniques is critical.

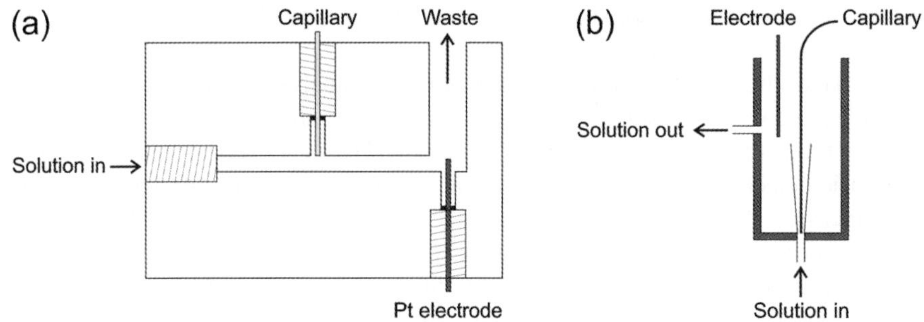

FIGURE 3.27

Schematic diagram of the firsts interfaces to couple FIA and CE. (a) Interface developed by Kuban et al. (b) Interface developed by Fang et al.

3.7.1 CE coupling to FIA

First online FIA–CE systems were developed almost simultaneously [79,80]. Both approaches provided the possibility of repeated multianalyte separation. In Figure 3.27(a) and (b) are shown these two original interfaces. Both designs consist of a flow-through channel into which one end of the CE separation capillary and an electrode are inserted. The interface shown in Figure 3.27(a) was fabricated by precise drilling in a piece of PMMA. In the other design (Figure 3.27(b)), the conical-flow-through channel was an Eppendorf pipette tip placed vertically in a supporting vial. In the latter it was important to keep the separation capillary as close as possible to the end of the conical flow inlet. These interfaces are currently widely used when coupling FIA/SIA to CE, sometimes with some modifications. For example, the original FIA–CE system was modified by replacing the peristaltic pump with gravity flow [81,82] obtaining better RSD values from repeated injections probably due to the avoidance of pulses by using gravity feed flow. Using a tubular Pt electrode can simplify the design of the interface to a 3-channel T-shape piece [83], this design was exploited in an FIA–CE coupled to ICP for chromium speciation analysis. CE has been also successfully coupled to electrospray ionization mass spectrometry [84] and AFS [85].

FIA–CE has also been coupled with contactless conductivity detection (C4D) for online analysis of metal cations (ammonium, potassium, calcium, magnesium, and sodium as complexes in aqueous 18-crown-6-ether-acetace electrolyte solution) [86]. In this system the ends of the separation capillary and the electrodes were placed opposite to each other into tubings which acted as flow-through channels.

3.7.2 CE coupling to SIA

In principle, the interfaces used for coupling SIA and CE have the same design as in FIA–CE systems. The only difference is the way SIA operates. SIA's main advantage over FIA is the possibility to drive the flow in two directions. Other benefits of SIA over FIA when coupling to CE are namely: the higher precision of the pressure generated by a syringe pump to pressurize the capillary, the lower flow rates available (μl/min), the possibility of using head column field amplified stacking by applying the pressure and electrokinetic injection at the same time and that SIA can be used for limited sample amounts.

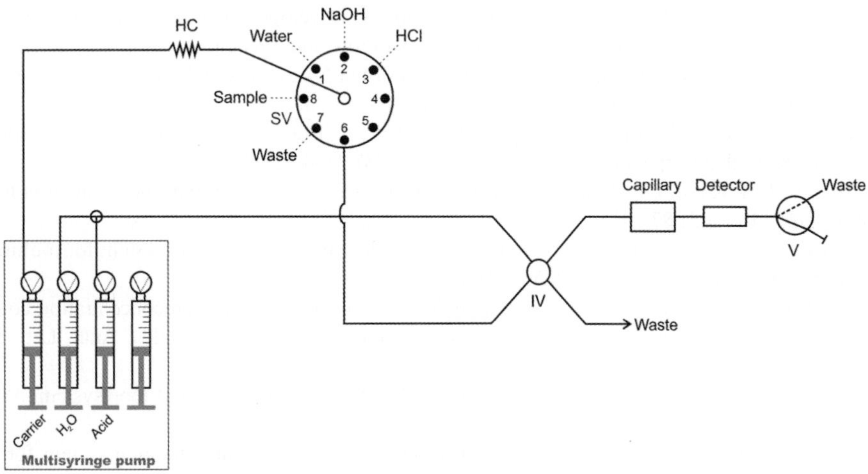

FIGURE 3.28

MSFIA–CE system. HC: holding coil; IV: injection valve; SV: selection valve; V: external solenoid valve.

SIA–CE has been successfully coupled to different detection techniques such as C4D [87], laser-induced fluorescence via a valve interface for online derivatization and analysis of amino acids and peptides [88] and ICP sector field mass spectrometer [89].

3.7.3 CE coupling to MSFIA

MSFIA allows background operations and thus, higher sample frequencies. The advantage of the employment of a robust and multichannel syringe pump allows the use of very fine sorbent material, and parallel operations, which are not possible with a single syringe pump (SIA) or multichannel peristaltic pump (FIA) with the same analytical performance and robustness. As an example an MSFIA–CE system for preconcentration, separation, and determination of nitrophenols [90] is shown in Figure 3.28. In this MSFIA–CE system, in-line sample acidification, analyte preconcentration, elution hydrodynamic injection, electrophoretic separation, detection, maintenance and conditioning of the SPE column and the separation capillary were successfully automated. A homemade photometric detection cell was used for on-capillary detection, using an optical fiber and a light emitting diode (LED).

References

[1] B. Karlberg, S. Thelander, Extraction based on the flow-injection principle. Part I. Description of the extraction system, Anal. Chim. Acta 98 (1978) 1–7.

[2] H.F. Bergamin, J.X. Medeiros, B.F. Reis, E.A.G. Zagatto, Solvent extraction in continuous flow injection analysis: determination of molybdenum in plant material, Anal. Chim. Acta 101 (1978) 9–16.

[3] M. Shamsipur, M.M. Zahedi, G. De Filippo, V. Lippolis, Development of a novel flow injection liquid-liquid microextraction method for the on-line separation and preconcentration for determination of zinc(II) using 5-(8-hydroxy-2-quinolinylmethyl)-2,8-dithia-5-aza-2,6-pyridinophane as a sensitive and selective fluorescent chemosensor, Talanta 85 (2011) 687–693.

[4] F. Cañete, A. Ríos, M.D. Luque de Castro, M. Valcárcel, Liquid-liquid extraction in continuous flow systems without phase separation, Anal. Chem. 60 (1988) 2354–2357.

[5] D.C. Shelly, T.M. Rossi, I.M. Warner, Multiple solvent extraction system with flow injection technology, Anal. Chem. 54 (1982) 87–91.

[6] T.M. Rossi, D.C. Shelly, I.M. Warner, Optimization of flow injection analysis system for multiple solvent extraction, Anal. Chem. 54 (1982) 2056–2061.

[7] J. Wang, E.H. Hansen, Flow injection on-line two-stage solvent extraction preconcentration coupled with Et-AAS for determination of bismuth in biological and environmental samples, Anal. Lett. 33 (2000) 2747–2766.

[8] R.H. Atallah, J. Ruzicka, G. Christian, Continuous solvent extraction in a closed-loop system, Anal. Chem. 59 (1987) 2909–2914.

[9] R.H. Atallah, G.D. Christian, S.D. Hartenstein, Continuous flow solvent extraction system for the determination of trace amounts of uranium in nuclear waste reprocessing solutions, Analyst 113 (1988) 463–469.

[10] K. Bäckström, L.G. Danielsson, Design of a continuous-flow two-step extraction sample work-up system for graphite furnace atomic absorption spectrometry, Anal. Chim. Acta 232 (1990) 301–315.

[11] R. Burakham, S. Lapanantnoppakhun, J. Jakmunne, K. Grudpan, Exploiting sequential injection analysis with lab-at-valve (LAV) approach for on-line liquid-liquid micro-extraction spectrophotometry, Talanta 68 (2005) 416–421.

[12] R.C. de Campos Costa, A.N. Araujo, Determination of Fe(III) and total Fe in wines by sequential injection analysis and flame atomic absorption spectrometry, Anal. Chim. Acta 438 (2001) 227–233.

[13] J. Wang, E.H. Hansen, Development of an automated sequential injection on-line solvent extraction-back extraction procedure as demonstrated for the determination of cadmium with detection by electrothermal atomic absorption spectrometry, Anal. Chim. Acta 456 (2002) 283–292.

[14] J. Wang, E.H. Hansen, FI/SI on-line solvent extraction/back extraction preconcentration coupled to direct injection nebulization inductively coupled plasma mass spectrometry for determination of copper and lead, J. Anal. At. Spectrom. 17 (2002) 1284–1289.

[15] A. Cladera, M. Miró, J.M. Estela, V. Cerdà, Multicomponent sequential injection analysis determination of nitro-phenols in waters by on-line liquid-liquid extraction and preconcentration, Anal. Chim. Acta 421 (2000) 155–166.

[16] M. Rezaee, Y. Assadi, M.R.M. Hosseini, E. Aghaee, F. Ahmadi, S. Berijani, Determination of organic compounds in water using dispersive liquid-liquid microextraction, J. Chromatogr. A 1116 (2006) 1–9.

[17] L. Kocúrová, I.S. Balogh, J. Skrliková, J. Posta, V. Andruch, A novel approach in dispersive liquid-liquid microextraction based on the use of an auxiliary solvent for adjustment of density. UV-vis spectrophotometric and graphite furnace atomic absorption spectrometric determination of gold based on ion pair formation, Talanta 82 (2010) 1958–1964.

[18] I.S. Balogh, L. Rusnáková, J. Skrlíková, M. Torok, V. Andruch, A spectrophotometric method for manganese determination in water samples based on ion pair formation and dispersive liquid-liquid microextraction, Int. J. Environ. Anal. Chem. (2011), http://dx.doi.org/10.1080/03067319.2010.537750.

[19] V. Andruch, C.C. Acebal, J. Skrilíková, H. Sklenárová, P. Solich, I.S. Balogh, F. Billes, L. Kocúrová, Automated on-line dispersive liquid-liquid microextraction based on a sequential injection system, Microchem. J. 100 (2012) 77–82.

[20] A.N. Anthemidis, K.I.G. Iannou, On-line sequential injection dispersive liquid-liquid microextraction system for flame atomic absorption spectrometric determination of copper and lead in water samples, Talanta 79 (2009) 86–91.

[21] A.N. Anthemidis, K.I.G. Ioannou, Development of a sequential injection dispersive liquid-liquid microextraction system for electrothermal atomic absorption spectrometry by using a hydrophobic sorbent material: determination of lead and cadmium in natural waters, Anal. Chim. Acta 668 (2010) 35–40.

[22] A.N. Anthemidis, K.I.G. Ioannou, Sequential injection dispersive liquid-liquid microextraction based on fatty alcohols and poly(etheretherketone)-turnings for metal determination by flame atomic absorption spectrometry, Talanta 84 (2011) 1215–1220.

[23] F. Maya, J.M. Estela, V. Cerdà, Completely automated in-syringe dispersive liquid-liquid microextraction using solvents lighter than water, Anal. Bioanal. Chem. 402 (2012) 1383–1388.

[24] F. Maya, B. Horstkotte, J.M. Estela, V. Cerdà, Lab in a syringe: fully automated dispersive liquid-liquid microextraction with integrated spectrophotometric detection, Anal. Bioanal. Chem. 404 (2012) 909–917.

[25] Global FIA: http://www.globalfia.com/.

[26] Mobitec: http://81.169.157.43/mobitec_de/index_content.htlm.

[27] C. Pons, R. Forteza, V. Cerdà, The use of anion-exchange disks in an optrode coupled to a multi-syringe flow-injection system for the determination and speciation analysis of iron in natural water samples, Talanta 66 (2005) 210–217.

[28] C. Pons, M. Miró, E. Becerra, J.M. Estela, V. Cerdà, An intelligent flow analyzer for the in-line concentration, speciation and monitoring of metals at trace levels, Talanta 62 (2004) 887–895.

[29] E. Rubí, R. Forteza, V. Cerdà, Preconcentration and colorimetric determination of Fe(III) by sequential injection analysis, Lab. Rob. Autom. 8 (1996) 149–156.

[30] E. Rubí, M.S. Jiménez, F. Bauzá, R. Forteza, V. Cerdà, Preconcentration and atomic absorption determination of iron by sequential injection analysis, Talanta 44 (1997) 553–562.

[31] M.I.G.S. Almeida, J.M. Estela, V. Cerdà, Multisyringe flow injection potentialities for hyphenation with different types of separation techniques, Anal. Lett. 44 (2011) 360–373.

[32] M. Miró, A. Cladera, J.M. Estela, V. Cerdà, Sequential injection spectrophotometric analysis of nitrite in natural waters using an on-line solid-phase extraction and preconcentration method, Analyst 125 (2000) 943–948.

[33] J. Avivar, L. Ferrer, M. Casas, V. Cerdà, Lab on valve-multisyringe flow injection system for fully automated uranium determination in environmental samples, Talanta 84 (2011) 1221–1227.

[34] M. Miró, W. Frenzel, V. Cerdà, J.M. Estela, Determination of ultratraces of nitrite by solid-phase preconcentration using a novel flow-through spectrophotometric optrode, Anal. Chim. Acta 437 (2001) 55–65.

[35] M. Miró, W. Frenzel, J.M. Estela, V. Cerdà, A novel flow-through disk-based solid-phase extraction diffuse reflectance optrode. Application to preconcentration and determination of trace levels of nitrite, Analyst 126 (2001) 1740–1746.

[36] L. Ferrer, G. de Armas, M. Miró, J.M. Estela, V. Cerdà, Flow-through optical fiber sensor for automatic sulfide determination in waters by multisyringe flow injection analysis using solid-phase reflectometry, Analyst 130 (2005) 644–651.

[37] Phenomenex: http://www.phenomenex.com/Phen/Products/onyx/index.htm.

[38] Chromolith: http://www.chromolith.com/.

[39] D. Kowalski, E. Pobozy, M. Trojanowicz, Flow-Injection preconcentration of chloramphenicol using molecularly imprinted polymer for HPLC determination in environmental samples, J. Autom. Methods Manage. Chem. (2011), http://dx.doi.org/10.1155/2011/143416.

[40] O. Kritsunankul, J. Jakmunee, Simultaneous determination of some food additives in soft drinks and other liquid foods by flow injection on-line dialysis coupled to high performance liquid chromatography, Talanta 84 (2011) 1342–1349.

[41] J.F.G. Jiménez, M.C. Valencia, L.F. Capitán-Vallvey, Parabens determination with a hybrid FIA/HPLC system with ultra-short monolithic column, J. Anal. Chem. 65 (2010) 188−194.

[42] J. Vichapong, R. Burakham, S. Srijaranai, K. Grudpan, Sequential injection-bead injection-lab-on-valve coupled to high-performance liquid chromatography for online renewable microsolid-phase extraction of carbamate residues in food and environmental samples, J. Sep. Sci. 34 (2011) 1574−1581.

[43] R. Burakham, S. Srijaranai, K. Grudpan, High-performance liquid chromatography with sequential injection for online precolumn derivatization of some heavy metals, J. Sep. Sci. 30 (2007) 2614−2619.

[44] J.P. Jesus, C.A. Suárez, J.R. Ferreira, M.F. Giné, Sequential injection analysis implementing multiple standard additions for as speciation by liquid chromatography and atomic fluorescence spectrometry (SIA-HPLC-AFS), Talanta 85 (2011) 1364−1368.

[45] P. Chantarateepra, W. Siangproh, S. Motomizu, O. Chailapakul, Automated online solid-phase extraction coupled with sequential injection-HPLC-EC system for the determination of sulfonamides in shrimp, Int. J. Electrochem. (2012), http://dx.doi.org/10.1155/2012/862823.

[46] H.M. Oliveira, M.A. Segundo, J.L.F.C. Lima, V. Cerdà, Multisyringe flow injection system for solid-phase extraction coupled to liquid chromatography using monolithic column for screening of phenolic pollutants, Talanta 77 (2009) 1466−1472.

[47] J.B. Quintana, M. Miró, J.M. Estela, V. Cerdà, Automated on-line renewable solid-phase extraction-liquid chromatography exploiting multisyringe flow injection-bead injection lab-on-valve analysis, Anal. Chem. 78 (2006) 2832−2840.

[48] P. Chocholous, P. Solich, D. Satinsky, An overview of sequential injection chromatography, Anal. Chim. Acta 600 (2007) 129−135.

[49] M. Fernández, H.M. González-San Miguel, J.M. Estela, V. Cerdà, Contribution of multi-commuted flow analysis combined with monolithic columns to low-pressure, high-performance chromatography, Trends Anal. Chem. 28 (2009) 336−346.

[50] J. Huclova, D. Satinsky, R. Karlicek, Coupling of monolithic columns with sequential injection technique. A new separation approach in flow methods, Anal. Chim. Acta 494 (2003) 133−140.

[51] D. Satinsky, J. Huclova, R.L.C. Ferreira, M.C.B.S.M. Montenegro, P. Solich, Determination of ambroxol hydrochloride, methylparaben and benzoic acid in pharmaceutical preparations based on sequential injection technique coupled with monolithic column, J. Pharm. Biomed. Anal. 40 (2006) 287−293.

[52] C.Z. Zacharis, G.A. Theodoris, A. Podgornik, A.N. Voulgaropoulos, Incorporation of a monolithic column into sequential injection system for drug-protein binding studies, J. Chromatogr. A 1121 (2006) 46−54.

[53] P. Chocholous, D. Satinsky, R. Sladkovsky, M. Pospisilova, P. Solich, Determination of pesticides fenoxycarb and permethrin by sequential injection chromatography using miniaturized monolithic column, Talanta 77 (2008) 566−570.

[54] M.A. Obando, J.M. Estela, V. Cerdà, Multisyringe chromatography (MSC) system for the on-line solid-phase extraction and determination of hydrochlorothiazide and losartan potassium in superficial water, groundwater and wastewater outlet samples, J. Pharm. Biomed. Anal. 48 (2008) 212−217.

[55] F. Maya, J.M. Estela, V. Cerdà, Multisyringe ion chromatography with chemiluminescence detection for the determination of oxalate in beer and urine samples, Microchim. Acta 173 (2011) 33−41.

[56] F. Maya, J.M. Estela, V. Cerdà, Interfacing on-line solid phase extraction with monolithic column multisyringe chromatography and chemiluminescence detection: an effective tool for fast, sensitive and selective determination of thiazide diuretics, Talanta 80 (2010) 1333−1340.

[57] C. Fernández, M.S. Larrechi, R. Forteza, V. Cerdà, M.P. Callao, Multisyringe chromatography (MSC) using a monolithic column for the determination of sulphonated azo dyes, Talanta 82 (2010) 137−142.

[58] C.A. Chávez, J.L. Guzmán-Mar, L. Hinojosa-Reyes, A. Hernández-Ramírez, L. Ferrer, V. Cerdà, Applicability of multisyringe chromatography coupled to on-line solid-phase extraction to the simultaneous determination of dicamba, 2,4-D, and atrazine, Anal Bioanal. Chem. 403 (2012) 2705−2714.

[59] E. Björklund, F. Maya, S.A. Bak, M. Hansen, J.M. Estela, V. Cerdà, Possibilities and limitations of the sequential injection chromatography technique for the determination of anticoccidial agents in water, pharmaceutical formulations and feed, Microchem. J. 98 (2011) 190–199.

[60] T. Hankemeier, S.P.J. van Leeuwen, J.J. Vreuls, U.A.T. Brinkman, Use of a presolvent to include volatile organic analytes in the application range of online solid-phase extraction-gas chromatography-mass spectrometry, J. Chromatogr. A 811 (1998) 117–133.

[61] T. Hyötyläinen, Principles, developments and applications of on-line coupling of extraction with chromatography, J. Chromatgr. A 1153 (2007) 14–28.

[62] T. Hyötyläinen, M.L. Riekkola, Sorbent and liquid-phase microextraction techniques and membrane-assisted extraction in combination with gas chromatographic analysis: a review, Anal. Chim. Acta 614 (2008) 27–37.

[63] T. Hyötyläinen, M.L. Riekkola, Approaches for on-line coupling of extraction and chromatography, Anal. Bioanal. Chem. 378 (2004) 1962–1981.

[64] A.J.H. Louter, U.A.T. Brinkman, R.T. Ghijsen, Fully automated water analyzer based on online solid phase extraction-gas chromatography, J. Microcolumn Sep. 5 (1993) 303–315.

[65] J.B. Quintana, W. Boonjob, M. Miró, V. Cerdà, Online coupling of bead injection lab-on-valve analysis to gas chromatography: application to the determination of trace levels of polychlorinated biphenyls in solid waste leachates, Anal. Chem. 81 (2009) 4822–4830.

[66] E.C. Goosens, R.G. Bunschoten, V. Engefen, D. De Jong, J.H.M. van der Berg, Determination of hexachlorocyclohexanes in ground water by coupled liquid-liquid extraction and capillary gas chromatography, J. High Resolut. Chromatogr. 13 (1990) 438–442.

[67] E. Fogelquist, M. Krysell, On-line liquid-liquid extraction in a segmented flow directly coupled on-column injection into gas chromatograph, Anal. Chem. 58 (1986) 1516–1520.

[68] J. Roeraade, Automated monitoring of organic trace components in water: I. Continuous flow extraction together with on-line capillary gas chromatography, J. Chromatogr. A 330 (1985) 263–274.

[69] G. Schulze, C.Y. Liu, M. Brodowski, O. Elsholz, W. Frenzel, J. Möller, Different approaches to the determination of ammonium ions at low levels by flow-injection analysis, Anal. Chim. Acta 214 (1988) 121–136.

[70] J. Junsomboon, J. Jakmunee, Flow injection conductometric system with gas diffusion separation for the determination of Kjeldahl nitrogen in milk and chicken meat, Anal. Chim. Acta 627 (2008) 232–238.

[71] X.L. Su, B.S. Yu, H.W. Tan, X.R. Yang, L.H. Nie, S.Z. Yao, Flow-injection determination of total ammonia and total carbon dioxide in blood based on gas-diffusion separation and with a bulk acoustic wave impedance sensor, J. Pharm. Biomed. Anal. 16 (1998) 759–769.

[72] H.L. Braz, D.T. Ito, J.A.F. da Silva, C.L. do Lago, J.J. Pedrotti, Trace levels determination of ammonium by flow injection analysis using gas-diffusion and capacitively coupled contactless conductivity detection, Electroanalysis 23 (2011) 2594–2600.

[73] J.N. Plant, K.S. Johnson, J.A. Needoba, L.J. Coletti, NH4-Digiscan: an in situ and laboratory ammonium analyzer for estuarine, coastal, and shelf waters, Limnol. Oceanogr. Methods 7 (2009) 144–156.

[74] M.T. Oms, A. Cerdà, A. Cladera, V. Cerdà, R. Forteza, Gas diffusion techniques coupled sequential injection analysis for selective determination of ammonium, Anal. Chim. Acta 318 (1996) 251–260.

[75] M.T. Oms, A. Cerdà, V. Cerdà, Preconcentration by flow reversal in conductometric sequential injection analysis of ammonium, Electroanalysis 8 (1996) 387–390.

[76] J. Klimundova, R. Forteza, V. Cerda, A multisyringe flow injection system coupled with a gas diffusion cell for ammonium determination Int, J. Environ. Anal. Chem. 83 (2003) 233–246.

[77] C. Henríquez, B. Horstkotte, V. Cerdà, Conductometric determination of ammonium by a multisyringe flow injection system applying gas diffusion, J. Environ. Anal. Chem. (2012), http://dx.doi.org/10.1080/03067319.2012.746322 1–17.

[78] P. Kubán, B. Karlberg, Flow/sequential injection sample treatment coupled to capillary electrophoresis. A review, Anal. Chim. Acta 648 (2009) 129–145.
[79] P. Kuban, A. Engström, J.C. Olsson, G. Thorsen, R. Tryzell, B. Karlberg, New interface for coupling flow-injection and capillary electrophoresis, Anal. Chim. Acta 337 (1997) 117–124.
[80] Z.L. Fang, Z.S. Liu, Q. Shen, Combination of flow injection with capillary electrophoresis. Part I. The basic system, Anal. Chim. Acta 346 (1997) 135–143.
[81] P. Kuban, P. Kuban, V. Kuban, Flow injection-capillary electrophoresis system with contactless conductivity detection and hydrostatic pressure generated flow. Application to the quantitative analysis of inorganic anions in water samples, Electrophoresis 24 (2003) 1935–1943.
[82] P. Kuban, P. Kuban, V. Kuban, Rapid speciation of Se(IV) and Se(VI) by flow injection-capillary electrophoresis system with contactless conductivity detection, Anal. Bioanal. Chem. 378 (2004) 378–382.
[83] M.F. Gine, A.P.G. Gervasio, A.F. Lavorante, C.E.S. Miranda, E. Carrilho, Interfacing flow injection with capillary electrophoresis and inductively coupled plasma mass spectrometry for Cr speciation in water samples, J. Anal. Atom. Spectrom. 17 (2002) 736–738.
[84] B. Santos, B.M. Simonet, B. Lendl, A. Rios, M. Valcarcel, Alternatives for coupling sequential injection systems to commercial capillary electrophoresis-mass spectrometry equipment, J. Chromatogr. A 1127 (2006) 278–285.
[85] D.D. Wang, F. Li, X.P. Yan, On-line hyphenation of flow injection, miniaturized capillary electrophoresis and atomic fluorescence spectrometry for high-throughput speciation analysis, J. Chromatogr. A 1117 (2006) 246–249.
[86] C. Sprung, H. Siren, S. Rovio, T. Työppönen, On-line flow injection-capillary electrophoresis instrument with contactless conductivity detection for sensitive cation analyses, Sep. Sci. Technol. 15 (2008) 3856–3872.
[87] T.D. Mai, S. Schmid, B. Müller, P.C. Hauser, Capillary electrophoresis with contactless conductivity detection coupled to a sequential injection analysis manifold for extended automated monitoring applications, Anal. Chim. Acta 665 (2010) 1–6.
[88] C.K. Zacharis, F.W.A. Tempels, G.A. Theodoris, A.N. Voulgaropoulos, W.J.M. Underberg, G.W. Somsen, G.J. De Jong, Coupling of sequential injection analysis and capillary electrophoresis-laser-induced fluorescence via a valve interface for on-line derivatization and analysis of amino acids and peptides, J. Chromatogr. A 1132 (2006) 297–303.
[89] C.A. Suárez, G.C.L. Araújo, M.F. Giné, M.H. Kakazu, J.E.S. Sarkis, Sequential injection analysis (SIA) for arsenic speciation by capillary electrophoresis hyphenated to inductively coupled plasma sector filed mass spectrometry (CE-ICP-SFMS), Spectrosc. Lett. 42 (2009) 376–382.
[90] B. Horstkotte, O. Elsholz, V. Cerdà, Multisyringe flow injection analysis coupled to capillary electrophoresis (MSFIA-CE) as a novel analytical tool applied to the preconcentration, separation and determination of nitrophenols, Talanta 76 (2008) 72–79.

CHAPTER

The Flow Workshop

4

CHAPTER OUTLINE

Flow Analysis. http://dx.doi.org/10.1016/B978-0-444-59596-6.00004-8

Frequently, and especially when developing new analytical flow methods, not all required components are commercially available. Very often, it is mandatory to build our own devices. This involves setting up our own suitable mechanical and electronic workshop. Although requiring an initial economical effort, this initial investment is quickly amortized, especially in research laboratories where innovative designs are self-evident. One of the greatest difficulties when developing or using these workshops is the lack of know-how, regarding electronics and mechanics.

Therefore the aim of this chapter is to provide guidance information and the basic knowledge to develop a workshop, e.g. main tools required, available materials, and how to manage them highlighting the advantages of these.

4.1 The flow workshop

Frequently, and especially when developing new analytical flow methods, not all required components are commercially available. A possible solution would be finding a mechanical and electronic workshop that builds for us our previously designed devices. However, very often when dealing with prototype designs, the production step becomes slow, tedious, and very expensive. Trial and error cycles are very common until the final version fulfilling all the required parameters of the new product is reached.

Another faster, more stimulating, and very important more economically affordable option consists on building our own devices. This involves setting up our own suitable mechanical and electronic workshop. Although requiring an initial economical effort, the initial investment is quickly amortized, especially in research laboratories where innovative designs are self-evident. Moreover, having our own workshop provides a huge flexibility to researchers, since designs can be easily modified and the desired material can be chosen depending on the chemicals to be used.

Some mechanical and electronic skills are necessary and especially useful for innovative research developments. Without intending to be an expert within the workshop, it is certainly important to know the basic language and "slang" used by the professionals with whom one must communicate to make them understand what you want. One of the greatest difficulties when developing or using these workshops is the lack of know-how, regarding electronics and mechanics.

Therefore, the aim of this chapter is to provide guidance information and basic knowledge to develop a workshop, e.g. main tools required, available materials, and how to manage them highlighting the advantages of these.

4.2 Mechanical workshop

A properly equipped mechanical workshop will allow us to build many passive devices selecting the appropriate materials to withstand the chemical reagents that will be used in the analytical method. Fortunately, most materials are usually soft and easy to handle, e.g. different types of plastics, aluminum, and brass, being rarely necessary to use materials difficult to handle, such as stainless steel.

4.2.1 Materials selection

Behavior of different materials in front of some chemicals is summarized in Table 4.1. This table can be used as a general guidance to avoid future problems in the laboratory, i.e. selecting the proper material to build our devices will prevent damages on it, e.g. erosion, dissolution, melting, and oxidation. Most common materials used in flow assemblies are described in detail below.

4.2.1.1 Teflon® PTFE

It is well known that polytetrafluoroethylene (PTFE Teflon) is one of the most resistant materials to chemicals including solvents. Teflon tubing and connectors have to date been the most widely used in flow assemblies on account of their high chemical resistance to virtually all types of reagents and solvents. Their greatest disadvantages are their low mechanical strength and high cost. While Teflon has so far been the norm, some routine applications can benefit from the use of more inexpensive alternative materials. Common uses are: tubing, reactors, gas diffusion membranes, and degassing membranes.

4.2.1.2 Teflon® FEP

Fluorinated ethylene propylene (FEP) is an effective alternative to PTFE Teflon. Sharing many of the properties of PTFE, e.g. can be heat-shaped, but being less permeable to gases and also more transparent. In fact, FEP is the most transparent of all resins in the Teflon family.

This material can withstand temperatures up to 50 °C and pH values from 0 to 14, has an oxygen permeability of 748 $cm^3/100\ in^2 \cdot 24\ h \cdot atm/mil$ at 25 °C and can be autoclaved. FEP is commonly used in tubing.

4.2.1.3 Teflon® PFA

Perfluoroalkoxy (PFA) is another effective alternative to PTFE Teflon. However, being more expensive than FEP, but withstanding higher temperatures and containing fewer impurities, is especially useful for delicate applications.

This material can withstand temperatures as high as 80 °C, has a good mechanical strength and an oxygen permeability of 881 $cm^3/100\ in^2 \cdot 24\ h \cdot atm/mil$ at 25 °C. PFA can be used over the pH range 0–14 and autoclaved. It is commonly used in tubing.

4.2.1.4 Kel-F®

Polychlorotrifluoroethylene (Kel-F/PCTFE/CTFE) shares the mechanical strength, heat resistance, and pH tolerance of PFA, but has an oxygen permeability of only 12 $cm^3/100\ in^2 \cdot 24\ h \cdot atm/mil$ at 25 °C. As previous described materials, it can be autoclaved. However, it is very expensive.

Kel-F can be sterilized with γ-rays, ethylene oxide, or heat and is only vulnerable to tetrahydrofuran (THF) and a few halogenated solvents. It is microwave transparent. Common uses of KEL-F are: connectors, valves, cells, and lab on valve (LOV) platforms.

4.2.1.5 Tefzel® ETFE

Ethylene-tetrafluoroethylene copolymer (Tefzel/ETFE) is structurally similar to Teflon, possessing also a high chemical resistance. This material has proved excellent for sealing surfaces and in applications involving aggressive solvents. Tefzel is also used in threaded materials, particularly in low-voltage lines and adapters. However, the polymer can be degraded or swollen by chlorinated solvents.

Tefzel can also withstand temperatures of up to 80 °C, has good mechanical strength, and can be used throughout the pH range. It has an oxygen permeability of 100 $cm^3/100\ in^2 \cdot 24\ h \cdot atm/mil$ at

25 °C and can be autoclaved. It is microwave transparent. However, it is very expensive. Tefzel is commonly used to build connectors, valves, and cells.

4.2.1.6 PVDF

Polyvinylidenefluoride (PVDF/Kynar/Hylar) has good mechanical and chemical resistance. It is microwave transparent and it is commonly used in connectors, valves, diffusion membranes, and screws.

4.2.1.7 TEFLON AF 2400

TEFLON AF 2400 is highly permeable to gases. It has a low refractive index, lower than water. This allows total reflexion using water as core in long path-length capillary cells. It is UV-transparent and very expensive. It is commonly used in tubing, membranes, windows, and optical fibers used as flow cells.

Table 4.1 Chemical Resistance of the Materials Common Used in Flow Assemblies

Chemical	PEEK	PPS	PTFE	Stainless Steel 316	FPM	EPDM	Perfluoro Elastomer	Teflon AF 2400	Kel-F
Acetaldehyde	A	A	A	A	C	A	A		A
Acetic acid 10%	A	A	A	A	C	A	A		A
Acetone	A	A	A	A	C	A	A		A
Ammonia water	A	B	A	A	A	A	A		A
Benzine	A	A	A	A	C	C	A		B
Caustic soda 50%	A	A	A	C	C	B	–		A
Chloroform	A	A	A	A	B	C	A	A	A
DMF	A	A	A	B	C	B		A	A
DMSO			A					A	A
Ethanol	A	A	A	A	B	A		A	A
Ethyl acetate	A	A	A	B	C	B		A	A
Formalin 37%	A	A	A	C	A	A	A	A	A
Hydrocarbons	A	A	A	A	A	C		A	A
Hydrochloric acid 20%	A 10	C	A	C	A	A	A		A
Hydrogen peroxide 30%	A	B	A	C	A	C	–		A
Nitric acid 10%	A	A	A	A	A	B	A		A
Phosphoric acid	A	A	A	C	A	B	–		A
Pure water	A	A	A	A	A	A	–		A
Seawater	A	A	A	C	A	A	–		A
Sodium hypochlorite 5%	A	A	A	C	A	B	A		A
Sulfuric acid <10%	A			B		A			A
Sulfuric acid 30%	C	A	A	C	A	B	A		A
Tetrahydrofuran	A	A	A	A	C	C	B		A
Toluene	A	A	A	A	A	C	A		B
Xylene	A	A	A	A	A	C	A		A

(Continued)

Chemical	Tefzel	PVDF	PET	PVC	PMMA	Polycarbonate	Delrin	Ultem	FEP	PFA
Acetaldehyde	A	C		C	C	C	A	A	A	A
Acetic acid 10%	A	A	A	C	A	B	C	A	A	A
Acetone	A	C	C	B	C	C	A	B	A	A
Ammonia water	A	A	A	C	A	C	C	A	A	A
Benzine	A	A	A	A	C	C	A	A	A	A
Caustic soda 50%	A	A		C	A	C	A		A	A
Chloroform	A	C	C	C	C	C	A	C	A	A
DMF	A	C		C	C		C		A	A
DMSO	A					B	C		A	A
Ethanol	A	B	A	C	C	C	A	A	A	A
Ethyl acetate	A	C	B	C	C	A	A	B	A	A
Formalin 37%	A	A		A	A	C	A		A	A
Hydrocarbons	A	A	A	C	C	B	A		A	A
Hydrochloric acid 20%	A	A	A	A	A	A		A	A	A
Hydrogen peroxide 30%	A	A		A	A	A	C	A	A	A
Nitric acid 10%	A	A	C	A	A	A	C	A	A	A
Phosphoric acid	A	B	A	A	A	A	C	A		
Pure water	A	A	A	A	A	A	A	A	A	A
Seawater	A	A		A	A	A	A	A	A	A
Sodium hypochlorite 5%	A	A		A	A	C	C		A	A
Sulfuric acid <10%	A	A	A	A	A	A		A	A	A
Sulfuric acid 30%	A	A	A	A	A	B	C	A	A	A
Tetrahydrofuran	A	B	C	C	C	C	A		A	A
Toluene	A	A	B	C	C	C	C	B	A	A
Xylene	A	A	B	C	C	C	A	B	A	A

Data from: http://www.coleparmer.com/Chemical-Resistance; http://nalgenecontainers.nuncbrand.com/PDF/Chemical%20Resistance%20Chart%20Detail.pdf.
A: Compatible B: Compatible under some conditions C: Not suitable for use.
Note: This table can be used as a general reference. Please check the compatibility of selected materials with chemicals under particular conditions before use.

4.2.1.8 PEEK™

Chemicals interacting with polyether ether ketone (PEEK™) are not pH-dependent. However, this material is to be avoided with nitric, sulfuric, and halogenated acids such as HF, HBr, and HI. It can be usually employed with HCl, and also with pure halogen gases. In addition, it should be used carefully with methylene chloride, THF, and dimethyl sulfoxide (DMSO) in order to avoid swelling.

PEEK possesses other interesting properties such as transparency to microwaves and a very low porosity; the latter makes it especially suitable as a biocompatible material by virtue of adsorbing only very small amounts of metals. By contrast, it is quite expensive and can be difficult to obtain and, especially, to shape.

PEEK can operate at temperatures up to 100 °C in tubing and 125 °C in connectors. It can be easily threaded, has an oxygen permeability of 14 $cm^3/100\ in^2 \cdot 24\ h \cdot atm/mil$ at 25 °C, and can be autoclaved.

Like Kel-F, PEEK can be sterilized with γ-rays, ethylene oxide, or heat. PEEK is commonly used in tubing, connectors, cells, and valves. It is also very used in microwave vessels in combination with Teflon. The Teflon inner vessel provides chemical resistance, whereas the external vessel made of PEEK provides mechanical robustness.

4.2.1.9 PMMA

Polymethyl methacrylate (PMMA/Perspex/Plexiglass) is colorless, transparent, but vulnerable to solvents and fragile. It is commonly used in connectors, cells, and LOV platforms.

4.2.1.10 Polycarbonate

Polycarbonate is a colorless material, transparent, but vulnerable to solvents and fragile. It is stronger than PMMA. It is commonly used to build connectors, cells, and LOV platforms.

4.2.1.11 PVC

Poly (vinylchloride) (PVC) is easily shaped, but vulnerable to solvents. It is elastic and cheap. PVC is used in connectors, cells, and screws.

4.2.1.12 Ultem

The Ultem resin family of amorphous thermoplastic polyetherimide (PEI) resins offers outstanding elevated thermal resistance, high strength and stiffness, and broad chemical resistance. Ultem is available in transparent and opaque custom colors, as well as glass-filled grades. Moreover, Ultem copolymers are available for even higher heat, chemical and elasticity needs. Ultem resins balance both mechanical properties and processability, offering exceptional flexibility and freedom of use. It can be used in LOV platforms instead of using PMMA when organics solvents have to be handled.

4.2.1.13 Delrin®

Acetal, Polyoxymethylene (Delrin/POM), is a robust polymer and an excellent choice for components that are to retain their mechanical strength over long periods (e.g. threaded connectors and adapters). Despite its mechanical strength, it cannot remain in contact with acid or alkaline solutions for a long time without degrading. Therefore, it should only be used with neutral aqueous solutions or organic solutions.

Delrin can withstand temperatures up to 60 °C and pH values from 6 to 8. It can be easily threaded and is amenable to autoclaving. It can be used in detectors, cell holders, diode heads, and screws.

4.2.1.14 UHMW PE

Ultra High Molecular Weight Polyethylene (UHMW PE) is widely used on account of its resistance to a number of chemicals, particularly in aqueous solutions. It is durable and highly suitable for operation at room temperature. However, it is to be avoided with organic and chlorinated solvents, and also with nitric acid. This material can withstand temperatures up to 50 °C and pH values of 0–14. It can also be autoclaved.

4.2.1.15 Polypropylene

Polypropylene (PP) is an all-purpose material resistant to common chemicals, spanning a wide pH range. This has made it the preferred choice for a number of uses. However, some chemicals can degrade, fade, or break it. Therefore, it is important to ensure that the chemicals to be used are compatible with it; this entails avoiding chlorinated and aromatic compounds, and also some organic solvents.

Polypropylene can be used at temperatures up to 40 °C and pH values of 0–14. It can also be autoclaved.

4.2.2 Tools

Devices are built from raw materials, such as plates, round, square, or hexagonal bars with different thickness, which need to be cut before being worked with precision. Useful tools for this purpose are described below.

For rough cutting of plastic plates a circular saw (tile cutter) is very useful; see Figure 4.1. This circular saw should have a fine and net cut. It is desirable to have a connection system that is coupled to a vacuum cleaner to absorb the dust and the possible chip formed.

Horizontal bandsaws (Figure 4.2) are more useful for cutting narrow rods and plates. When dealing with small pieces, it is better to use a band saw such as that shown in Figure 4.3.

In both cases the appropriate type of saw must be selected depending on the type of material to be cut: plastic, metal, glass, or ceramic. For the latter, it is recommended to use a cutting fluid to prevent overheating of the saw and to lengthen the saw life.

Sander, polisher, and grinder are very useful tools for the finishing touches (Figure 4.4).

A digital caliper is also very helpful when precision measures are required (Figure 4.5).

Another frequent operation when developing devices for the laboratory is the realization of threaded, whether male using the dies or taps (Figure 4.6). In FIA standard threaded taps and dies are used i.e. ¼″-28, while the standard for optical fiber is ¼″-36.

FIGURE 4.1

Tile cutter. (For color version of this figure, the reader is referred to the online version of this book.)

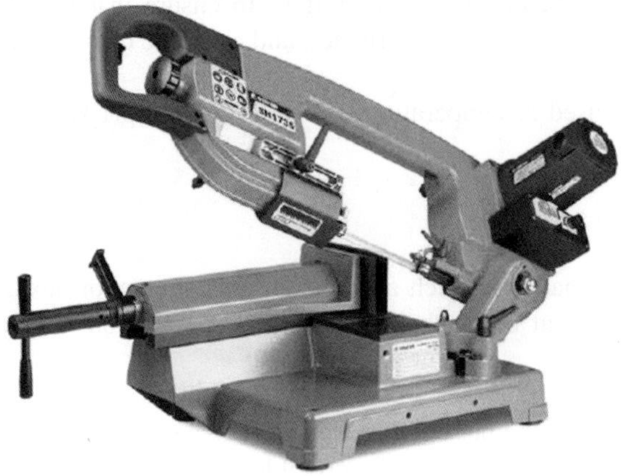

FIGURE 4.2

Horizontal bandsaw. (For color version of this figure, the reader is referred to the online version of this book.)

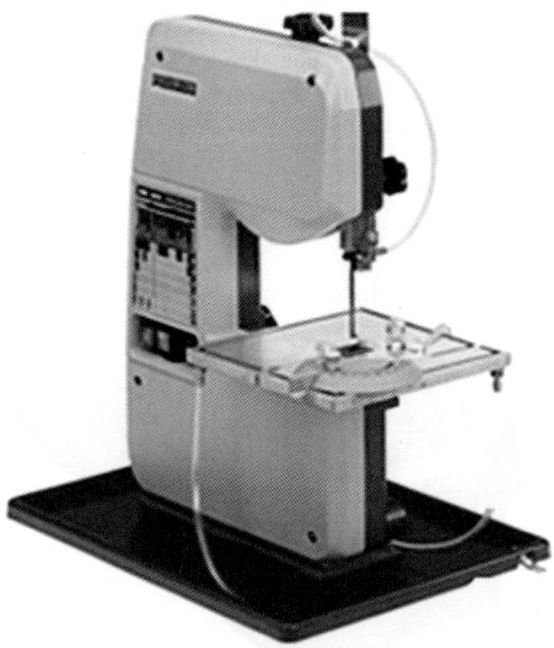

FIGURE 4.3

Bandsaw suitable for small pieces. (For color version of this figure, the reader is referred to the online version of this book.)

FIGURE 4.4

Sanding, polishing, and grinding machine. (For color version of this figure, the reader is referred to the online version of this book.)

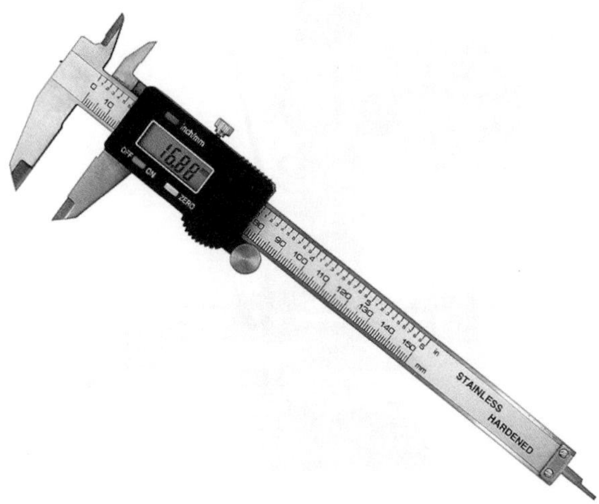

FIGURE 4.5

Digital caliper. (For color version of this figure, the reader is referred to the online version of this book.)

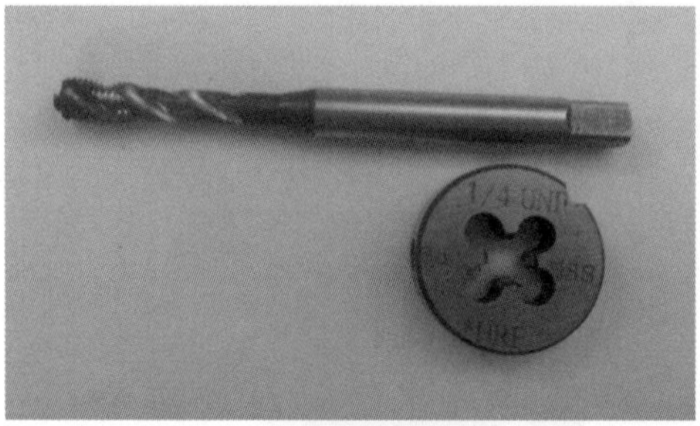

FIGURE 4.6

Tap and die used to thread. (For color version of this figure, the reader is referred to the online version of this book.)

Special attention has to be paid to flow spectrophotometric cuvettes, since they have metric six (M6) thread, with similar measures to ¼″-28. A common mistake is trying to thread an FIA screw to a spectrophotometric cuvette what leads to liquid leakage and thread forced when trying to avoid the latter.

Although the computer numerical control (CNC) milling machine controlled by computer described below is widely used, the accurate manual drilling–milling machine shown in Figure 4.7 is also frequently required.

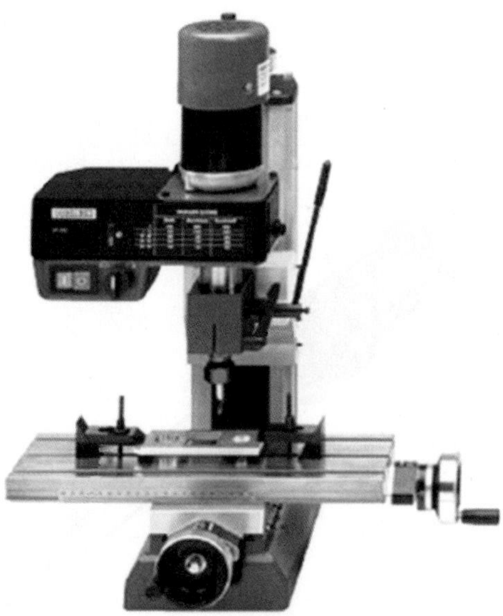

FIGURE 4.7

Manual drilling and milling machine. (For color version of this figure, the reader is referred to the online version of this book.)

4.2.2.1 The lathe

Along with the mill, the lathe is also a basic and useful tool, frequently used to build devices precisely. When a large number of replicates have to be built, it is almost mandatory to use computer controlled systems. By this, the desired device is easily reproduced. Computer-controlled CNC mills with a software CAD/CAM that allows the piece design, programming, and executing the movement of the cutting tool are very useful.

Below are summarized the main steps of the sequence to follow when using these computer-controlled machines:

- Indicate the gross size of the piece.
- Draw the profile of the element to build.
- Program the tool work.
- Tool installation, i.e. setting the zeros of the *x*, *y* and *z* axis. The final result depends fundamentally on the precision of the installation. A wrong setting will ruin the piece dimensions.
- Program execution.
- Check that the dimensions of the first piece are correct.
- Adjust the position of the tool if necessary to achieve the correct dimensions in the next pieces.
- Once the desired piece is achieved, repeat the program execution as many times as the number of pieces desired.

Figure 4.8 shows some of these steps applied to the development of typical screws one used in HPLC and other typical flow screw. Figure 4.9 shows the lathe EMCO Compact 5.

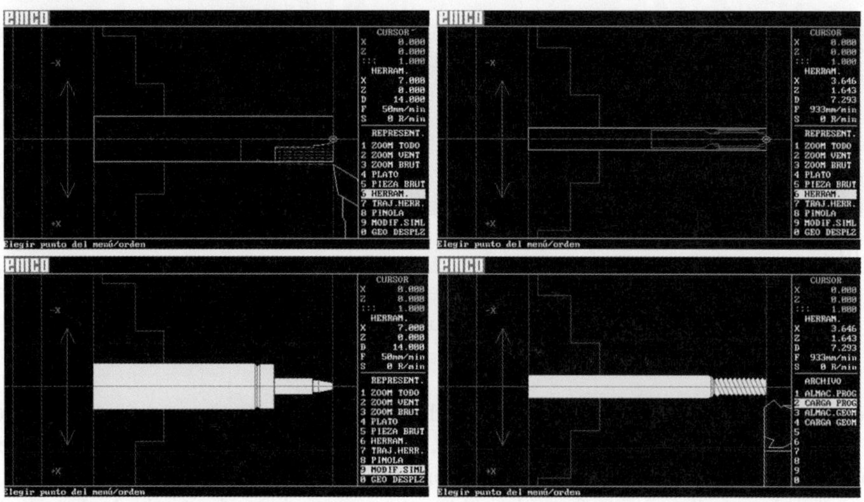

FIGURE 4.8

EMCO Compact 5 lathe software used to design and build screws. (For color version of this figure, the reader is referred to the online version of this book.)

FIGURE 4.9

EMCO Lathe Concept Turn 55, with automatic tools turret of eight positions with a maximum lathing diameter of 52 mm (2.1″) and a maximum lathing length of 215 mm (8.5″). The raw piece is held by the spindle that draws a circular movement while the tool moves toward the rotation axis (*z*) or perpendicular to it (*x*). (For color version of this figure, the reader is referred to the online version of this book.)

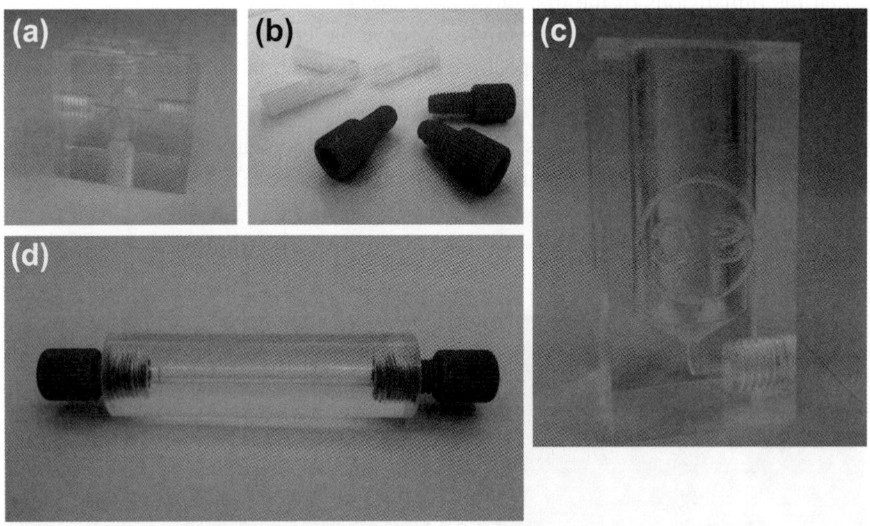

FIGURE 4.10

Some components made with the lathe. (a) 6-way connector. (b) Screws. (c) Mixing chamber. (d) Microcolumn. (For color version of this figure, the reader is referred to the online version of this book.)

Figure 4.10 shows some pieces built with the lathe. It has to be taken into account that some parts of the piece have to be finished manually, e.g. threads, unless having a lathe with the appropriate decoder these cannot be built automatically. Threads are commonly finished with the manual milling machine with a motor with two directions of rotation. Also, some experience is required when dealing with some materials, such as methacrylate, since the heat generated by the friction with the tool during processing can lead to melt the material and sticking the piece to the tool, being able to destroy it. Therefore, it is important to work by stages, extracting the chips formed between them.

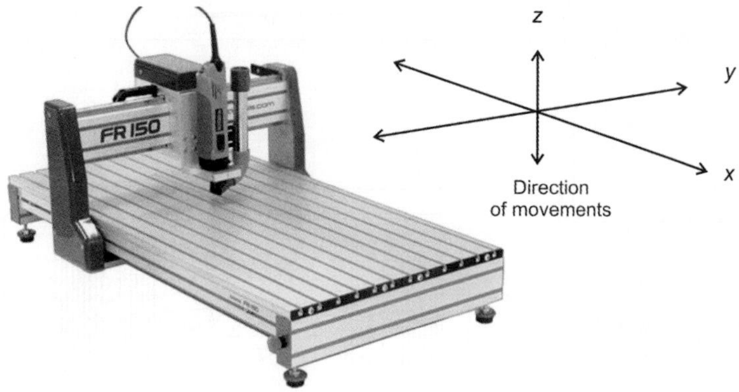

FIGURE 4.11

Milling machine Alarsis FR150. Useful working area: 860 × 460 cm. Useful height *z*-axis: 80 mm. High-speed shift (10,000 mm/min). Resolution 0.02 mm. (For color version of this figure, the reader is referred to the online version of this book.)

4.2.2.2 The milling machine

Unlike with the lathe, when working with a milling machine, the swivel is the tool and not the piece, i.e. staying the piece in static position. In flow systems, the pieces required are small and low in height. Thus, it is convenient to select a milling machine with a large path in the axis x and y, not being so important the axis z.

CNC milling machines (Figure 4.11) are really useful for the development of electronic printed circuit boards, and for the development of laboratory equipment. Different complementary programs have to be used depending on what is willed to build, e.g. VCarve for laboratory equipment and Eagle for electronic printed circuit boards.

4.2.2.3 Machining laboratory equipment

For this purpose, two programs are combined, i.e. VCarve is used to design the pieces and generate the G-Code that will be used by the MACH3 program to build the piece with the milling machine.

VCarve program works by steps:

- Defining the size of the starting raw material.
- Draw of the piece to be built. Figure 4.12 shows the design and drawing made exploiting VCarve to build a chip on valve.
- Tools selection and work programming. Figure 4.13 shows a screenshot of the tools programming step.
- G-Code files generation. Figure 4.14 shows the piece simulation, once this simulation is the one desired, G-Code files to be used by the MACH3 program will be generated and stored.

VCarve is not able to control the milling machine. This is made by the MACH3 program, using the files generated by the VCarve program.

First and foremost, the primordial step to be carried out with the MACH3 is to find the correct zero-axis position, as when working with the lathe. With this purpose, the tool, generally the milling cutter,

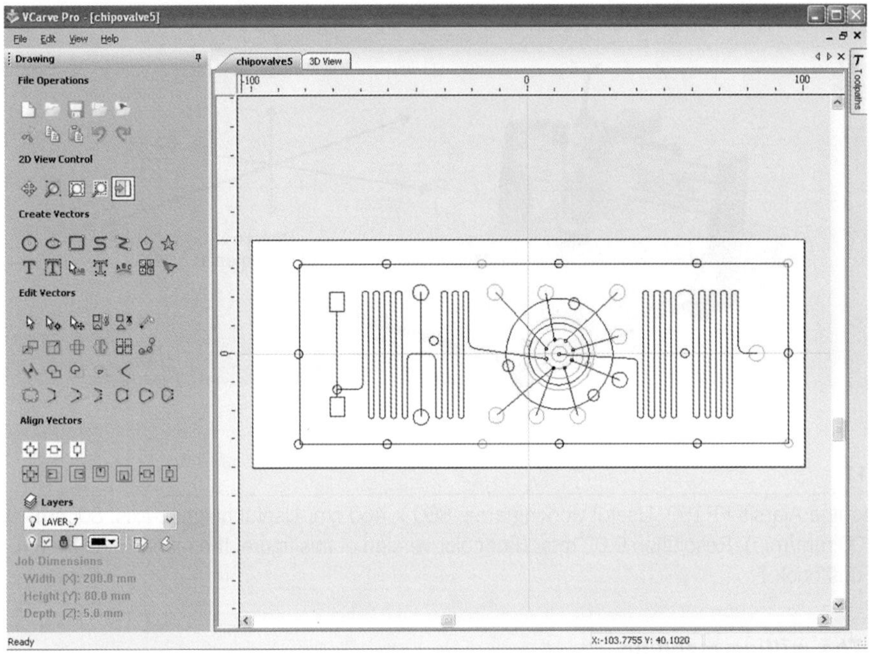

FIGURE 4.12

Chip on valve design and drawing exploiting VCarve. (For color version of this figure, the reader is referred to the online version of this book.)

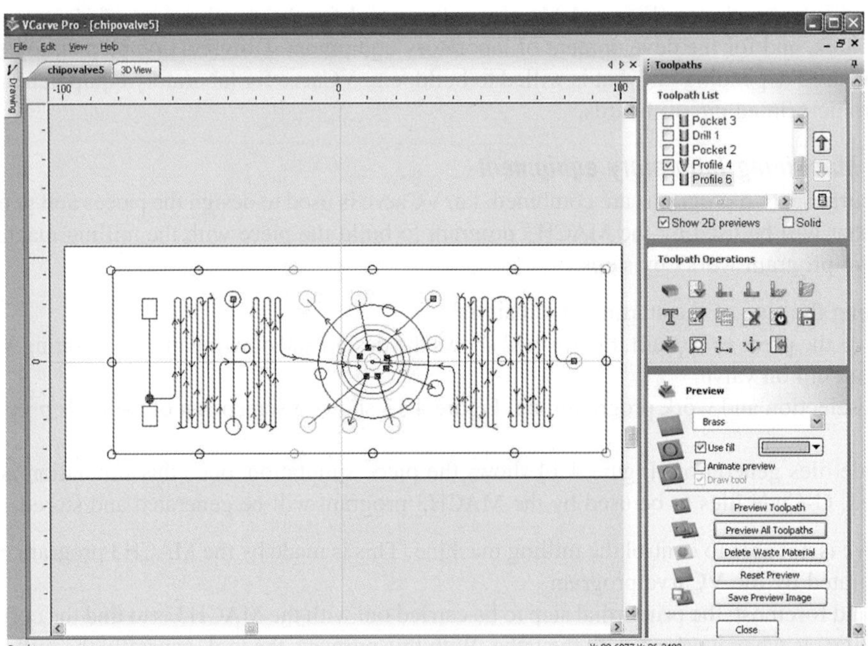

FIGURE 4.13

Tools programming by VCarve. (For color version of this figure, the reader is referred to the online version of this book.)

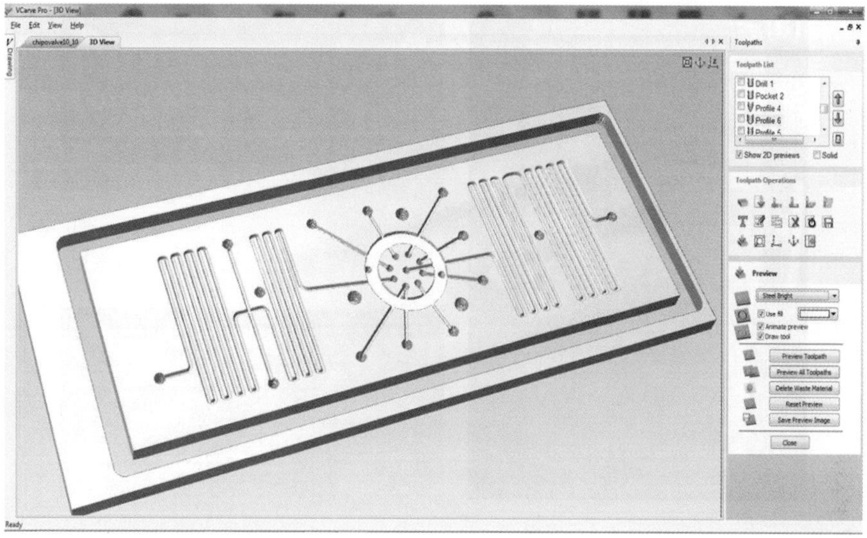

FIGURE 4.14

Piece simulated by VCarve. (For color version of this figure, the reader is referred to the online version of this book.)

position is manually controlled, searching the middle of x and y axis and positioning the tool touching the piece surface to set the zero of the z axis.

Once the tool has been correctly positioned, the program generated by VCarve can be executed. Working speed can be previously selected to assure material and tool integrity. Figure 4.15 shows a screenshot of the computer monitor when working with MACH3.

Figures 4.16 and 4.17 show some devices built by the combined use of VCarve and MACH3. It is important to remind when dealing with certain materials, e.g. methacrylate that holes and any other

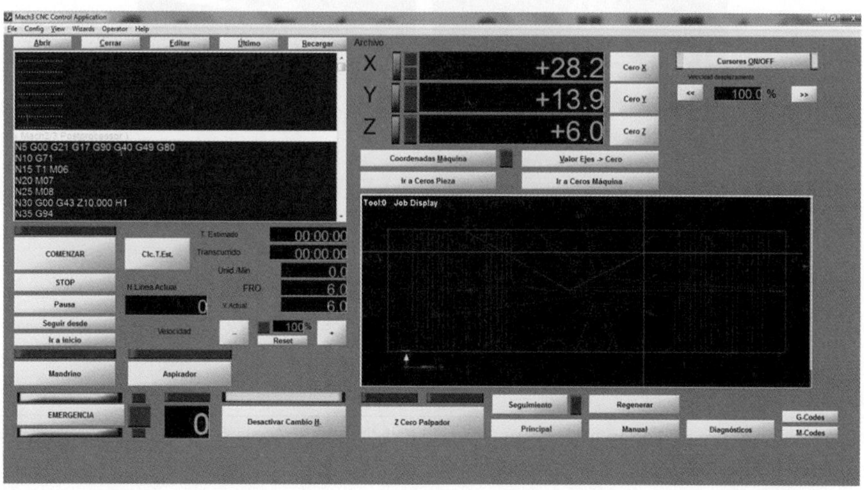

FIGURE 4.15

Screenshot of MACH3. (For color version of this figure, the reader is referred to the online version of this book.)

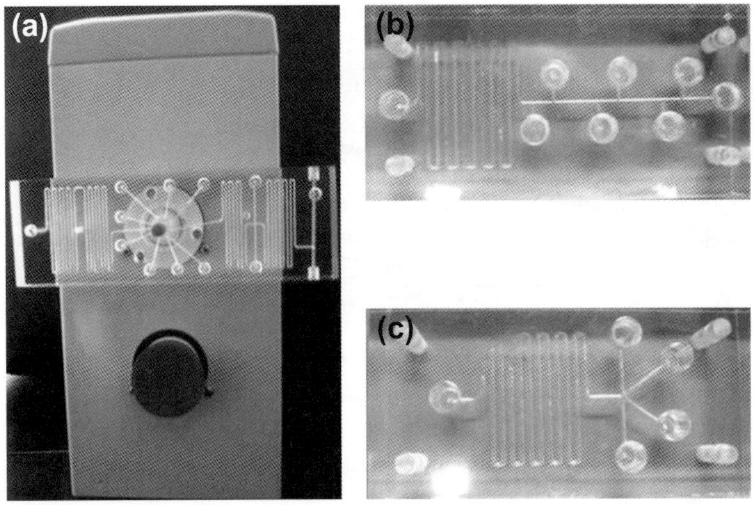

FIGURE 4.16

Some methacrylate chips built by combined use of VCarve and MACH3. (a) Chip on valve mounted atop the selection valve. (b) Chip designed for cross-injection analysis. (c) Chip especially used in kinetic methods. (For color version of this figure, the reader is referred to the online version of this book.)

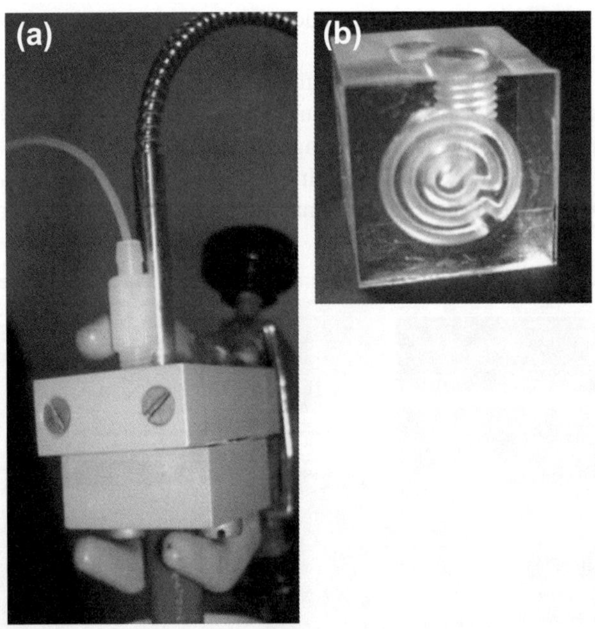

FIGURE 4.17

Some pieces made with the milling machine. (a) Optrode. (b) Chemiluminescence cell. (For color version of this figure, the reader is referred to the online version of this book.)

work made with the milling cutters and drills cannot be made at once but in several steps to avoid material melting and so work disabling, together with blockage and breakage of the tool. In general, other plastics don't pose this difficulty, but polycarbonate.

4.3 Electronics workshop

It is possible to find on the market many electronic devices necessary for flow system assemblies, whether constructed, e.g. detectors, power supply, input converter cards and digital output cards, relays, or in kit form. However, frequently it is mandatory to build your own electronics, e.g. to adapt the analog outputs from one device to the other input.

But no panic, current electronics works generally with concepts and circuits that do not require a great knowledge in the field. Thus, for example, designing a signal amplification adapter requires only a power source, an operational amplifier chip, e.g. the classic μA741, and two resistors whose relationship gives the value of the gain (Figure 4.18).

How to build an electronic device is summarized in the following steps:

- Appropriate design selection based on bibliography or internet sources.
- Check its availability in the market.
- If it is not commercially available, then select and acquire the circuit components.
- In case of having a solderless breadboard (Figure 4.19), mount on it the desired circuit and check it works properly.
- Once verified the correct performance of the circuit, design the final version of the printed electronic circuit and mount the components on it.

A more detailed description of the elements and processes required to build these electronics devices is given below.

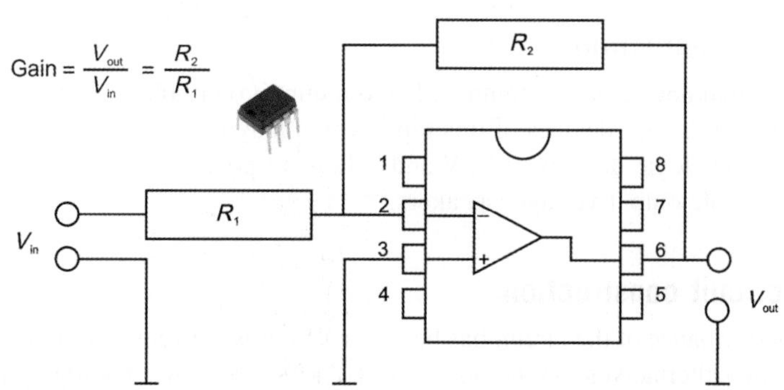

FIGURE 4.18

Amplifier with a gain = R_2/R_1.

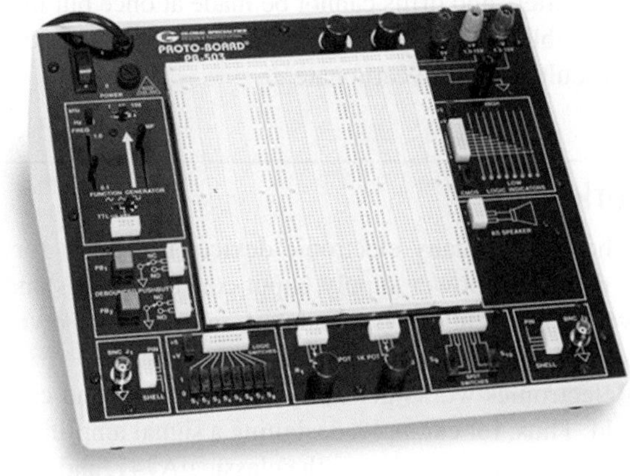

FIGURE 4.19

Solderless breadboards. (For color version of this figure, the reader is referred to the online version of this book.)

4.3.1 Electronic workshop's tools

The electronic workshop requires little, basic, very common and cheap instrumentation. Main elements are: soldering iron, multimeter, oscilloscope, adjustable voltage supply, and function generator.

4.3.2 Circuit selection

There is extensive literature on electronic circuits from edited collections on a professional level, and other aimed at amateur hobby. Internet is nowadays a powerful tool, which offers varied sources, providing also information from manufacturers of all types of components, with descriptions not only of their products features but also of their applications. As mentioned above, frequently and hopefully the desired circuit will be commercially available, otherwise the following steps can be followed to build it.

4.3.3 Prototype construction

Before building the final device, it is recommended to mount it on a solderless breadboard. Currently, on the market there are various devices of this type, which may include power sources required, e.g. typically +5 V for TTL logic devices, +12 V and −12 V to power-integrated circuits, the function generator with adjustable output voltage, speakers.

4.3.4 Printed circuit construction

Once the proper performance of the circuit has been verified, it is convenient to build a more definitive system. There are two alternatives: on the one hand it can be made by a photographic system, on the other hand with the milling machine. The first requires fewer tools, while the second does not require chemicals. In both cases an image of the printed circuit to be developed is needed. There are different ways of obtaining it. However, we will focus on the most comfortable one, based on the use of commercially available software.

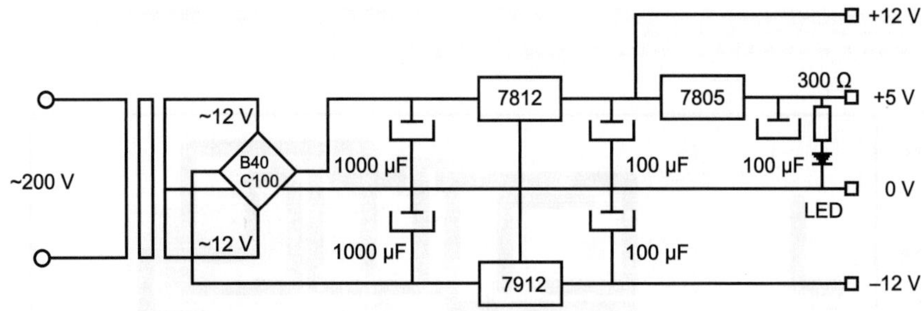

FIGURE 4.20

Schematic depiction of a stabilized power supply.

4.3.4.1 Printed circuit design exploiting the program Eagle

Eagle software not only provides an image of the printed circuit, but also exploiting a plug-in (pcb-G-Code) generates the G-Code required to manufacture the printed circuit using a milling machine exploiting the program MACH3.

First of all, the electronic scheme has to be edited, selecting at the same time the components to be used. By this, the size of all the components will be taken into account for the plate construction.

As an example, the construction of a stabilized power supply very useful for the development of various devices used in flow techniques, e.g. as source of a Hamamatsu photomultiplier used in chemiluminescence detectors, is presented. Figure 4.20 shows a schematic depiction of the cited electronic circuit, able to supply +5, +12 and −12 V.

Figure 4.21 shows the result obtained using the schematic editor of the software Eagle to design the above-mentioned power supply.

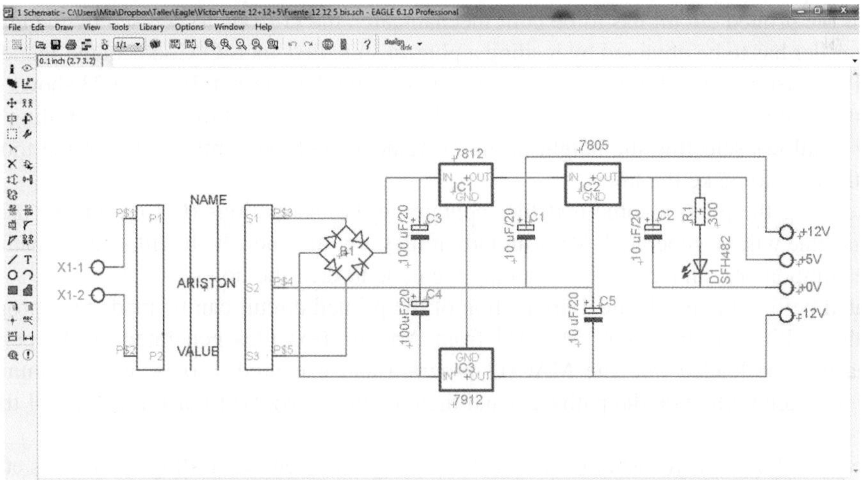

FIGURE 4.21

Schematic depiction of the power supply provided by Eagle editor. (For color version of this figure, the reader is referred to the online version of this book.)

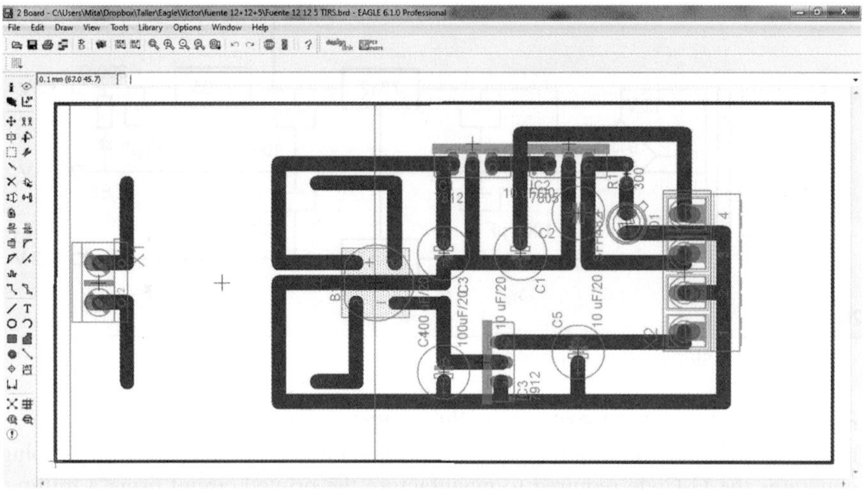

FIGURE 4.22

Electronic board obtained with Eagle. (For color version of this figure, the reader is referred to the online version of this book.)

Then exploiting the board editor, the step two of the process, i.e. printed circuit design, is carried out. The user can distribute the components as desired and interconnections between these are also easily made. Figure 4.22 shows the plate at this stage. This can be printed and used to make the printed circuit by the photographic system. This design can also be sent to plate manufacturer companies.

Then if the printed circuit has to be built mechanically, the G-Code to be used by the milling machine controller has to be obtained by loading a plug-in so-called "pcb-G-Code" in the Eagle program as an ULP extension. This plug-in can be freely obtained on the internet. Figure 4.23 shows the simulation provided by the G-Code to check how will be the printed circuit made by the milling machine. This plug-in allows selecting the height in the movements and the depth to etch of the tool and the depth of the drill to make the holes.

By executing the plug-in some text files containing the instructions (G-Code) to be used by the milling machine will be generated. Most important files are: the one which will etch the pathways between the components, and the one containing the code to drill the holes.

As stated before, the mechanical production of the printed circuit can be made by a program that controls the milling machine, e.g. MACH3 from Artsoft. Text files generated by the pcb-G-Code plug-in have to be loaded into the MACH3 program and executed. Firstly, it is recommended to load the file *.etch.txt to etch the pathways and then the file *.drill.txt to drill the holes of the printed circuit.

Figure 4.24 shows a screenshot of the MACH3 program when the milling machine is etching the pathways between the components on the blank printed circuit to be used in a multipumping system.

Once all mechanical and electronics components have been produced, these have to be put together to finally assembly the devices.

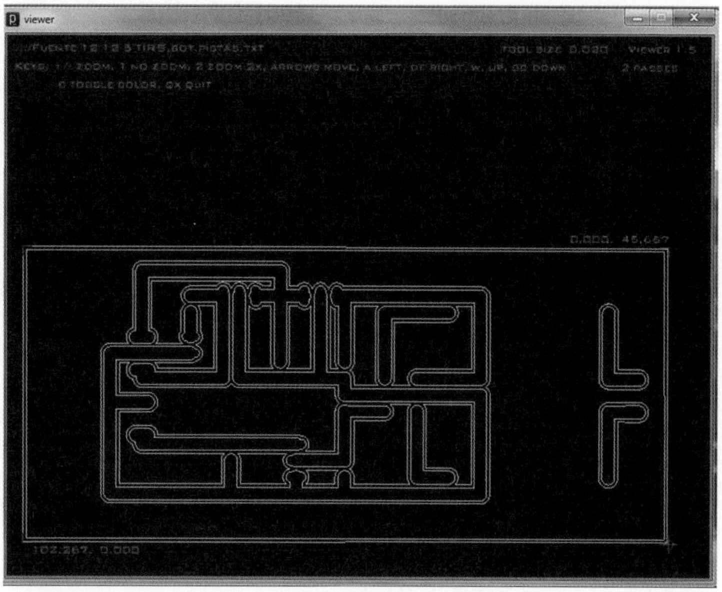

FIGURE 4.23

Printed circuit simulation. (For color version of this figure, the reader is referred to the online version of this book.)

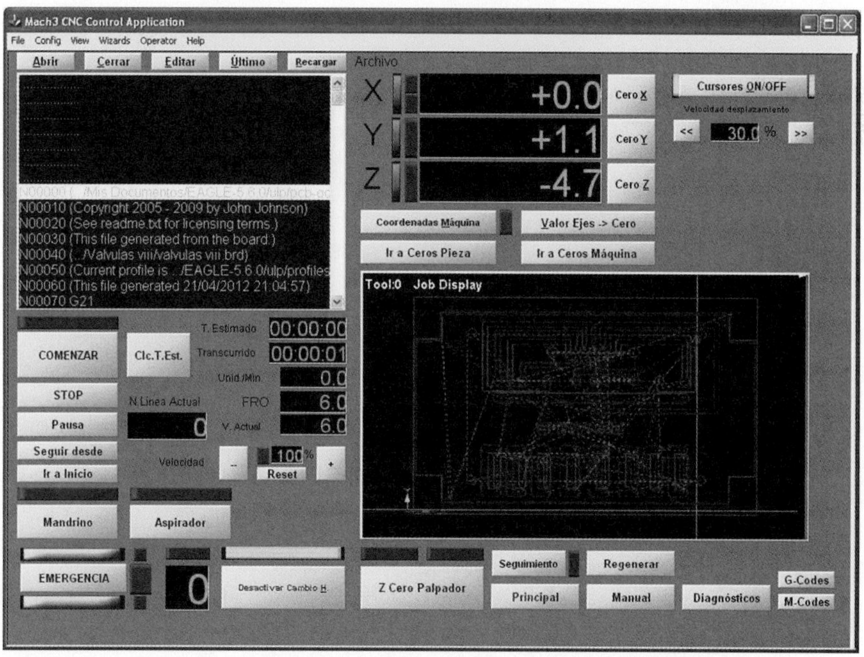

FIGURE 4.24

MACH3 screenshot. (For color version of this figure, the reader is referred to the online version of this book.)

CHAPTER

Instrumentation: Detectors, Accessories and Interfaces

5

CHAPTER OUTLINE

Flow Analysis. http://dx.doi.org/10.1016/B978-0-444-59596-6.00005-X

In the previous chapter how to build useful devices commonly used in flow assemblies was described. In the present chapter most relevant elements that compose the instrumentation commonly used in flow systems, including detectors, accessories, with special emphasis on the two most frequently employed in practice, namely: electrochemical and optical detectors together with most used interfaces are presented.

This chapter also deals with various types of auxiliary devices used in flow systems including power sources for general purposes and light sources, electronic circuits for controlling solenoid valves, peripheral instruments such as inductively coupled plasma mass spectrometry (ICP-MS) or gas chromatography (GC), and amplifiers. The devices described here have been of great help in constructing our own flow systems.

5.1 Optical detectors

The detection system constitutes an essential element of a flow assembly. In most cases, detectors commonly found in the laboratory are somehow adapted by using a suitable flow cell. However, an increasing number of detectors are being designed specifically to be used in flow systems. These range from quite simple models for routine applications to highly complex ones applied in research.

Optical detectors are the most widely implemented in flow systems for various reasons, but particularly for their robustness and high reproducibility. Optical detectors can be classified in light emission or absorption detectors. Despite both frequently use the same type of light measuring system the latter requires an additional light excitation source.

Many of the components of optical detectors are common to all. For this reason, we have classified them in terms of the spectral region they operate in, and in whether they rely on atomic or molecular absorption or emission.

5.1.1 Light sources

Light sources are used by absorption detectors. These differ essentially in the spectral region they encompass and also they can be of the continuous or discrete type, i.e. they span a wide spectral region or emit largely monochromatic light. This difference usually has an impact on cost.

5.1.1.1 Discrete light sources

A number of highly stable, inexpensive sources emitting discrete light are currently available. The greatest difficulty is finding one that operates reliably at the desired wavelength. Light-emitting diodes (LEDs) and gas or diode lasers possibly constitute the most interesting types of ultraviolet, visible and infrared light sources.

5.1.1.1.1 Light-emitting diodes

LEDs are simple, usually inexpensive semiconductors available in standard formats. They feature a high electrical yield since, unlike filament lamps their power is largely used to emit light with little heating.

LEDs are being increasingly favored as light sources for two main reasons: firstly because these require very low power consumption and secondly there are LEDs which operate at increasingly low wavelengths verging on the UV region. In fact, recently developed LEDs afford wavelengths as low as 245 nm with a strong light intensity and very low power consumption which makes them especially suitable for incorporation into portable equipment.

LEDs are very stable and usually powered by very simple electronic devices. Since the amount of light emitted is proportional to the current intensity passing through them, they are usually powered by current-based sources rather than the more usual voltage-based sources. While old LEDs require 10–15 mA in order to supply high enough light intensities, recently developed superbright diodes afford similar intensities with much lower current intensities, e.g. below 1 mA in some cases.

LEDs of different colors can be encapsulated in a single element to construct an RGB (red green, blue) LED. By separately controlling the intensity of each integrated LED, specific colors can be achieved. This is especially interesting in flow techniques since the schlieren effect caused by differences in refractive index can be corrected.

Commercial LEDs span a wide range of wavelengths. Figure 5.1 shows the emission spectra of a selection of LEDs. However there are still some spectral gaps that are rather difficult to bridge. LEDs can vary widely in cost, from very low, i.e. ca. 1 euro for IR diodes to more than 200 euros for UV diodes. Recently, a white LED has been developed to circumvent some shortcomings. It uses a blue LED containing a phosphor to convert a portion of emitted light into a band spanning most of the visible spectrum. By using a white LED the schlieren effect can be corrected and it can be used in

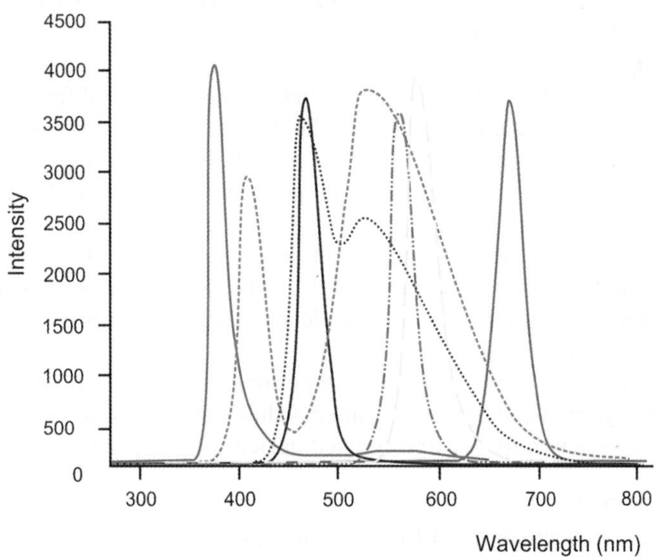

FIGURE 5.1

Emission spectra of a selection of LEDs. (For color version of this figure, the reader is referred to the online version of this book.)

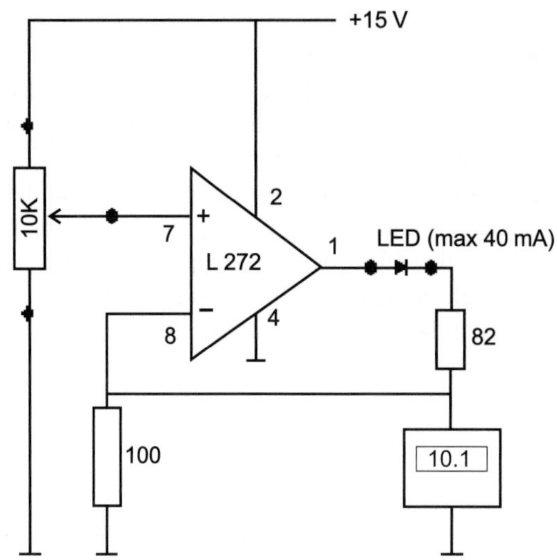

FIGURE 5.2

Electronic scheme of a very simple light source for laser and light-emitting diodes.

replacement of wolframium lamps providing advantages such as lower consumption, higher stability, lower heating and longer life.

Figure 5.2 shows the electronic scheme of a very simple light source manufactured in our workshop [1] which can power LEDs and laser diodes with current intensities up to 40–60 mA.

Delivered current intensity is adjusted via a potentiometer and can be read on an optional, low-cost digital display consisting of a fixed-scale voltmeter measuring the difference between the two ends of a 100 Ω resistor. By accurately adjusting the on-screen scale, one can have the current intensity supplied to the LED displayed on it. Care should always be exercised not to surpass the manufacturer's recommended maximum level in powering LEDs in order to avoid rendering them unusable.

Figure 5.3 shows a homemade constant-current power source for LEDs and laser diodes. The screen displays the intensity supplied to the LEDs in mA. Sources have been designed to power one, two or four LEDs simultaneously but independently. Those having two or four separate outputs are especially useful for RGB LEDs.

In addition, LEDs can be readily exchanged as they are inserted in a standard plastic head that can be screwed to various types of cell holders. Figure 5.4 shows an LED head mounted on a cell holder (Figure 5.4(a)) directly screwed to a charge-coupled device (CCD) spectrophotometer.

5.1.1.1.2 Laser diodes

Laser diodes constitute an alternative to LEDs inasmuch as they span a similar wavelength range, i.e. from IR to violet. However, laser diodes dissipate a much higher fraction of power as heat than LEDs, and are more expensive than these, ranging from 6 € (red, IR) to 600 € or more (violet). Despite they were formerly attractive replacements for LEDs their high cost has made them less competitive, especially after the inception of superbright LEDs.

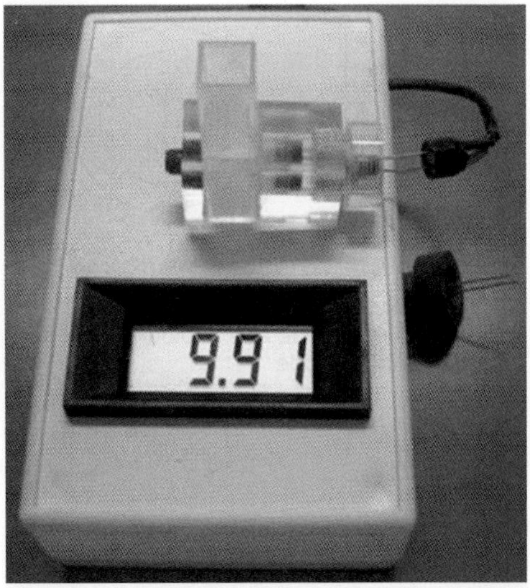

FIGURE 5.3

Homemade constant-current power source for LEDs and laser diodes powering a diode mounted on a threaded cell holder head.

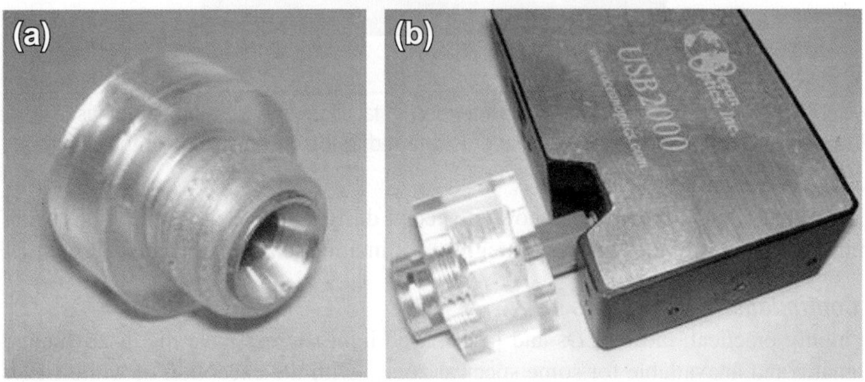

FIGURE 5.4

(a) LED head. (b) Threaded head directly connected to a CCD spectrophotometer. (For color version of this figure, the reader is referred to the online version of this book.)

Laser diodes can use the same types of power source as LEDs provided these can deliver an adequate current intensity, ca. 40 mA. Most recent models include a diode to read and help stabilize the emitted light intensity (Figure 5.5).

There are various commercially available circuits for powering laser diodes that range from devices generating continuous laser light to others allowing light modulation which is of special interest for fluorimetric applications. These sources are powered by low-voltage batteries or by some of the

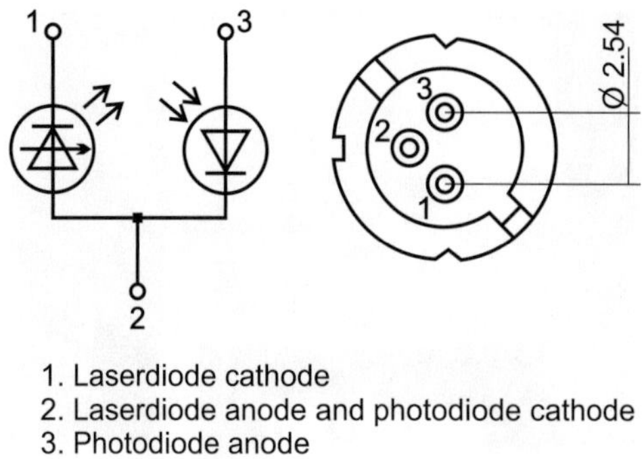

FIGURE 5.5

Scheme of a laser diode integrating a photodiode for stabilizing emitted light.

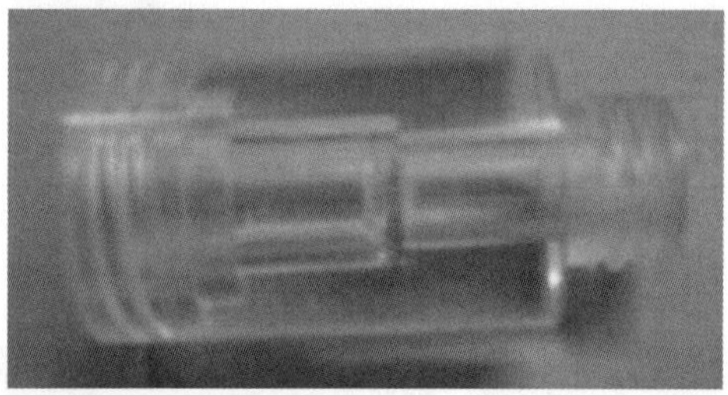

FIGURE 5.6

Sciware Systems collimator with an acrylic head intended to facilitate the incorporation of laser diodes into cell holders. (For color version of this figure, the reader is referred to the online version of this book.)

previous described circuits. Figure 5.6 shows a special designed head produced by Sciware Systems for accommodating laser diodes that can integrate collimating lenses exchangeable with LEDs.

5.1.1.2 Continuous light sources

Although highly practical, most LEDs and laser-based light sources have the disadvantage of being monochromatic and unavailable for some spectral zones, with the exception of white LEDs. Continuous light sources provide a wider spectral region, allowing corrected measures by a second wavelength when exploiting a detector capable of simultaneously measuring two wavelengths.

5.1.1.2.1 Visible light sources

The tungsten filament lamp is the most common source of visible light. This lamp requires a stable-voltage power source (Figure 5.7) and waiting for light stabilization. Such a source can power low-intensity tungsten lamps of the type used in electric torches.

5.1.1.2.2 Ultraviolet light sources

Most commonly used UV light source is the deuterium lamp. There are commercially available visible-deuterium sources with an SubMiniature version A (SMA) fiber-optic output connector

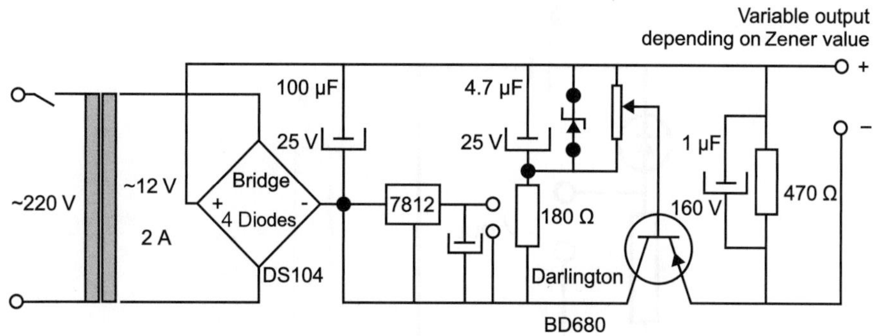

FIGURE 5.7

Stabilized power source for a tungsten lamp allowing the output voltage, the maximum level of which depends on that of the Zener, to be selected.

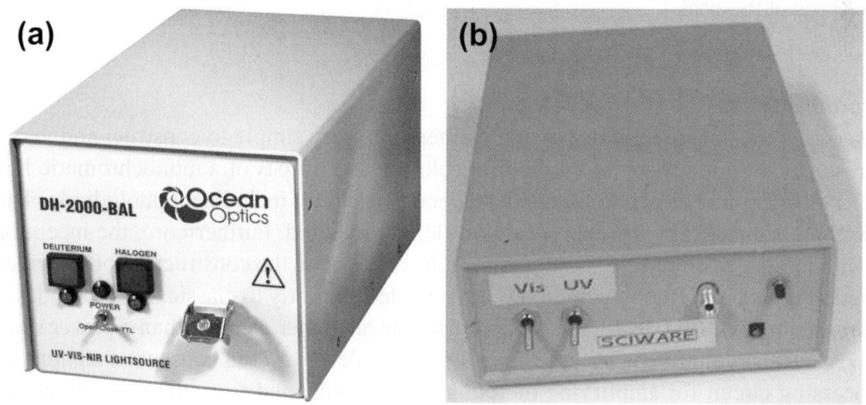

FIGURE 5.8

(a) Commercial UV–visible light source. (b) Low-power UV–visible light source. (For color version of this figure, the reader is referred to the online version of this book.)

(Figure 5.8(a)). Some recently developed visible-deuterium lamps (see Figure 5.8(b)) are powered by radio frequency sources and can operate in a flashing mode, which saves energy and stabilization time. In addition, they use voltages low enough to be used in field measurements.

Another frequently used UV light source, particularly in fluorimetry, is the xenon lamp, which can be of the pulsating type and synchronized with detector readings.

5.1.2 Photometric detectors

Flow techniques use a wide range of photometric detectors from simple colorimeters to CCD and diode array spectrophotometers. While photometers typically available in the laboratory are often adapted to flow systems by using an appropriate cell, a number of specially designed detectors are also currently available.

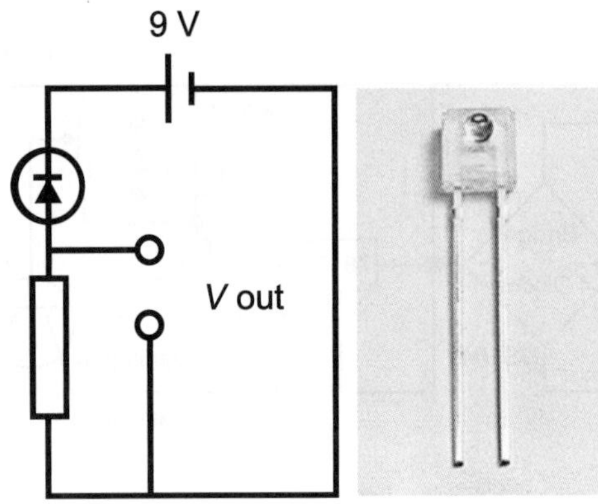

FIGURE 5.9

Photometric detector using a Hamamatsu photodiode. (For color version of this figure, the reader is referred to the online version of this book.)

5.1.2.1 Colorimeters

Colorimeters, do not contain light dispersion element, are very simple to construct and very inexpensive as a result. The simplest configuration of a colorimeter consists of a monochromatic light source e.g. an LED or laser diode, a cell holder and an electronic circuit including a photodiode. The specific LED to be used is dictated by the analytical wavelength required. Furthermore, the inception of photodiodes with integrated circuits, e.g. Hamamatsu, has facilitated the construction of extremely simple colorimeters (Figure 5.9). These photodiodes are operated similarly to classical photodiodes; however, they feature a gain from 1300 to 13,000 without needing any other element than a power source, e.g. a battery, and a current/voltage converter resistor. For example, the S7183 consists of a photodiode and a signal processing circuit for amplifying the photocurrent generated from the photodiode up to 1300 times. The S7183 can be used the same way as a reverse-biased photodiode, and in most cases, it delivers sufficient output voltage by just connecting a load resistor. The output is a few volts and can thus be directly measured with a meter or an output to an A/D converter.

Performance can be enhanced by incorporating an operational amplifier (Figure 5.10). Thus, a stabilized reference voltage can be used in conjunction with a summation circuit in order to alter the baseline of the graphical output at will.

When choosing a photometric detector, it is important to check that the intensity of the light source can be adjusted and that its sensitivity will be adequate for the intended purpose. Often, the chosen detector becomes saturated or is supplied with an inadequate current intensity, resulting both in distorted analytical signals.

5.1.2.2 Spectrophotometers

Most often, flow techniques are coupled to common spectrophotometers available in the laboratory, which are operated at a single wavelength. As with colorimeters, this can result in distorted signals by effect of changes in refractive index in the liquids circulating through the flow cell, i.e. the schlieren effect.

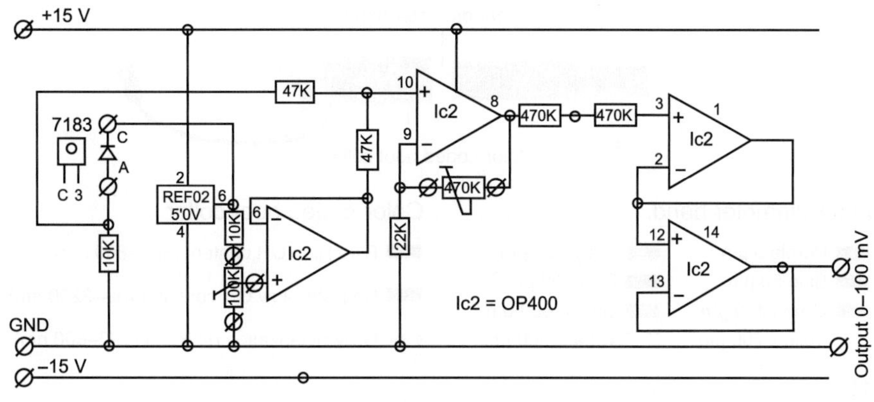

FIGURE 5.10

Photometric detector capable of baseline adjustment.

Frequently, the schlieren effect can be corrected in flow injection analysis (FIA) by injecting the sample into a carrier with the same refractive index, and since reagents will be continuously injected into liquids of the same index, no double peak will appear. This effect, however, is very difficult to suppress in sequential injection analysis (SIA) as this technique relies on the stacking of sample and reagent plugs and this causes schlieren effect at each interface change. The most efficient way of avoiding it is by using a detector capable of simultaneously measuring two wavelengths, e.g. a diode array or CCD spectrophotometer. One of the monitored wavelengths should be that of the analytical signal and the other anywhere the target species exhibits no absorption. Since the effect of changes in refractive index is not specific, it can be partly or completely suppressed by subtracting the absorbances measured at the two previous wavelengths.

CCD detectors are of special interest in this context since these can be used with optical fibers not requiring mechanically moved components, being their wavelengths highly reproducible. Also, they can obtain the whole spectra within few milliseconds. This allows obtaining spectra at each point along a flow profile facilitating application of multivariate chemometric techniques. It also exposes changes in refractive index which reflect in horizontal shifts in the spectrum, as fluids pass through the detector.

These detectors have a very small footprint being readily integrated with liquid drivers. Also, some can be directly powered on a computer USB (universal serial bus) port, which enormously facilitates their use in portable equipment.

5.1.2.3 Accessories for photometric detectors

5.1.2.3.1 Optical fibers

Optical fibers have gained widespread use in flow techniques. One of their greatest advantages in this context is that they allow cells to be used in a clear environment since both the light source and the detector can be placed at a fairly long distance from them. Although several forms are currently in use, the most widely accepted are those based on SMA 96 connectors with a ¼-inch thread and 36 threads/inch.

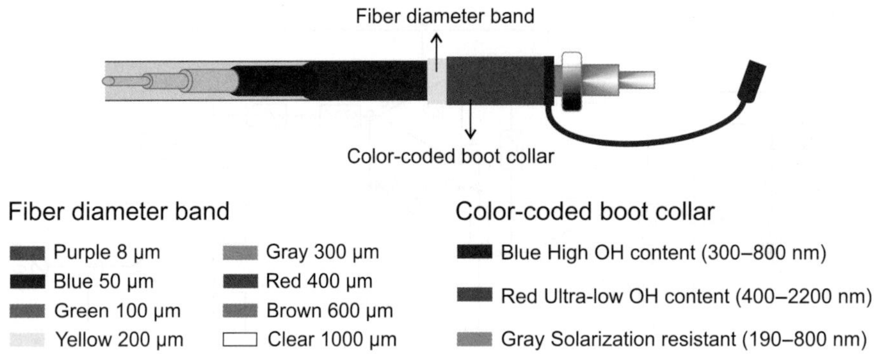

FIGURE 5.11

Optical fiber coding. (For color version of this figure, the reader is referred to the online version of this book.)

Optical fibers can be plastic, i.e. methacrylate, or inorganic, i.e. glass for the visible spectral region and quartz for the UV region and range from 8 to 1000 μm in diameter. Some manufacturers use color codes to facilitate their identification (Figure 5.11) [2].

Optical fibers consist of a core of a given material and a coating layer of a different material with a lower refractive index. This results in total reflection of light circulating within the fiber and avoids leaking through its outer walls.

Because AF Teflon has a lower refractive index than water, it can be used for this purpose. Thus, any light impinging on water will reflect on Teflon walls and the water–Teflon combination will act as an optical fiber. This allows spectrophotometric and fluorimetric cells of a long, not necessarily aligned light path, to be constructed such as liquid waveguide capillary cells (LWCCs).

5.1.2.3.2 Cells

Flow cells are available in a variety of shapes and materials. Glass (visible) and quartz (UV–visible) microcells with their inlet and outlet at the top have been the most widely used to date (Figure 5.12). The light path most frequently used is 10 mm and the observation hole is at a height of 8.5 or, more often, 15 mm from the base. The hole diameter can be 1, 1.5 or 2 mm corresponding to cell volumes of 8, 18 and 32 μL, respectively. A very small hole results in increased resolution and low void volumes, at the expense of a reduced amount of energy reaching the detector which can restrict its use. Attention!!! Connectors typically used in these cells are metric 6, very similar to ¼-inch 28 threads/inch FIA screws. A common mistake is using FIA screws with spectrophotometric cells leading to liquid leaks.

Z-cells and their analogs (Figures 5.13) are amenable to coupling with optical fibers. The light path is of 10 mm and light goes in and out via a detached window consisting of a borosilicate, quartz or sapphire disc. The cell body is plastic, e.g. methacrylate, polyether ether ketone or Teflon. Measurements with these cells are frequently distorted by effect of the presence of entrapped bubbles. The connectors used are typically ¼-inch 28 threads/inches for liquids and ¼-inch 36 threads/inch for optical fibers.

LWCCs [3,4] are versatile and can enhance the sensitivity and improve limits of detection of optical instrumentation, by detecting as much of the optical signal as possible while minimizing background noise. LWCCs are based on the use of a capillary with a lower refractive index than the liquid core contained in it, so the light introduced into the liquid core of the capillary is totally internally

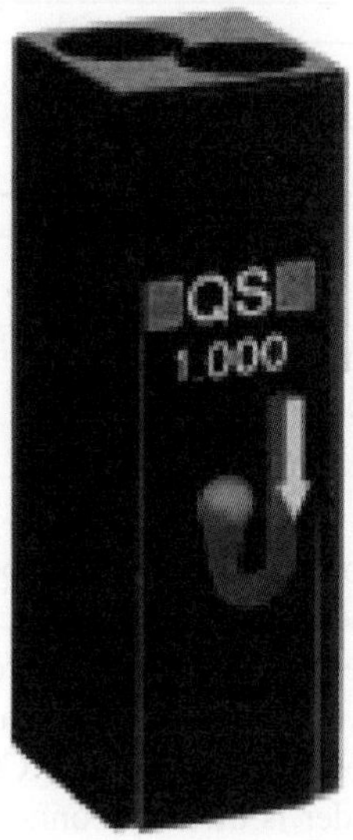

FIGURE 5.12

Typical 1-cm path length flow cell. (For color version of this figure, the reader is referred to the online version of this book.)

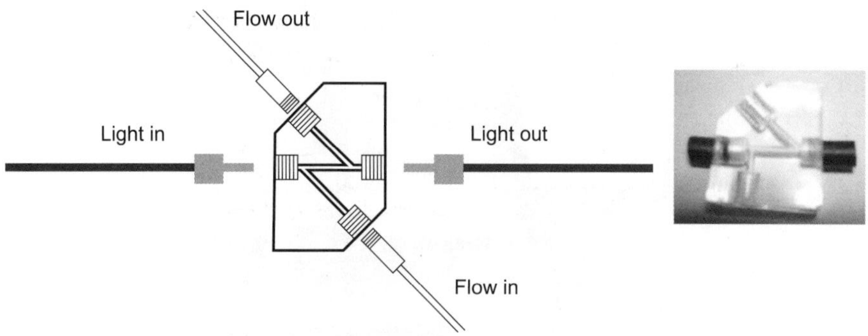

FIGURE 5.13

Z-cells. (For color version of this figure, the reader is referred to the online version of this book.)

reflected down the capillary towards the detector (Figure 5.14). Moreover these capillaries are flexible allowing miniaturization of cells ranging from 10 to 500 cm of length. Thus, taking into account that the absorbance is directly proportional to the path length, an improvement of two orders of magnitude in the signal regarding to a 1 cm path length cell can be accomplished. The internal diameter of these capillaries can vary from μm up to mm. Results of increasing the internal diameter are lower backpressure problems but higher internal volumes.

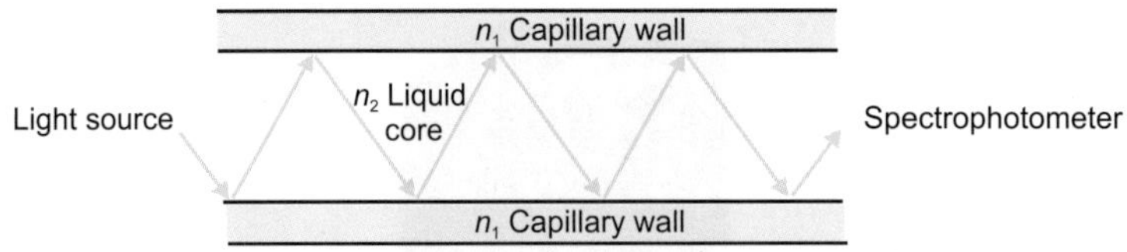

FIGURE 5.14

Schematic depiction of the LWCCs operation. *n*: refractive index, being $n_1 < n_2$. (For color version of this figure, the reader is referred to the online version of this book.)

5.1.2.3.3 Cell holders

Traditional flow cells require a holder for placement. The type of holder to be used is dictated by the particular detector. With conventional spectrophotometers, the holder can be one of the models used in batch mode. Fiber-optic spectrophotometers, however, require a modified cell holder, which is available in various forms. Figure 5.15 shows a cell holder to which a light emitting head powered by a current-based source and an optical fiber connected to a CCD spectrophotometer are screwed on each side. In Figure 5.4(b), the cell holder is directly connected to the spectrophotometer, avoiding the need for an extra optical fiber.

Alternatively, a stand-alone light source with fiber-optic output can be used to transfer light to a cell holder furnished with a fiber-optic input in order to capture light and drive it, with or without scatter, to the detector.

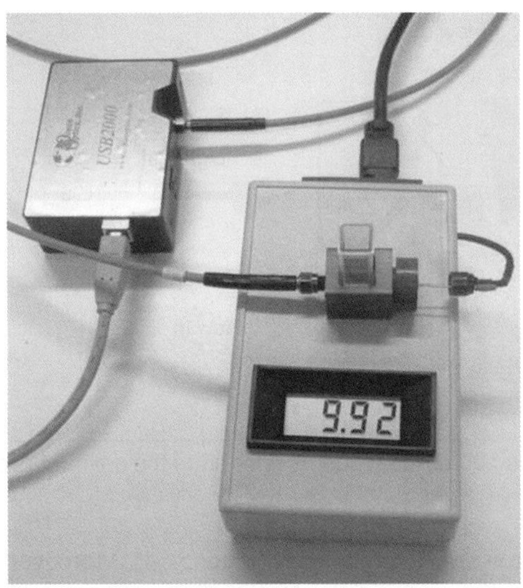

FIGURE 5.15

Self-contained spectrophotometric detection system consisting of a current-based power source, an LED screwed to a cell holder and an optical fiber connecting the holder to a CCD spectrophotometer. (For color version of this figure, the reader is referred to the online version of this book.)

5.1.2.3.4 Connectivity of accessories

As shown above, optical detection systems can be assembled in a variety of ways. Easily switching between available choices requires standardizing connectors. On the one hand, standard threads for connection to liquid drivers are usually ¼-inch, 28 threads/inch, which is the typical format in FIA. On the other hand, standard threads for connection of optical fibers are typically those of SMA 96 (¼-inch, 36 threads/inches). Additionally standardizing head threads can greatly facilitate connection of various components. Figure 5.16 shows some accessories with the European metric thread 14 fabricated in our workshop.

5.1.3 Atomic Absorption Spectroscopy (AAS) detectors

While some molecular spectrophotometers have been specially designed for use in flow techniques, no dedicated atomic absorption instrument for this purpose has to date been developed. Rather, classical batch atomic absorption spectrophotometers are normally used and attention is paid to ensure appropriate outgoing flow-rates.

Figure 5.17 shows an SIA system for the AAS determination of iron by preconcentration of the metal on a Chelex 100 column and elution with nitric acid [5]. The system uses a Crison Compact titrator to operate both burettes in such a way that the nebulizer will continuously receive liquid at a constant flow-rate. Another interesting example of coupling ASS detection and flow techniques is a multisyringe flow injection analysis (MSFIA) system with cold-vapor AAS developed to determine mercury [6].

5.1.4 Luminescence detectors

Luminescence detectors most widely used in flow techniques are those commonly available in the laboratory, which are adapted by using an appropriate flow cell.

5.1.4.1 Fluorimetric detectors

Controlling a fluorimeter via computer software allows the stopped-flow technique to be implemented and making possible to obtain variable-angle spectra for chemometric processing exploiting a multivariate analysis technique.

Although a number of dedicated flow cells and cell holders for fluorimetric measurements exist, none tested so far in the authors' laboratory has proved as sensitive as those used in classical photomultiplier fluorimeters.

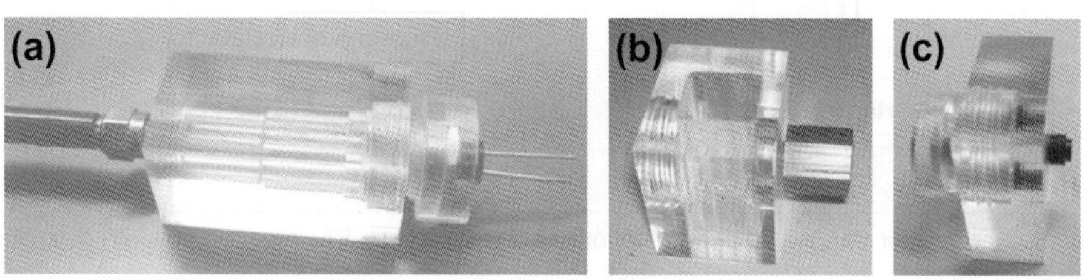

FIGURE 5.16

(a) Collimator for transferring light emitted by an LED head to an optical fiber screwed to it. (b) Cell holder. (c) LED head. All these accessories have the European metric thread 14. (For color version of this figure, the reader is referred to the online version of this book.)

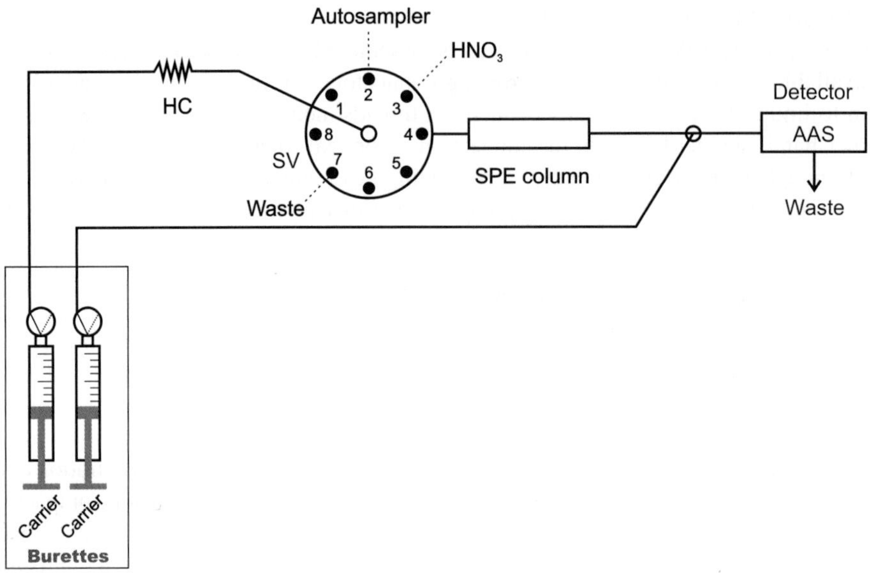

FIGURE 5.17

Schematic depiction of an SIA system for the AAS determination of iron. HC: holding coil, SV: selection valve.

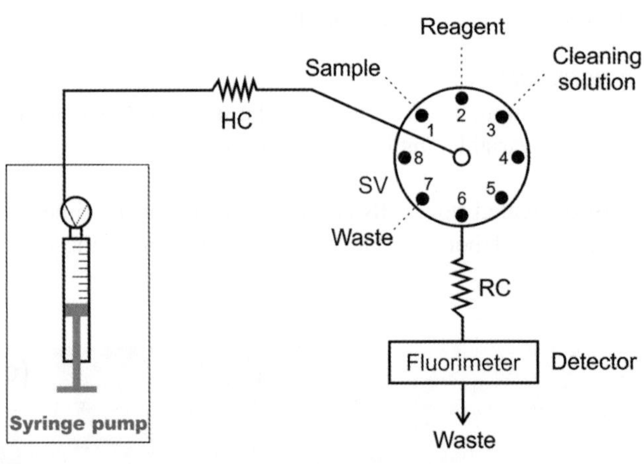

FIGURE 5.18

Fluorimetric SIA system for aluminum determination. HC: holding coil, RC: reaction coil, SV: selection valve.

As example, Figure 5.18 shows an SIA system with a fluorimetric detector for the determination of aluminum with sodium 8-hydroxyquinoline sulfate [7]. Another example is an in-syringe dispersive liquid–liquid microextraction system for aluminum determination in seawater exploiting a special designed fluorescence detector and flow cell [8]. The detection cell comprised a glass tube of 3 mm internal diameter (i.d.) used as detection flow cell with an LED placed on top of it. A photomultiplier tube was mounted perpendicularly to the glass tube and served for detecting the fluorescence emission.

5.1.4.2 Chemiluminescence detectors

Dedicated chemiluminescence detectors are scant. However, they can be easily constructed by using a Hamamatsu integrated photomultiplier. A photomultiplier (Figure 5.19) contains a resistor network intended to power dinodes, an integrated low/high voltage converter and a current/voltage converter amplifier in addition to an output for gain control via an external potentiometer.

As shown in Figure 5.20, a stabilized symmetric source of ± 12 V is sufficient to power the photomultiplier obtaining a full-scale output of 10 V which can be directly connected to an A/D converter. By changing the voltage of the gain regulator, gain levels from 1 to 1000 can be set at will.

The detector devised in our workshop consists of two separate boxes. One contains the photomultiplier and flow cell, and the other the whole circuitry required to operate the detector including an RS232C interface to connect to a PC serial port. The instrument is supplied with a dedicated DLL file intended to facilitate its use in flow techniques by running the software AutoAnalysis.

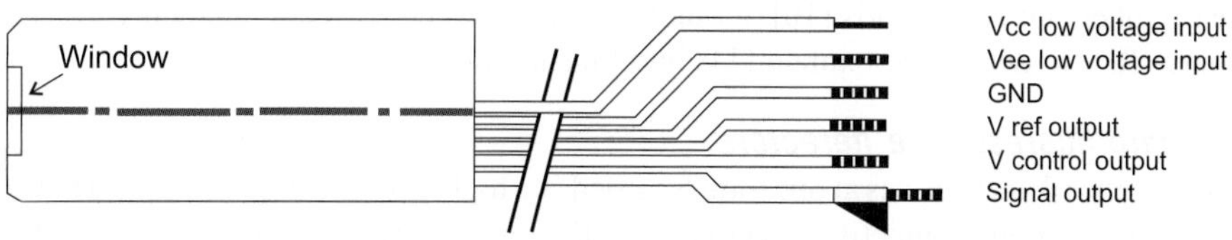

FIGURE 5.19

Hamamatsu photomultiplier integrating a low/high voltage converter, a resistor network and a current/voltage converter, featuring a full-scale output of 10 V.

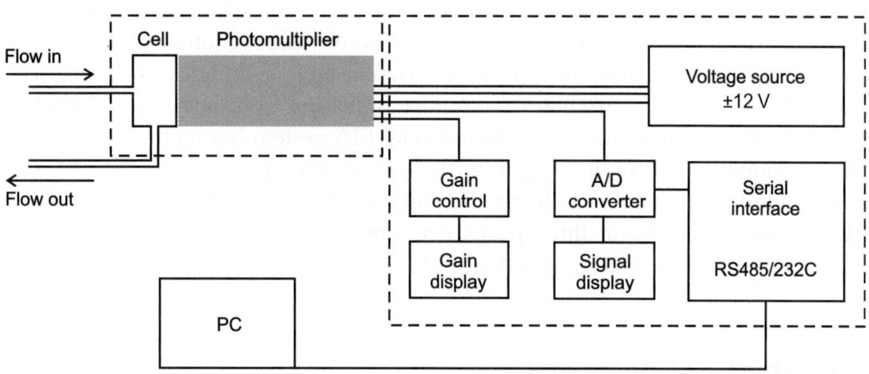

FIGURE 5.20

Block diagram of the Sciware Systems chemiluminescence detector.

Table 5.1 Comparison of MSFIA, a Commercial FIA and SIA Systems for As Determination

Parameter	FIA	SIA	MSFIA
Detection limit (μg/l)	0.05	0.67	0.07
Linear range (μg/l)	0.1–8	2.5–70	0.25–12
% RSD		1.9	4.9
Injection throughput (samples/h)	45	33	113
Analytical throughput (samples/h)		6	36
$NaBH_4$ concentration (%)	1.2	0.12	0.24
$NaBH_4$ consumption (ml/injection)	4.7	0.5	0.3
HCl, KI consumption (ml/injection)	11.3	0.5	0.6
Injected volume (ml)	5.7	0.5	0.6
Sample volume required (ml/injection)	11.5	0.5	0.6

Flow systems with chemiluminescence detection have been used exploiting classical reactions such as that of luminol with hydrogen peroxide in the presence of an appropriate metal as catalyst, and also for sulfide [9], phenolic compounds [10] or other chemiluminometric reagents such as tris (2,2′-bipyridylruthenium(III)) to determine oxalate [11] and thiazide diuretics [12].

5.1.4.3 Atomic fluorescence detectors

Atomic fluorescence detection was at one time regarded as a highly promising technique. However, it was soon superseded by more competitive alternatives. But, it has recently regained some of its former popularity as an effective tool for determining mercury [13,14] and hydride-forming species [15,16]. Some commercial flow analyzers currently exist such as the PSA Analytical system [17] that uses borohydride ion as a reductant in order to generate hydrogen by decomposition in an acidic medium, the resulting gas is employed to produce a small flame that ultimately causes the target hydride to decompose.

An effective way of saving reagents, particularly sodium borohydride which is very expensive, is by using flow techniques such as MSFIA [18]. Since MSFIA is a batch flow technique, the flame is fed from a hydrogen cylinder, which allows the amount of borohydride required to be further reduced.

Table 5.1 compares the performance of three different flow techniques exploiting hydride generation atomic fluorescence spectrometry for As determination. As can be seen, MSFIA [18] provides detection limits similar to those of a commercial FIA system but with a much lower reagents and sample consumption and higher injection throughput. Comparing with the SIA system a 10 times lower limit of detection is achieved exploiting MSFIA together with a higher sample throughput. The ability to obtain three peaks per sample with a single syringe load in MSFIA results in 2.5−3 times higher throughput than in FIA and SIA.

5.2 Electrochemical detectors

Electrochemical methods are commonly used in routine analysis of environmental, clinical and industrial samples. These are an attractive alternative to spectrometric techniques such as flame atomic absorption spectrometry (FAAS), electrothermal atomic absorption spectrometry (ETAAS), inductively coupled plasma-optical emission spectrometry (ICP-OES), inductively coupled plasma-mass spectrometry (ICP-MS) and atomic fluorescence spectrometry (AFS) for trace analysis. Its widespread use is devoted to its low cost, small size, ease of manipulation and high capacity to online coupling, what leads to high sample throughput and portability [19–23].

Flow cells for electrochemical detectors are less commercially available than are equivalent cells for optical detectors, so often these must be constructed by users.

Electrochemical detectors have some limitations such as:

- The strong influence of the sample matrix on the results, which can distort them so much leading to spurious conclusions.
- The great ease with which electrodes can be fouled or poisoned upon direct contact with the sample. This leads to increasingly inaccurate responses and irreproducibility in results.
- The range of elements that can be determined with some electrochemical techniques is usually much narrower than that for optical detectors.

As with optical detection, electrochemical methods can be implemented with equipment commonly available in the laboratory including conductimeters, pH-meters, potentiometers or coulometers. Especially useful in this context are potentiostats which facilitate application of an accurately known, constant voltage to the sensing electrode; and galvanostats which allow constant electrolysis currents to be applied.

One of the undeniable advantages of electrochemical methods is the commercial availability of very complex equipment allowing a wide variety of techniques to be implemented. The sole difficulty in this respect is to find an appropriate flow cell for each application.

5.2.1 Conductimetric detectors

Conductimetric determinations in flow systems can be performed by using ordinary conductimeters. These, however, are difficult to interface to a computer when the whole system is to be automated and possess a low data refresh rate.

The greatest disadvantage of conductimetric detectors is their absolute lack of selectivity, which must frequently be achieved by using a previous pretreatment such as ionic chromatography, capillary electrophoresis or gas diffusion, among others [20,24].

Figure 5.21 shows some conductimetric-cells (C-cells) which have been coupled to different flow techniques. Figure 5.21(a) shows a C-cell made of methacrylate with two perpendicular channels [25]. Solutions flow across one of the channels and in the other one there are two copper wires whose separation, dictated by the channel width, work as electrodes. Commercial C-cells designed for ionic chromatography can also be coupled to FIA systems [26]. Figure 5.21(b) shows a special designed flow C-cell made of methacrylate and encompassing three holes two perpendicular between them to let the solution flow and the third hole is used as the electrode holder, allowing variable distances between the electrodes [27].

Tubular C-cells shown in Figure 5.21(c) can also be used in flow systems [28]. These are built with two concentric steel cylinders of different size.

Capacitive coupling contactless conductivity detectors (C4D) which avoid contact with the solutions using high frequencies (>1 MHz) have gained popularity due to their applications in capillary electrophoresis and ionic chromatography [20,21,29,30]. These have also been coupled to flow

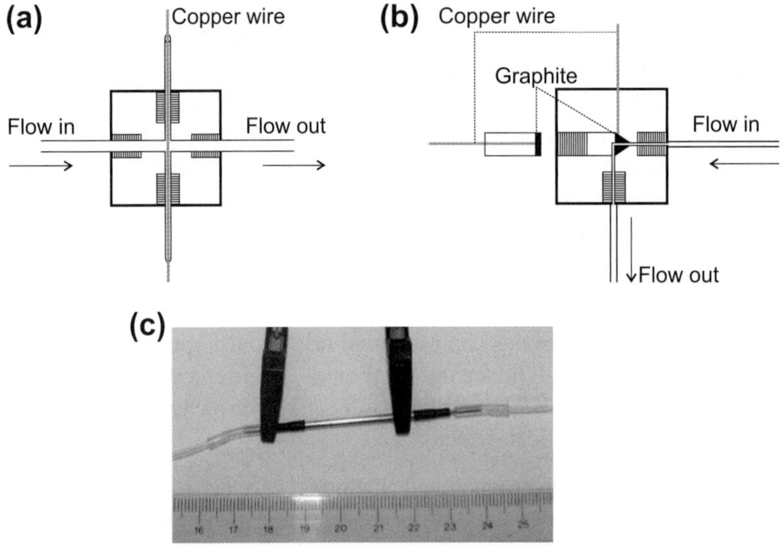

FIGURE 5.21

(a) C-cell made of methacrylate with two perpendicular channels. (b) Special designed flow C-cell made of methacrylate and encompassing three holes allowing variable distances between the electrodes. (c) Tubular C-cell. (For color version of this figure, the reader is referred to the online version of this book.)

techniques, in particular to FIA and SIA, and applied to volatile species previously isolated by gas diffusion [31–33].

5.2.2 Potentiometric detectors

Potentiometric detectors are among electrochemical detectors the most widely favored in flow techniques on account of their easy operation, direct applicability to underivatized analytes and extremely wide dynamic concentration ranges, which often span several orders of magnitude. Their recordings consist of peaks the potentials of which are proportional to the logarithmic concentration of analyte provided an ionic strength buffer (ISAS or TISAB) is used as carrier.

The measuring system should comprise an indicator electrode and a reference electrode. Also, a grounding device should be used to suppress stray potentials that may arise in flow systems. The grounding device can be readily constructed by inserting a piece of an injection needle between two previously threaded methacrylate blocks (Figure 5.22) and using an electrical wire to connect the needle to the mains ground. Detector geometry plays a major role in potentiometric flow measurements.

The main advantage of coupling potentiometric detectors to flow systems with fast recording of transient signals is the possibility of kinetic discrimination between ionic interferences what leads to improvements in membrane electrodes sensitivity.

There are also flow systems exploiting two or more electrodes, which allow differential or multicomponent determinations. Figure 5.23 shows an SIA system for the sequential determination of chloride and pH in water. The system uses the two analog inputs for selective electrodes on the back of the

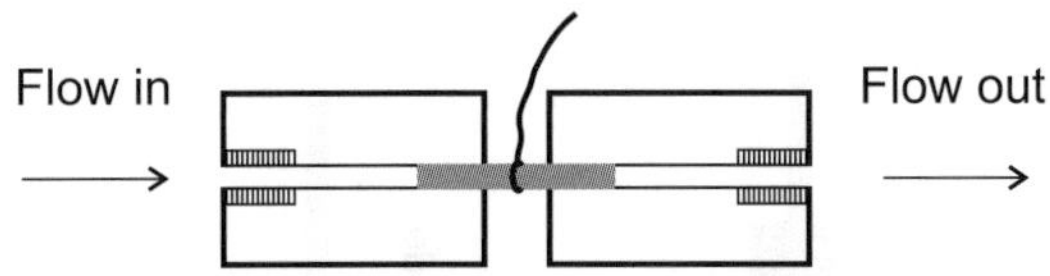

FIGURE 5.22

Grounding device used to suppress stray potentials.

Crison Compact system [34]. A similar system has been used for the sequential determination of chloride and fluoride [35]. The two systems are arranged differently owing to the disparate responsiveness of their electrodes. Thus, while the chloride and fluoride electrode provide a rapid response, the glass electrode is slow especially in response to pH in low-buffered water. For this reason, the system for the sequential determination of chloride and fluoride has both electrodes inserted in the same channel in order to simultaneously obtain their responses. The sole precaution to be adopted here is that the buffer used to adjust the ionic strength and pH of fluoride (TISAB) should contain KNO_3 instead of NaCl.

The slow response of the glass electrode entails inserting it in a channel another than that for chloride. Usually, a relative large volume of water sample is passed through the channel accommodating the glass electrode in order to ensure that any residue of the previous sample has been completely removed, after which the electrode reading is allowed to level off for 3 min. Meanwhile, the switching valve is actuated in order to propel another water sample, containing the ISAS, to the chloride electrode. Because of the chloride electrode is highly responsive, it provides typical peaks that can be used to determine chloride concentrations from a conventional calibration curve, i.e. a plot of peak height or peak potential as a function of the logarithmic concentration of chloride.

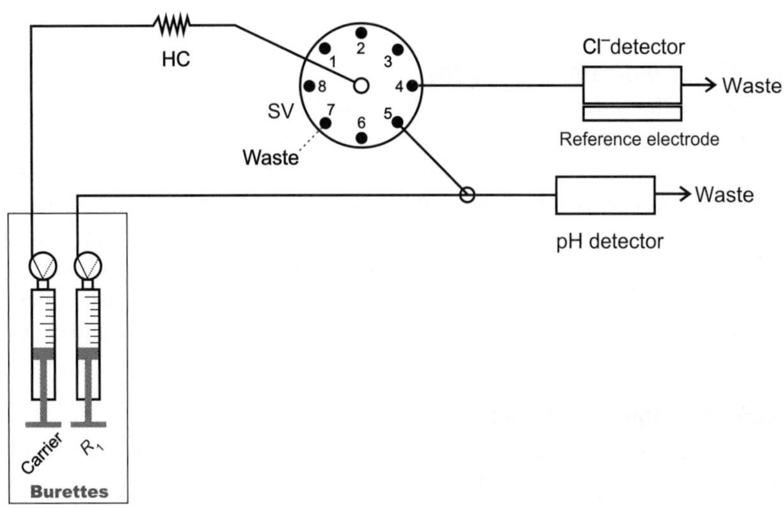

FIGURE 5.23

SIA system for the sequential determination of chloride and pH in water. HC: holding coil, R: reagent, SV: selection valve.

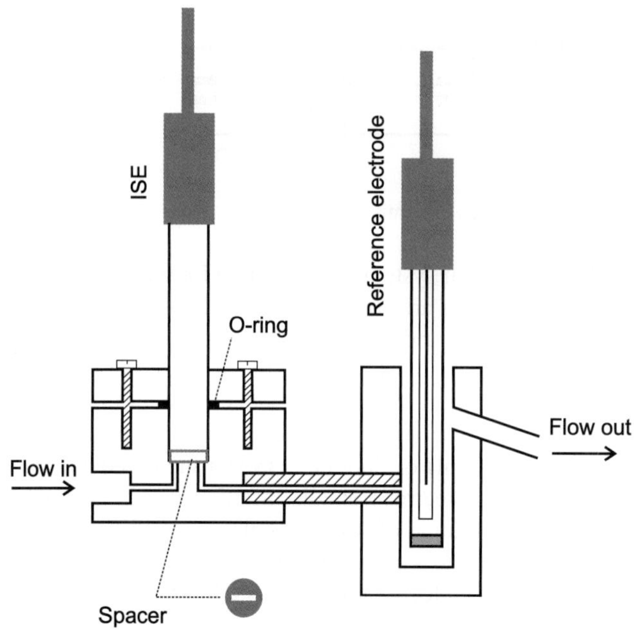

FIGURE 5.24

Adapters for conventional selective and reference electrodes. ISE: ion selective electrode.

By the time the chloride peak completes, the glass electrode will be ready for making a reliable measurement of pH.

5.2.2.1 Conventional indicator electrodes

Flow techniques use both conventional selective electrodes designed for batch potentiometric measurements and tubular electrodes specially developed for flow systems.

Figure 5.24 shows a device intended to facilitate the use of conventional electrodes as flow sensors. The potentiometric circuit is closed by using a reference electrode; in some cases, the electrode must be of the double salt bridge type to ensure compatibility with the analyte, e.g. chloride. A simple solution involves connecting the waste tube to the output of the indicator electrode in a beaker and inserting the reference electrode in the waste liquid. Alternatively, one can use a methacrylate well to screw the output of the selective electrode on one side and to evacuate the liquid excess via an overflow trough on the other.

5.2.2.2 Tubular indicator electrodes

An advantage of tubular electrodes specially designed for flow systems is that they don't cause distortion in the flow which results in decreased dispersion.

Tubular electrodes are available in various configurations [20,36–39]. Figure 5.25 shows some potentiometric flow cells designed for polycrystalline membrane electrodes, e.g. for chlorides, and plastic membrane electrodes, which are deposited on the inner surface of a hole drilled in an epoxy resin/graphite cylinder.

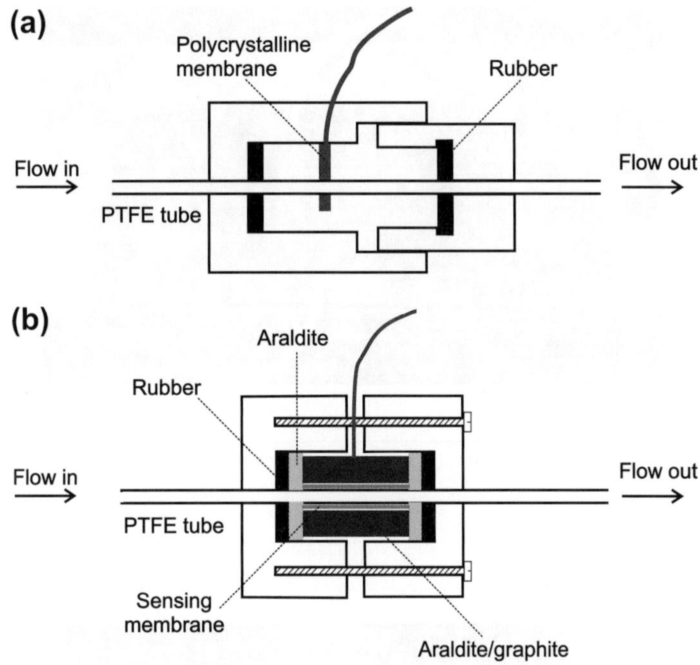

FIGURE 5.25

(a) Potentiometric flow cell for polycrystalline electrodes. (b) Potentiometric flow cell for plastic membrane electrodes.

As shown above, potentials of several electrodes can be simultaneously acquired. Figure 5.26 shows a potentiometric system with eight high-impedance inputs allowing up to eight selective electrodes and a reference electrode to be connected. The impedance adapter board is connected to an A/D converter that is in turn interfaced to a computer via its serial port.

5.2.3 Voltammetric and amperometric detectors

As noted earlier, one of the shortcomings of flow techniques is the lack of flow cells which would allow the instruments currently available to be coupled to flow systems. However, especially common among available cells are amperometric and coulometric cells for solid electrodes. Some have been designed for HPLC detectors, not being all useful for flow systems owing to the high pressure required to pass liquids through them.

Hydrodynamic characteristics of voltammetric measurements significantly depend on the flow-through detector design [20,40].

While some commercial cells for dropping mercury electrodes have existed in the past, most have been withdrawn from the market. Figure 5.27 shows a very simple methacrylate flow cell to connect to the capillaries of an Amel 473 polarograph, which has the advantage of being computer controlled.

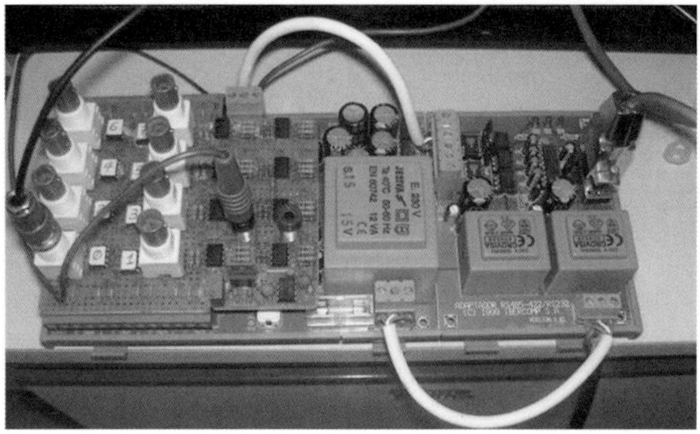

FIGURE 5.26

Multiplexed 8-channel potentiometric system. (For color version of this figure, the reader is referred to the online version of this book.)

FIGURE 5.27

Flow cell for use with a static mercury drop electrode in polarographic and voltammetric methods. (For color version of this figure, the reader is referred to the online version of this book.)

Figure 5.28 depicts an MSFIA system with a flow cell connected to the mercury capillary of an Amel 473 polarograph. The unknown sample to be analyzed for cadmium determination is treated with bisulfite to remove dissolved oxygen and propelled by the carrier into the flow cell, where it enters and leaves via different sides. The waste is poured into a beaker accommodating the reference electrode and a Pt auxiliary electrode. The carrier contains sodium chloride, acting the outgoing liquid

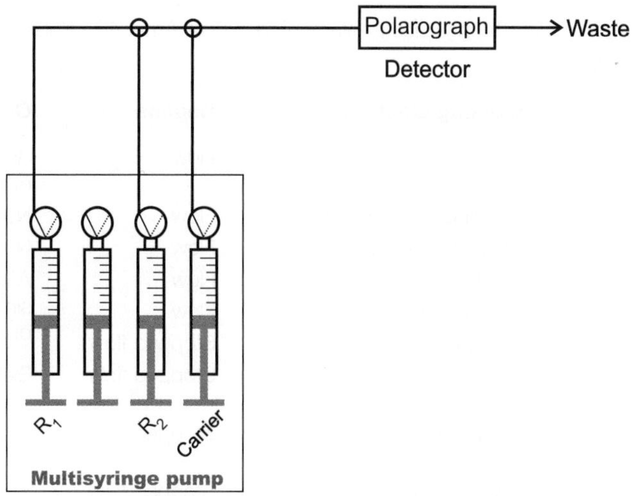

FIGURE 5.28

MSFIA system coupled to an Amel 473 polarograph equipped with a flow cell for the mercury working electrode. R: reagents.

column as a salt bridge between the three electrodes. Measurements can be made on a hanging mercury drop (voltammetry) or on multiple drops (polarography). In any case, the underside of the cell has a threaded hole into which a closed Teflon loop containing mercury can be screwed in order to evacuate the used mercury.

This system is governed via AutoAnalysis software that allows the implementation of a number of polarographic and voltammetric techniques ranging in complexity from amperometry with three electrodes to anodic or cathodic differential pulse stripping voltammetry.

Table 5.2 lists the flow techniques afforded by this system. Many are still at a developmental stage and not yet being used in practice.

Stripping voltammetry is widely used in flow approaches to determine trace levels of transition metal ions [20]. Anodic stripping voltammetry is the most exploited stripping technique for mono and multielemental determinations of metals [19,22,41,42]. Its cathodic variant and the adsorptive cathodic stripping voltammetry allow organic compounds and organometallics complexes determinations [41].

Automation of stripping voltammetry by flow techniques provides several advantages, e.g. constant renovation of the solution over the working electrode (WE) with a small volume of sample, forcing the mass transfer to the surface, what leads to higher preconcentration efficiencies and thus to an improved method sensitivity. Moreover, all the steps of the process, e.g. conditioning, cleaning and preparation of the WE, analyte preconcentration, balance, stripping and signal acquisition as well as time and potential control, can be automated.

FIA systems provide a lower sample and reagents consumption and a simpler operation in comparison to continuous flow systems [43–50]. However, lower preconcentration efficiencies are achieved because of lower contact time between the sample and the WE and the sample

Table 5.2 Electroanalytical Flow Techniques Implemented on the Amel 473 Polarograph and Controlled via the Software AutoAnalysis

Technique	Working Electrode	Regime	Comments
Amperometry	Solid	Flow	Subject to memory effects
	Hanging mercury drop	Flow	No memory effects
	Dropping mercury	Flow	No memory effects
Programmed pulse amperometry	Solid	Flow	Affords the simultaneous detection of several elements
	Hanging mercury drop	Flow	
	Dropping mercury	Stopped flow	
Differential pulse polarography	Dropping mercury	Stopped flow	Extremely sensitive
DC voltammetry	Solid, hanging mercury drop or mercury film	Stopped flow	
Differential pulse polarography	Solid, hanging mercury drop or mercury film	Stopped flow	
Square-wave voltammetry	Solid, hanging mercury drop or mercury film	Stopped flow	Affords very rapid scans
Linear stripping voltammetry	Hanging mercury drop or mercury film	Stopped flow	
Differential pulse stripping voltammetry	Hanging mercury drop or mercury film	Stopped flow	Extremely sensitive
Square-wave stripping voltammetry	Hanging mercury drop or mercury film	Stopped flow	

dispersion inside the flow system [51] together with signal fluctuations due to peristaltic pump pulses [21]. SIA overcomes some of these drawbacks, e.g. by using a piston pump signal fluctuations are considerably reduced. Moreover the way SIA operates allowing flow reversal permits passing the sample as many times as desired through the WE improving the preconcentration efficiency. Also SIA flexibility enables the automation of more pretreatments steps devoted to sample cleaning by eliminating mostly organic molecules that can be adsorbed on the electrode rising the capacitive current and so the residual current [52–63].

Hanging mercury drop electrodes and mercury drop film electrodes (MFE) are the most commonly used electrodes. Nonetheless, the toxicity of mercury has motivated the search of less toxic materials with similar electrochemical properties [64]. Glassy carbon electrodes have been coupled to flow systems [22,65,66], but its surface dirties or is modified too easily requiring cleaning and polishing protocols which frequently involve disassembling the flow cell. Moreover, noisy and low defined voltamperograms are obtained. Bismuth film electrodes are more environmentally friendly and provide high resolution peaks and sensitivity [67]. Also its high mechanical resistance makes them more adequate than MFE for coupling to flow systems [22,64,67–70]. Screen printed electrodes are very easy to build and low-cost consisting in printing with carbon ink the electrode on an inert surface containing the three electrodes (working, counter and reference) concealed in

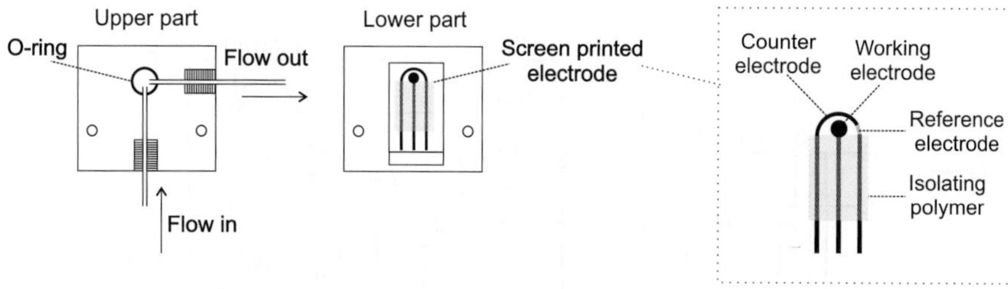

FIGURE 5.29

Homemade BiF-screen printed electrode that has been successfully coupled to an MSFIA system for Cd determination.

a very small surface what leads to miniaturization. Figure 5.29 shows a homemade BiF-screen printed electrode that has been successfully coupled to an MSFIA system for Cd determination in natural water samples [71]. This cell allows an easy and reproducible replacement of the BiF-screen printed electrode and with a very small volume minimizes the dead volume reducing reagents and sample consumption.

5.3 Auxiliar electronic devices

5.3.1 Stabilized voltage power sources

Stabilized voltage power sources constitute the most widely used type of source for electronic circuits. Although most supply a constant voltage, other having a stabilized variable voltage are especially useful in the laboratory.

5.3.1.1 Stabilized fixed-voltage sources

Stabilized fixed-voltage sources can be constructed very easily by using voltage regulators. Figure 4.20 in Chapter 4 shows an example of a very useful source with analog and digital integrated circuits. It can deliver the voltages typically needed by operational amplifiers (+12, −12 V) and TTL logic gates (+5 V). This circuitry is used in large number of laboratory instruments, e.g. chemiluminescence detectors.

5.3.1.2 General-purpose stabilized variable-voltage sources

Tungsten filament lamps have to date been the most widely used light sources for the visible spectrum and part of the infrared region. These lamps require a power source capable of supplying a fairly high current since most of the energy they receive is dissipated as heat.

Sources capable of powering tungsten lamps in a stable manner allowing the amount of light emitted to be adjusted to the sensitivity requirements of a given type of detector are required. Sometimes the intensity of the power source may be inadequate or extremely excessive, e.g. for modern CCD detectors furnished with fiber optics, detector integration times and/or fiber diameters have to be adjusted, among other factors. Figure 5.7 shows a very simple circuit where stabilization relies on the voltage of a reference Zener diode and a potentiometer which allows an appropriate portion of the supplied voltage to be selected. We have used various types of tungsten lamps ranging from

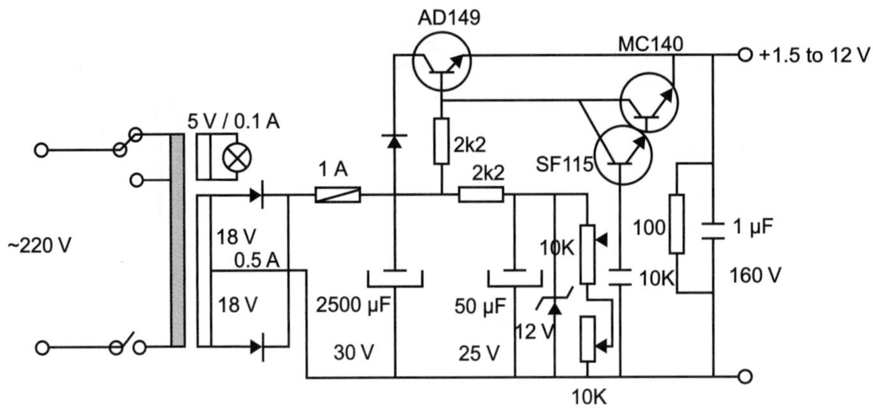

FIGURE 5.30

Stabilized variable-voltage power source.

those for electric torches to 12 V models for car lights, the required voltage being obtained by using an appropriate Zener. It is very important to place the Darlington transistor BD680 on an efficient heat sink in order to avoid excessive heating by effect of the high current consumption of the lamps. The circuit shown in Figure 5.7 includes a 12 V output that can be used to power a fan in order to cool the transistor.

Figure 5.30 depicts a general-purpose source capable of supplying a variable potential between 1.5 and 12 V at a maximum current intensity of 1 A. This source is very useful for rapid testing of solenoids, and also for powering lamps and a wide range of small accessories with modest current requirements. The AD149 transistor used is somewhat outdated, but can be replaced with an equivalent, more modern version serving the same purpose. This is also the case with the MC 140 and SF115 transistors in the Darlington circuit, which can be replaced with Darlington BD680 integrated transistors.

5.3.1.3 Constant-current power sources

With laser diodes and LEDs, which facilitate the construction of very stable, low-cost, high-power monochromatic light sources, constant-current power sources are preferred to stabilize voltage sources. Such sources can be readily constructed by using an LM317 or LM350 integrated circuit, which affords stable currents of up to 1 A. Moreover, the advent of white LEDs, which span the whole visible spectrum, allowed the problem arising from instability in emitted light provided by tungsten lamps to be solved and also enabled the development of improved portable photometers equipped with low-power elements.

Figure 5.31 shows a circuit capable of powering, in an independent manner, up to three laser or LEDs in addition to a tungsten lamp, spanning the whole visible spectrum. Its combination with a CCD spectrophotometer allows the schlieren effect arising from changes in the refractive index to be readily suppressed. This is an especially useful type of source by virtue of its high stability and low consumption, which make it suitable for field measurements.

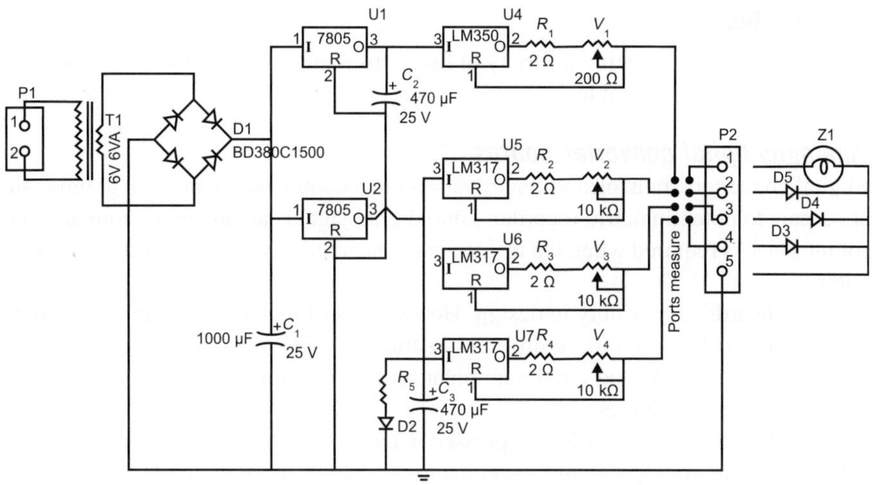

FIGURE 5.31

Constant-current source for independently powering three laser diodes or LEDs and a low-voltage, low-power tungsten lamp.

FIGURE 5.32

Picture of the power source shown in Figure 5.31. (For color version of this figure, the reader is referred to the online version of this book.)

As can be seen in Figure 5.32, it has three white polarized connectors for laser diodes and LEDs at the front part and another connector for tungsten lamps on the left position. It has adjustable 20-turn potentiometers which allow the current intensity supplied to each element to be individually adjusted. If needed, the three diode outputs can be used to power an RGB LED, which delivers the three primary colors (red, green and blue). The tungsten lamp can be replaced by a white LED.

5.3.2 Interface boards

A large number of boards performing a variety of functions required to complete the operation of automatic systems are currently available.

5.3.2.1 Analogic/digital converter boards

Analogic/digital converters transform voltages varying in a continuous manner, e.g. those supplied by instrument outputs to potentiometric recorders, into digital signals amenable to computer processing. This type of interface is required when the instrument concerned has no dedicated ports for connection to a computer.

A/D converter boards vary widely in design. However, most include a multiplexed input affording connection to up to eight or even more different instruments. Most simple models have a fixed full-scale value, e.g. 2.5, 5 or 10 V, and can be unipolar to detect voltages above 0 V or bipolar to detect both positive and negative voltages.

Resolution of A/D converter boards is expressed in bits. Most inexpensive models are 8-bit boards; this means that they can provide a resolution equal to the full-scale value divided by 2^8. Thus, an 8-bit board with a full-scale voltage of 10 V will have a resolution of 10/256 V = 0.039 mV. Modern boards are usually 12, 14 or 16-bit models. More expensive models feature software-adjustable gain and allow their full-scale levels to be selected among a range of available values.

5.3.2.2 Digital/analogic converter boards

These boards perform the opposite function of A/D boards. Computers can only process digital data, being able to deliver only two voltage levels, e.g. 0 or 5 V if they operate at the TTL level. That is, computers are able to start and stop a continuous motor but not to change its rotation speed. When a computer delivers a digital datum to a converter board, this transforms it into a voltage proportional to the magnitude of the datum within a continuous range of variation, usually from zero to the full-scale value. As A/D interfaces, these boards have a specific resolution; however, unlike them, they usually have only one or two analog outputs.

D/A converter boards are used, for example, to construct modern polarographs by replacing the classical voltage splitting scanning potentiometer using a cursor to deliver the potential required by the WE with the output of a D/A converter which is subjected to none of the restrictions of mechanical devices resulting in increased reproducibility.

5.3.2.3 Digital input/output boards

Controlling some auxiliary devices in flow assemblies entails using appropriate relays. If such devices additionally require computer control, the simplest choice is a digital I/O board for insertion into a PC slot. This type of board usually contains a large number of outputs capable of triggering low-power reed relays and also of optocoupled digital inputs capable of detecting whether a voltage exists at their gates or not.

In fact, multicommutated flow injection analysis systems use a digital I/O board to control fast-switching solenoid valves, and so did early MSFIA systems to control independent solenoid valves.

5.3.2.4 Communication boards

One of the essential requirements when automating a flow system is connecting the instruments and/or apparatuses involved to appropriate controllers, usually microprocessors or personal computers.

Instruments equipped with two-way communication ports can greatly facilitate their integration in automatic systems. To check if a given instrument has the required ports for a specific purpose, following considerations should be taken into account:

1. The type of communication protocol available. RS232C serial ports were the norm; lately, however, they have been largely superseded by USB ports. In fact, most computers currently on the market include several USB ports but no RS232C port. The HP-IB interface (GP-IB, IEEE485), which is more difficult to program, has been designed for insertion into a PC slot. This precludes its use in portable computers and also in many current PCs, which don't have an appropriate slot. In fact, most modern instruments are equipped with USB interfaces instead. By contrast, the HP-IB interface has the advantage of affording connection of multiple instruments to a single computer.
2. Required communication protocols are also important. That is the software to set up the connection and a list of the commands accepted by the instrument. Since manufacturers usually supply users with closed software not amenable to alteration, communication protocols are especially important when users themselves have to develop the software needed to control the body of instruments and apparatuses integrated in their flow systems.

5.3.2.4.1 Parallel interfaces

Parallel Centronics interface was a communication interface used mainly with printers and the standard for computers, even though it has been replaced by the USB interface. In some exceptional cases, it is used to communicate with and/or control other devices, e.g. EMCO lathes, Fischer Technic interfaces.

Parallel IEEE488 interface is a universal standard developed in closed cooperation with Hewlett–Packard, hence its designation: HP-IB (Hewlett Packard Interface Bus). Other manufacturers call it GP-IB (General Purpose Interface Bus). This type of interface is very common in physical and electronic instruments, and also in HP instruments for chemical measurements.

5.3.2.4.2 Serial interfaces

Even though Serial RS232/485 interface continues to be widely used, it has never been standardized as regards connectors and wiring, being available in various forms including modem, null-modem and configurations involving more than three wires. This makes it rather difficult to program and troubleshoot. Thus, it is often very difficult to ascertain whether faulty performance of a system including this interface is due to the interface itself, a defective connection or a software bug. Moreover one of its greatest hindrances at present is its rapid disappearance as standard port in computers in favor of USB interfaces. However, this problem can be solved by using an appropriate adapter to connect RS232C devices to USB ports.

USB is a very fast serial communication standard greatly improving on the RS232C interface. Main advantages of USB are that it can be connected while the system is On and that it can operate as a common bus, allowing several USB connectors to be used on the same computer.

It employs a stacked star topology allowing simultaneous operation of up to 127 devices. At the root or apex of the layers, is the host controller that controls all traffic flowing through the bus. This topology allows many devices to be connected to a single logical bus avoiding that the devices which are further down suffer delay. Unlike other architectures, USB is not a bus for storage and shipment, so that there is no delay in sending a data packet to lower layers.

USB 2.0 allows the connection to low consumption devices, i.e. with a maximum consumption of 100 mA per port. However, devices allowing four ports per USB output permit from 100 mA up to 500 mA per port. USB 3.0 allows 1 A per port.

5.3.2.4.3 Ethernet interface

The Ethernet interface is especially common for connecting some equipment to the Internet via a computer. Ethernet is a family of computer networking technologies for local area networks.

Ethernet standards comprise several wiring and signaling variants of the OSI physical layer in use with Ethernet. The original 10BASE5 Ethernet used coaxial cable as a shared medium. Later the coaxial cables were replaced by twisted pair and fiber optic links in conjunction with hubs or switches. Data rates were periodically increased from the original 10 megabits per second to 100 gigabits per second.

Systems communicating over Ethernet divide a stream of data into shorter pieces called frames. Each frame contains source and destination addresses and error-checking data so that damaged data can be detected and re-transmitted.

5.3.2.5 Multifunction boards

Some commercially available boards integrate several of the above-described functions. For example the Metrabyte DAS8 board includes a 14-bit A/D converter with a multiplexed 8-way input, a D/A converter output, three digital inputs, four digital outputs, and two timers.

5.3.2.6 Boards for stand-alone systems

Most of the previous described boards either come with a computer or are inserted in a PC slot. Current PCs come with various types of slots (ISA, AGP, PCI) which are only compatible with the respective matching board types. As a result, many boards used in old computers cannot be employed in newer ones. Care should therefore be exercised to avoid purchasing an outdated board model unusable with current PCs. Another disadvantage of these boards is that relocating them, even in the same computer, entails opening its case.

An effective alternative is to construct instruments with a built-in interface. The Ibercomp RS485 series of boards can be very useful for this purpose [72]. This series includes A/D converter, D/A converter and digital I/O boards that can be mounted on instruments and connected to a single RS485/RS232C interface. The RS485 portion of the interface can support several boards and the RS232C portion ensures compatibility with computer serial ports.

The circuit shown in Figure 5.26 uses this type of board. The board on the left is a potentiometric system that affords connection of up to eight selective electrodes and a reference one. Each electrode is furnished with an impedance adapter the output of which is connected to the multiplexer input of the Ibercomp A/D board—the one under the impedance adapter. The A/D board is connected to the RS485 communication interface, which is equipped with two small transformers. In addition, the A/D board accommodates the RS232C interface, which allows direct connection of the multiplexed potentiometer to the computer.

5.3.3 Solenoid valve and pump controllers

Fast-switching three-way solenoid valves constitute the technical basis of the multicommutated flow analysis technique and also are common elements of multisyringe flow injection analysis systems. We have therefore devised basic circuits to facilitate their control. Solenoid valves are also employed in the MPFS technique.

Figure 1.20 in Chapter 1 is a block diagram of a system consisting of an industrial digital I/O board with eight output relays and an interface delivering the electrical power required by the valves which is supplied by an external 12 V source.

MSFIA systems have this auxiliary circuitry integrated via a connecting strip at the back. The strip has four 12 V outputs each allowing several valves to be simultaneously connected provided their total consumption does not surpass the manufacturer's specifications.

The channel connected to the common one when a valve is switched on should be that used for the shortest time in order to avoid overheating of the solenoid and deformation of the valve's internal components as a result. Overheating can be avoided by using auxiliary devices to reduce the electrical power supplied to the valve to the minimum level required in order to keep it operating after it is activated. Such device is manufactured by NResearch as a voltage splitter consisting of a solenoid coil and a resistor connected in parallel to a capacitor (Figure 5.33(a)). Initially the capacitor is uncharged. On application of current, the transistor starts to conduct it and the 12 (−0.6) V delivered is directly used to power the solenoid coil and actuate the valve as a result. As the capacitor is charged, the applied voltage, V+, is split between the coil and the resistance (2*R*), thereby reducing the capacitor voltage albeit to an extent still allowing the valve to continue operating. Diode D1 is used to protect the system from overvoltages when the valve is switched off. This circuit has the disadvantage that its components rely on the resistance *R* of the coil. Table 5.3 lists resistance and capacitance values recommended for various valve models. Figure 5.33(b) shows an alternative system based on this type of circuit and used to protect four NResearch valves. As can be seen, it comprises four capacitors in addition to the resistors and diodes required.

Another solution is to supply the nominal voltage by using a succession of pulses in such a way that the time during which the voltage is zero is shorter than that taken to switch the valve Off. In

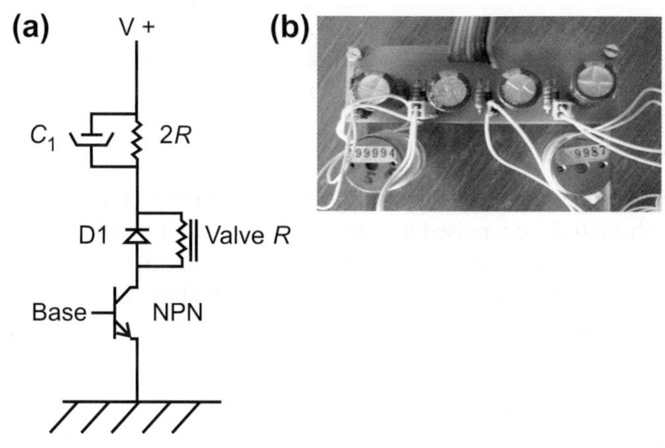

FIGURE 5.33

(a) Valve protection circuit from NResearch. (b) Electronic device for protecting up to four valves based on the circuit shown in (a). (For color version of this figure, the reader is referred to the online version of this book.)

Table 5.3 Resistance (R, Ω) and Capacitance (C, μF) Values to be Used in the Protective Circuit of Figure 5.33(b)

			Coil			Experimental	
Element	**Manufacturer**	**Model**	***R***	**2*R***	***C***	***R***	***C***
Valve	NResearch	T225K031	93	180	556	120	470
Valve	NResearch	T161K031	125	250	232	120	470
Valve	Takasago	MTV-3-N1/AUKG	33.5	68	1764	33/0.5 W	4700
Valve	Takasago	STV-3-1/4/UKG	63.4	126		68/0.5 W	4700
Pump	Biochem	120SP12-25	3X.8	74	1764	120	150
Pump	Biochem	090SP12-8	51.1	102			

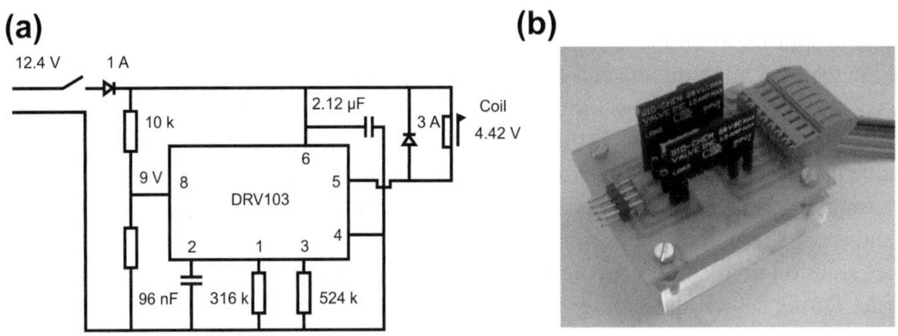

FIGURE 5.34

(a) Pulse-driven valve protection circuit. (b) Picture of the system based on the circuit described in (a) affording connection of up to four protected solenoid valves to the back of a multisyringe burette. (For color version of this figure, the reader is referred to the online version of this book.)

this way, the effective voltage can be reduced to nearly one-half the nominal value and the power dissipated by the valve substantially decreased as a result. Figure 5.34(a) depicts the DRV103 integrated circuit, which was specially designed for this purpose, and Figure 5.34(b) a system based on it. The use of these devices considerably reduces the temperature reached by valves, thereby substantially prolonging their useful lifetimes.

5.3.4 Detectors controlled as peripherals

Usually detectors are controlled by a DLL when exploiting AutoAnalysis. However, big instruments such as ICP-MS or GC, which are governed by their own software, can be controlled as peripheral instruments of the MSFIA burette if they have an events input executable by an external trigger. The synchronism between them can be performed through the 12 V digital output of the MSFIA

FIGURE 5.35

Relay used to interface MSFIA and instruments such as ICP-MS or GC. (For color version of this figure, the reader is referred to the online version of this book.)

burette and exploiting a relay (Figure 5.35) which can trigger the events input, e.g. of the ICP-MS. While the instrument executes its own software, AutoAnalysis can work simultaneously carrying out other steps of the method, e.g. clean up, preconcentration, what leads to higher injection throughputs.

5.3.5 Amplifiers

Amplifiers are also required to construct flow systems, either as part of the detectors used or to adapt the output of some instrument, e.g. a recorder, to the input of another device, e.g. a data acquisition system.

5.3.5.1 General-purpose amplifiers

Commercially available integrated circuits facilitate the construction of amplifiers for specific needs. Figure 5.36 shows an inverter amplifier circuit, i.e. one where the output signal is inverted with respect to the input signal, the gain of which is given by the ratio of the two resistances ($G = R_2/R_1$).

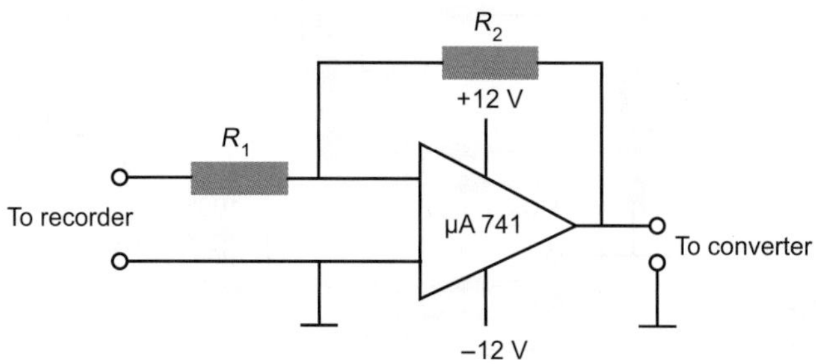

FIGURE 5.36

Typical voltage adapter amplifier.

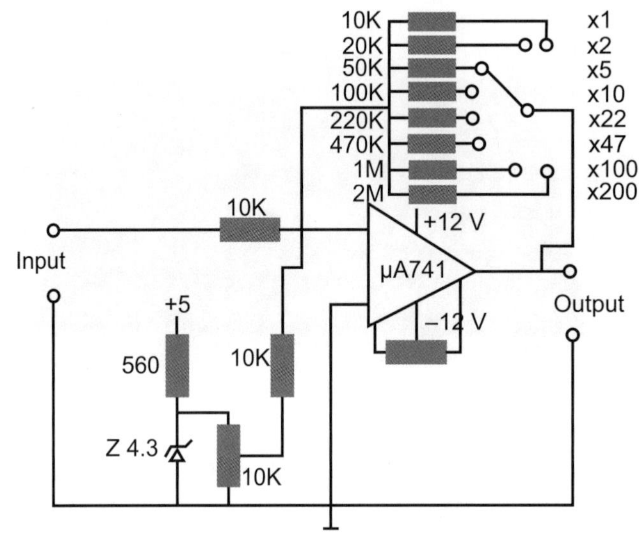

FIGURE 5.37

Adjustable gain amplifier affording baseline shifting.

Thus, in order to connect a recorder with a full-scale voltage of 10 mV to an A/D converter with one of 5 V an amplifier with a gain $G = 5000\ \text{mV}/10\ \text{mV} = 500$ will be needed. Thus, if $R_1 = 1\ \text{k}\Omega$, R_2 will have to be 500 kΩ. Amplifiers can be powered using the source shown in Figure 4.20 in Chapter 4.

Figure 5.37 depicts an amplifier the gain of which can be adjusted by changing the position of a switch to select the feedback resistance of the operational amplifier. The system includes an additional circuit consisting of a Zener intended to supply a reference voltage. A 10 K potentiometer is

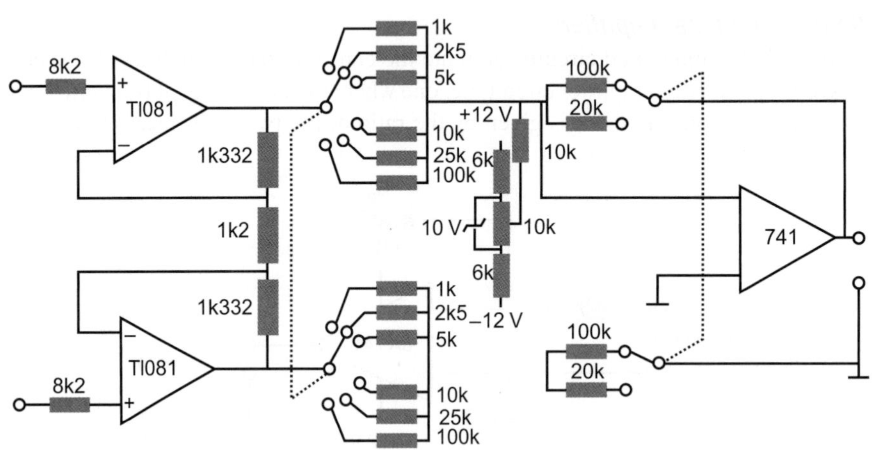

FIGURE 5.38

Adjustable gain differential instrumental amplifier.

used to select a fraction of the voltage in order to alter the baseline onset. This device can also be powered by the source shown in Figure 4.20 in Chapter 4, which supplies all the voltages required.

5.3.5.2 Differential amplifiers

In some cases, reducing input noise is easier by using differential amplifiers such as those employed to construct thermometric detectors. Figure 5.38 shows an example of such amplifier.

References

[1] Sciware Systems: http://www.sciware-sl.com.

[2] Ocean Optics: http://www.Oceanoptics.com.

[3] L.J. Gimbert, P.J. Worsfold, Environmental applications of liquid-waveguide-capillary cells coupled with spectroscopic detection, Trends Anal. Chem. 26 (2007) 914–930.

[4] F. Maya, J.M. Estela, V. Cerdà, Multisyringe flow injection analysis hyphenated with liquid core waveguides for the development of cleaner spectroscopic analytical methods: improved determination of chloride in waters, Anal. Bioanal. Chem. 394 (2009) 1577–1583.

[5] E. Rubí, M.S. Jiménez, F. Bauzá de Mirabó, R. Forteza, V. Cerdà, Preconcentration and atomic absorption determination of iron by sequential injection analysis, Talanta 44 (1997) 553–562.

[6] L.O. Leal, O. Elsholz, R. Forteza, V. Cerdà, Determination of mercury by multisyringe flow injection system with cold-vapor atomic absorption spectrometry, Anal. Chim. Acta 573–574 (2006) 399–405.

[7] C. Brach-Papa, B. Coulomb, J.L. Boudenne, V. Cerdà, F. Theraulaz, Spectrofluorimetric determination of aluminium in drinking waters by sequential injection analysis, Anal. Chim. Acta 457 (2002) 311–318.

[8] R. Suárez, B. Horstkotte, C.M. Duarte, V. Cerdà, Fully-automated fluorimetric determination of aluminum in seawater by in-syringe dispersive liquid-liquid microextraction using lumogallion, Anal. Chem. 84 (2012) 9462–9469.

[9] F. Maya, J.M. Estela, V. Cerdà, Improving the chemiluminescence-based determination of sulphide in complex environmental samples by using a new, automated multi-syringe flow injection analysis system coupled to a gas diffusion unit, Anal. Chim. Acta 601 (2007) 87–94.

[10] A.R.T.S. Araujo, F. Maya, M.L.M.F.S. Saraiva, J.L.F.C. Lima, J.M. Estela, V. Cerdà, Flow system for the automatic screening of the effect of phenolic compounds on the luminol-hydrogen peroxide-peroxidase chemiluminescence system, Luminescence 26 (2011) 571–578.

[11] F. Maya, J.M. Estela, V. Cerdà, Multisyringe ion chromatography with chemiluminescence detection for the determination of oxalate in beer and urine samples, Microchim. Acta 173 (2011) 33–41.

[12] F. Maya, J.M. Estela, V. Cerdà, Interfacing on-line solid phase extraction with monolithic column multisyringe chromatography and chemiluminescence detection: an effective tool for fast, sensitive and selective determination of thiazide diuretics, Talanta 80 (2010) 1333–1340.

[13] J.L. Guzmán-Mar, L. Hinojosa-Reyes, A.M. Serra, A. Hernández-Ramírez, V. Cerdà, Applicability of multisyringe chromatography coupled to cold-vapor atomic fluorescence spectrometry for mercury speciation analysis, Anal. Chim. Acta 708 (2011) 11–18.

[14] A.M. Serra, J.M. Estela, V. Cerdà, An MSFIA speciation based on anion-exchange membrane, Talanta 78 (2009) 790–794.

[15] A.M. Serra, J.M. Estela, V. Cerdà, A MSFIA system for selenium speciation by atomic fluorescence spectrometry, J. Anal. At. Spectrom. 27 (2012) 1858–1862.

[16] L.O. Leal, L. Ferrer, R. Forteza, V. Cerdà, Multicommutation flow techniques in the hydride generation-atomic fluorescence determination of arsenic, Trends Anal. Chem. 30 (2011) 761–770.

[17] PSA Analytical. http://www.psanalytical.com/.

[18] N.V. Semenova, L.O. Leal, R. Forteza, V. Cerdà, Multisyringe flow-injection systems for total inorganic arsenic determination by hydride generation-atomic fluorescence spectrometry, Anal. Chim. Acta 455 (2002) 277–285.
[19] M. Trojanowicz, M. Szewczynska, M. Wcislo, Electroanalytical flow measurements – recent advances, Electroanalysis 15 (2003) 347–365.
[20] M. Trojanowicz, Recent developments in electrochemical flow detections – a review: part I. Flow analysis and capillary electrophoresis, Anal. Chim. Acta 653 (2009) 36–58.
[21] M. Trojanowicz, Advances in Flow Analysis, Wiley-VHC, Weinheim, 2008, pp. 672.
[22] A. Economou, Recent developments in on-line electrochemical stripping analysis – an overview of the last 12 years, Anal. Chim. Acta 683 (2010) 38–51.
[23] J. Buffe, M.L. Tercier-Waeber, Voltammetric environmental trace-metal analysis and speciation: from laboratory to in situ measurements, TrAC Trends Anal. Chem. 24 (2005) 172–191.
[24] M. Trojanowicz, Recent developments in electrochemical flow detections – a review: part II. Liquid chromatography, Anal. Chim. Acta 688 (2011) 8–35.
[25] J. Junsomboon, J. Jakmunee, Flow injection conductometric system with gas diffusion separation for the determination of Kjedahl nitrogen in milk and chicken meat, Anal. Chim. Acta 627 (2008) 232–238.
[26] L.N. Moskvin, G.L. Grigor'ev, A.G. Papsueva, Flow-through conductometric analysis for microconcentrations of ammonia in natural water with chromatomembrane separation into deionized water, Russ. J. Appl. Chem. 74 (2001) 976–980.
[27] M.T. Oms, A. Cerdà, V. Cerdà, Preconcentration by flow reversal in conductometric sequential injection analysis of ammonium, Electroanalysis 8 (1996) 387–390.
[28] C. Henríquez, B. Horstkotte, V. Cerdà, Conductometric determination of ammonium by a multisyringe flow injection system applying gas diffusion, Int. J. Environ. Anal. Chem. (2012) 1–17.
[29] J.A. Fracassi da Silva, N. Guzman, C.L. do Lago, Contactless conductivity detection for capillary electrophoresis: hardware improvements and optimization of the input-signal amplitude and frequency, J. Chromatogr. A 942 (2002) 249–258.
[30] P. Kuban, P.C. Hauser, Contactless conductivity detection in capillary electrophoresis: a review, Electroanalysis 16 (2004) 2009–2021.
[31] Z. Hohercakova, F. Opekar, A contactless conductivity detection cell for flow injection analysis: determination of total inorganic carbón, Anal. Chim. Acta 551 (2005) 132–136.
[32] H.L. Braz, D.T. Ito, J.A.F. da Silva, C.L. do Lago, J.J. Pedrotti, Trace levels determination of ammonium by flow injection analysis using gas-diffusion and capacitively coupled contactless conductivity detection, Electroanalysis 23 (2011) 2594–2600.
[33] K. Sereenonchai, S. Teerasong, S. Chan-Eam, P. Saetear, N. Choengchan, K. Uraisin, N. Amornthammarong, S. Motomizu, D. Nacapricha, A low-cost method for determination of calcium carbonate in cement by membraneless vaporization with capacitively coupled contactless conductivity detection, Talanta 81 (2010) 1040–1044.
[34] J. Alpízar, A. Crespí, A. Cladera, R. Forteza, V. Cerdà, Simultaneous determination of chloride and pH in waste waters by sequential injection analysis, Lab. Rob. Autom. 8 (1996) 165–170.
[35] J. Alpízar, A. Crespí, A. Cladera, R. Forteza, V. Cerdà, Simultaneous determination of chloride and fluoride ions in waters by sequential injection analysis, Electroanalysis 8 (1996) 1051–1054.
[36] I.M.P.L.V.O. Ferreira, J.L.F.C. Lima, Tubular potentiometric detector for flow injection based on homogeneous crystalline membranes sensitive to copper, cadmium and lead, Analyst 119 (1994) 209–212.
[37] C.M.C.M. Couto, J.L.F.C. Lima, M.C.B.S.M. Montenegro, Increased sensitivity potentiometric detectors for FIA, Anal. Sci. 13 (1997) 403–408.
[38] J.A. Rodriguez, E. Barrado, M. Vega, J.L.F.C. Lima, Speciation of inorganic arsenic in waters by potentiometric flow analysis with on-line preconcentration, Electroanalysis 17 (2005) 504–511.

[39] J.R. Santos, R.A.S. Lapa, J.L.F.C. Lima, Development of a tubular fluoride potentiometric detector for flow analysis: evaluation and analytical applications, Anal. Chim. Acta 583 (2007) 429–436.

[40] K. Toth, K. Stulik, W. Kutner, Z. Feher, E. Linder, Electrochemical detection in liquid flow analytical techniques: characterization and classification (IUPAC technical report), Pure Appl. Chem. 76 (2004) 119–1138.

[41] R. Pérez-Olmos, J.C. Soto, N. Zarate, M.C.B.S.M. Montenegro, Sequential injection analysis using electrochemical detection: a review, Anal. Chim. Acta 554 (2005) 1–16.

[42] M.S.M. Quintino, L. Angnes, Batch injection analysis: an almost unexplored powerful tool, Electroanalysis 16 (2004) 513–523.

[43] P. Daneshgar, P. Norouzi, M.R. Ganjali, Rapid determination of bisacodyl in flow injection system combination by a novel sensitive adsorptive square-wave voltammetry, Sens. Actuators B Chem. 136 (2009) 66–72.

[44] P. Daneshgar, P. Norouzi, A.A. Moosavi-Movahedi, M.R. Ganjali, E. Haghshenas, F. Dousty, M. Farhadi, Fabrication of carbon nanotube and dysprosium nanowire modified electrodes as a sensor for determination of curcumin, J. Appl. Electrochem. 39 (2009) 1983–1992.

[45] H.P.A. Nouws, C. Delerue-Matos, A.A. Barros, J.A. Rodrigues, A. Santos-Silva, F. Borges, Square-wave adsorptive-stripping voltammetric detection in the quality control of fluoxetine, Anal. Lett. 40 (2007) 1131–1146.

[46] P. Norouzi, R. Dinarvand, M.R. Ganjal, A.S.E. Meibodi, Application of adsorptive stripping voltammetry for the nano-level detection of tramadol in biological fluids and tablets using fast fourier transform continuous cyclic voltammetry at an Au microelectrode in a flowing system, Anal. Lett. 40 (2007) 2252–2270.

[47] H.P.A. Nouws, C. Delerue-Matos, A.A. Barros, J.A. Rodrigues, Electroanalytical determination of paroxetine in pharmaceuticals, J. Pharm. Biomed. Anal 42 (2006) 341–346.

[48] J.A. Rodrigues, A.A. Barros, P.J. Almeida, P.G. Rodrigues, A.G. Fogg, Flow injection square wave cathodic stripping voltammetric determination at a hanging mercury drop electrode of rapidly reduced compounds: determination of diacetyl in wine as 2,3-dimethylquinoxaline, Anal. Chim. Acta 449 (2001) 119–127.

[49] P. Norouzi, M.R. Ganjali, P. Matloobi, Sub-second adsorption for sub-nanomolar monitoring of metoclopramide by fast stripping continuous cyclic voltammetry, Electrochem. Commun. 7 (2005) 333–338.

[50] W. Yantasee, C. Timchalk, Y.H. Lin, Microanalyzer for biomonitoring lead (Pb) in blood and urine, Anal. Bioanal. Chem. 387 (2007) 335–341.

[51] M.D. Luque de Castro, I. Izquierdo, Flow injection stripping analysis, Electroanalysis 3 (1991) 457–467.

[52] P.G. Rodrigues, J.A. Rodrigues, A.A. Barros, R.A.S. Lapa, J.L.F.C. Lima, J.M. Machado Cruz, A.A. Ferreira, Automatic flow system with voltammetric detection for diacetyl monitoring during brewing process, J. Agric. Food Chem. 50 (2002) 3647–3653.

[53] N.K. Djane, K. Ndung'u, G. Johansson, L. Mathiasson, S. Armalis, Supported liquid membrane coupled on-line to potentiometric stripping analysis at a mercury-coated reticulated vitreous carbon electrode for trace metal determinations in urine, Analyst 132 (1998) 393–396.

[54] N. Parthasarathy, M. Pelletier, M.L. Tercier-Waeber, J. Buffle, On-line coupling of flow through voltammetric microcell to hollow fiber permeation liquid membrane device for subnanomolar trace metal speciation measurements, Electroanalysis 13 (2001) 1305–1314.

[55] S. Armalis, I. Krikŝĉiuniene, E. Kubiliene, N.K. Djane, K. Ndungu, L. Mathiasson, Stripping analysis of trace metals at flow-through reticulated vitreous carbon electrode after the preconcentration by supported liquid membrane technique, Int. J. Environ. Anal. Chem. 74 (1999) 233–242.

[56] D.J. Whitworth, P.E. Achterberg, M. Nimmo, P.J. Worsfold, Validation and in situ application of an automated dissolved nickel monitor for estuarine studies, Anal. Chim. Acta 371 (1998) 235–246.

[57] A. Biscombe, M. Nimmo, M. Gledhill, E.P. Achterberg, An automated monitor to determine trace metal particle/dissolved interactions in natural waters, Anal. Chim. Acta 521 (2004) 69–76.
[58] C. Billing, D.R. Groot, J.F. van Staden, Determination of arsenic in gold samples using matrix exchange differential pulse stripping voltammetry, Anal. Chim. Acta 453 (2002) 201–208.
[59] M.L. Tercier-Waeber, J. Buffle, Submersible online oxygen removal system coupled to an in situ voltammetric probe for trace element monitoring in freshwater, Environ. Sci. Technol. 34 (2000) 4018–4024.
[60] J. Wang, B. Tian, J. Wang, J. Lu, C. Olsen, C. Yarnitzky, K. Olsen, D. Hammerstrom, W. Bennett, Stripping analysis into the 21st century: faster, smaller, cheaper, simpler and better, Anal. Chim. Acta 385 (1999) 429–435.
[61] J. Wang, J. Wang, J. Lu, B. Tian, D. MacDonald, K. Olsen, Flow probe for in situ electrochemical monitoring of trace chromium, Analyst 124 (1999) 349–352.
[62] L.B.O. dos Santos, J.C. Masini, Square wave adsorptive cathodic stripping voltammetry automated by sequential injection analysis: potentialities and limitations exemplified by the determination of methyl parathion in water samples, Anal. Chim. Acta 606 (2008) 209–216.
[63] E. Muñoz, S. Palmero, M.A. García-García, A continuous flow system design for simultaneous determination of heavy metals in river water samples, Talanta 57 (2002) 985–992.
[64] A. Economou, A. Voulgaropoulos, On-line stripping voltammetry of trace metals at a flow-through bismuth-film electrode by means of a hybrid flow-injection/sequential-injection system, Talanta 71 (2007) 758–765.
[65] P. Manisankar, G. Selvanathan, S. Viswanathan, H. Gurumallesh Prabu, Electrochemical determination of some organic pollutants using wall-jet electrode, Electroanalysis 14 (2002) 1722–1727.
[66] J.F. van Staden, M.C. Matoetoe, Simultaneous determination of copper, lead, cadmium and zinc using differential pulse anodic stripping voltammetry in a flow system, Anal. Chim. Acta 411 (2000) 201–207.
[67] J. Wang, J.M. Lu, S.B. Hocevar, P.A.M. Farias, B. Ogorevc, Bismuth-coated carbon electrodes for anodic stripping voltammetry, Anal. Chem. 72 (2000) 3218–3222.
[68] D. Monticelli, E. Ciceri, C. Dossi, Optimization and validation of an automated voltammetric stripping technique for ultratrace metal analysis, Anal. Chim. Acta 594 (2007) 192–198.
[69] W. Siriangkhawut, S. Pencharee, K. Grudpan, J. Jakmunee, Sequential injection monosegmented flow voltammetric determination of cadmium and lead using a bismuth film working electrode, Talanta 79 (2009) 1118–1124.
[70] Y. Wang, Z. Liu, G. Yao, P. Zhu, X. Hu, Q. Xu, C. Yang, Determination of cadmium with a sequential injection lab-on-valve by anodic stripping voltammetry using a nafion coated bismuth film electrode, Talanta 80 (2010) 1959–1963.
[71] C. Henríquez, L.M. Laglera, M.J. Alpizar, J. Calvo, F. Arduini, V. Cerdà, Cadmium determination in natural water samples with an automatic multisyringe flow injection system coupled to a flow-through screen printed electrode, Talanta 96 (2012) 140–146.
[72] Ibercomp. http://www.ibercomp.es.

CHAPTER

Software

6

CHAPTER OUTLINE

Computer advances have played a major role in flow techniques development. On the one hand, one of the greatest hindrances to the development of flow techniques other than segmented flow analysis (SFA) and flow injection analysis (FIA) was the lack of interfaces to control the analytical equipment via a computer. Most flow techniques require computer control, but FIA which has the invaluable advantage of using very simple assemblies that are easy to control manually. This provided popularity and widespread success to this technique, immediately after its inception. By contrast, the need for computer control in SIA and other multicommutated flow techniques together with the lack of experience in connecting computers to instruments at the time it was developed delayed its success.

On the other hand, advances in information technology have provided most salient flow techniques' benefits, such as fully automation, autonomy, precision and accuracy. Therefore, advances in more recent techniques such as SIA, multicommutated flow analysis (MCFIA), multisyringe flow injection analysis (MSFIA) and lab on valve (LOV), and their combinations, have relied on the availability of appropriate dedicated software for their control.

6.1 Software for flow techniques

Early software for flow techniques was largely specific for a particular problem, system or technique. Subsequent developments led to more general software which, however, was not

Flow Analysis. http://dx.doi.org/10.1016/B978-0-444-59596-6.00006-1

made commercially available as it had a reduced prospective user base. In addition, further development was impeded by the restrictions of the MS DOS operating system, which included a limited ability to manage large amounts of computer memory and the lack of multitasking capabilities.

Especially prominent among the early software used with flow techniques was Flowtek, which was developed by van Staden and coworkers [1]; FIALab, which was devised by Ruzicka's group [2]; and DARRAY [3], which was developed by our research group. Some of these computer programs continue to be readily available and used at present as they greatly facilitate the implementation of flow techniques. There are in the market also more general software, e.g. Labview, which can be used to control instruments, data acquisition and processing, being however more general and complicated than specific software.

In this chapter we will mainly focus on the description of a software specially designed to flow techniques "AutoAnalysis", developed by our research group [4].

6.2 Technical aspects

Flow analysis techniques use a diverse set of instruments from different manufacturers and with different communication interfaces. Additionally, each manufacturer uses different protocols to implement such communication. These may even vary within the same manufacturer.

Analytical methods also vary from one another, since different techniques and instrumentation are used for each method. Despite this, such methods can be formalized as a sequence of steps, i.e. laboratory unit operations, operating different instruments and apparatuses. Furthermore data processing is very similar for all methods.

All these considerations should be taken into account in order to develop or choose a software. The best choice would be a software which operates the same way. To fulfill all these requirements, the application should develop analytical methods, acquire and process data independently of the hardware. AutoAnalysis was designed following this model, i.e. consisting of three layers, namely, hardware (layer 1), instruments and communication channels (layer 2) and application (layer 3).

The application is developed on the Microsoft Windows platform (Win32) and the intermediate layer is implemented as dynamic link libraries (DLLs), moreover, a communication interface between it and the main application is defined. The bottom layer (layer 1) consists of the devices that are connected directly to the computer, interface cards and device drivers, manufacturer libraries for the control of the equipment and programming the connection interfaces.

The middle layer (layer 2) consists of communication channels and tools, i.e. it is subdivided into two parts. However, there are not two sublayers, since each part is directly connected to the upper and lower layer. Communication channels are defined to provide the application with enough information of the actual scheme of hardware connections.

The middle layer (layer 2) prevents the application (layer 3) to communicate directly with the hardware (layer 1). This is one of the objectives of the layer 2 achieving independence of the hardware configurations from the application layer.

Finally there is the analytical application layer of concern. It includes the connected hardware configurations, methods and data acquisition and processing, being the final user interface.

6.3 AutoAnalysis

AutoAnalysis is a software which affords high flexibility in automating analytical methods. Thus, with an appropriate choice of instrumental modules and the required DLLs, AutoAnalysis allows the automatic or semiautomatic implementation of a host of flow techniques including FIA, SIA, MCFIA, MPFS, MSFIA, LOV, capillary electrophoresis (CE) and high-performance liquid chromatography (HPLC).

Although it was primarily conceived for flow techniques, AutoAnalysis also affords the automation of other laboratory processes such as maintaining the pH of a medium by addition of appropriate amounts of an acid or base (pHStat) or controlling bioreactors. In fact, this software provides a wide range of solutions only limited by the user's instrumentation.

One of the greatest advantages of AutoAnalysis is that it allows a variety of problems to be solved by using a single software suite, thereby avoiding the need to learn and use a different program for each instrument, providing an open environment. Since AutoAnalysis has been developed for the Win32 platform, it can be run simultaneously with other programs such as word processors, spreadsheets or Internet browsers.

AutoAnalysis accepts conditional variables and commands; in fact, it has its own programming language, which facilitates the development of intelligent routines. Thus, AutoAnalysis is capable of taking a decision by itself based on analytical results gathered at a given point without intervention of the analyst.

The software is written in Delphi and Visual C++ and offers a window-based graphical surface. It runs on all 32-bit operation systems from Microsoft Corporation® (MS-Windows XP). The minimum hardware requirements for AutoAnalysis include a computer with a Pentium microprocessor running Windows 95 or higher, 16 MB of RAM (32 MB is recommended) and 10 MB of free hard disk space.

6.3.1 Description of the AutoAnalysis software suite

AutoAnalysis consists of a core program and a body of DLLs that facilitate connection to various types of instruments and their control. The distinctive feature of this software based on DLLs at 32 bits is the possibility of using a single and versatile application without further modification for whatever instrumentation and detection system needed. It involves a basic protocol which allows the implementation of specific and individual DLLs, addressing the configuration of the assembled flow analyzer. Thus being a very flexible tool, easy to handle by nonspecialized users.

The core program is hardware independent. This allows the same user interface and facilities to be used with any type of assembly and instruments. Instruments are connected to a communication channel (a DLL) that is managed by the software in a user transparent manner. A single channel can be shared by several instruments at once, each being governed by a specific DLL.

The software automatically manages individual registered user accounts, which include personal configurations, preferred methods and outputs.

These system specifications and user control facilities allow the computer code required to implement a given method to be automatically loaded in memory, thereby saving large amounts of system resources (RAM, hard disk).

6.3.2 Execution of AutoAnalysis

AutoAnalysis operates in the following sequence of steps:

1. Configuring the connected hardware: the user chooses the equipment and connections to be used and selects the most appropriate settings for the channels and instruments that are to come into play.
2. Designing the analytical method to be used with the connected hardware: the programming tools of the software are used to develop the most appropriate method via the sequence of operations and decisions.
3. Implementing the previously developed methods to perform the required operations and acquire the target data.
4. Processing the previously acquired data by subjection to a mathematical treatment such as curve smoothing (e.g. with the Savitzky–Golay algorithm), derivation, calibration or spectral refining. Data can be also exported to be processed with other software (e.g. Excel spreadsheet).

6.3.2.1 Hardware configuration

The establishment of a communication protocol or "channel", e.g. RS232, RS485, USB, IEEE488, incorporation and configuration of the connected hardware, e.g. syringe pump, spectrometer, etc., are carried out via individual DLLs. Since these are installed and loaded as required for each system forming an individual configuration, modification of the instrumentation assembly requires a minimum of adaptation effort (Figure 6.1). Program versatility is only limited by the availability of the

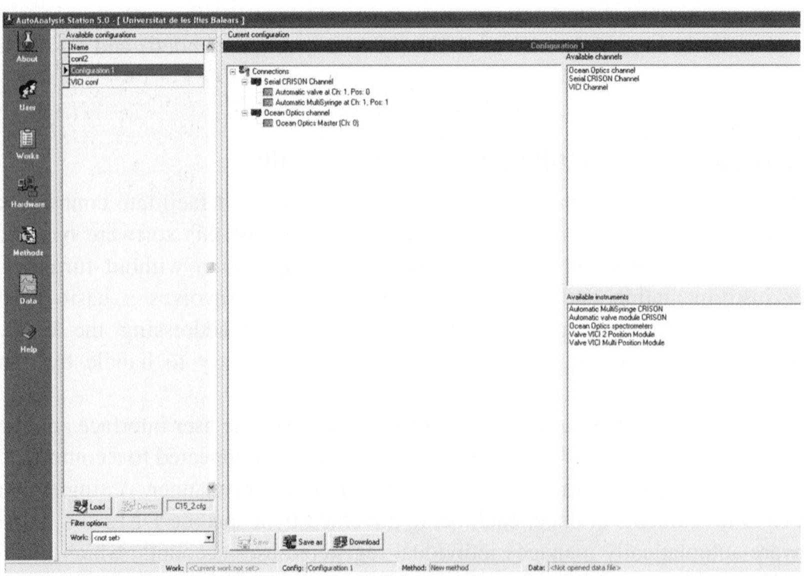

FIGURE 6.1

Hardware window where channels and instruments are loaded to establish a configuration. (For color version of this figure, the reader is referred to the online version of this book.)

required DLLs. Up-to-now, DLLs for seven communication channels and 30 devices are available including atomic fluorescence, molecular fluorescence, spectrometric and electrochemical detectors, autosampler, syringe-, peristaltic-, and solenoid-micropumps, valve modules, and I/O, A/D, or D/A PC digital cards for the connection and communication with other devices.

The Hardware option allows the settings for the body of equipment components connected to the computer to be loaded, saved, deleted or changed (see Figure 6.1). In fact, the configuration menu only displays the DLLs for the communication channels and instruments employed by the users in their particular applications. The software will only load into memory the DLLs explicitly selected by the user for a specific purpose, avoiding the computer's memory overloading.

In configuring one's system, it is very useful to use a detailed block diagram for the process including the channels to be used to connect the analytical equipment to the computer. First, the communication channels required are loaded. Then the instruments needed are implemented. Finally, the instruments concerned are linked with the communication channels present in the configuration tree.

Figure 6.1 shows the AutoAnalysis configuration for an SIA system with spectrophotometric detection. The branches of the tree in the "current configuration" pane bear a Crison serial communication channel and an Ocean Optics USB communication channel. Once the required communication channels are chosen, their operational settings (communication rate, word length, number of stop bits, control parity) can be selected in terms of the instruments to be employed.

Once all channels have been selected and configured, the miniaturized Ocean Optics CCD spectrophotometer is linked to its communication channel, and so are the Crison valve module and burette to theirs, e.g. first the valve (position 0) and then the burette (position 1). After these elements being chained, they can be individually configured and operational characteristics such as the type of valve used (e.g. one selection valve and one injection valve), number of ports, syringe volumes, burette steps, can be selected.

After loading the required configuration for the connected instruments, they can be addressed in each method and corresponding command forms become available in the editor window in addition to the basic and former described functions.

Data processing is the same, independently of the instruments or configuration loaded. At the same time, data processing and method development are centralized, providing ease of use for the users by minimizing the time and efforts required to add a new instrument to the system.

6.3.2.2 Development of the analytical method

The analytical method is created by sequencing functions, e.g. wait, variable and conditionals or instructions for the instruments by their selection from a pull-down menu and assigning specified commands to them (Figure 6.2). This provides a variability of programming functions for method creation including procedures, loops, indexing, basic calculations, variables, user input, waiting steps, event marking, comments, conditional enquiries, and online data processing. Further functions enable method test execution, definition of detection thresholds, manual data processing, such as calibration, data smoothing, spectral information processing, peak adjustment, and data export. Furthermore, this software allows the use of conditionals which enables the development of smart systems.

AutoAnalysis also allows the development of procedures. Procedures are prior-created instruction protocols, which can be loaded by all methods within the same hardware configuration. Main

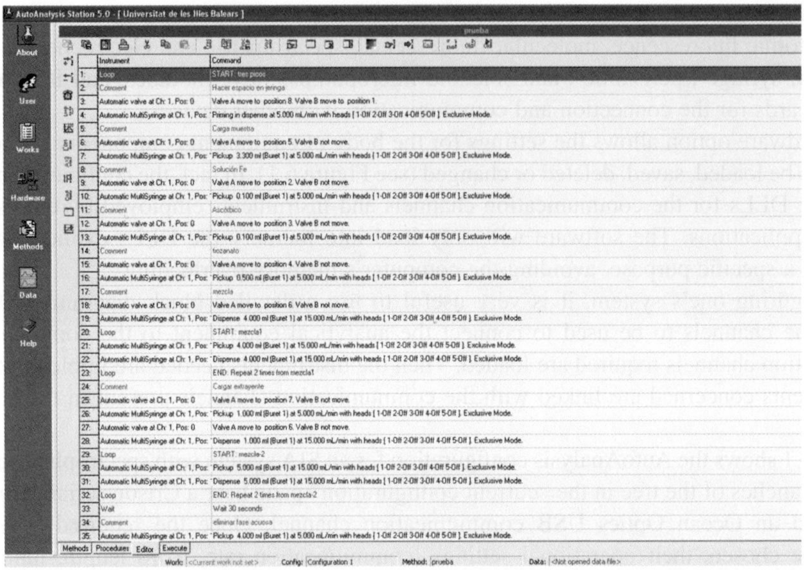

FIGURE 6.2

Method edition window. (For color version of this figure, the reader is referred to the online version of this book.)

advantages are higher method clearness by apparent shorting, module-like programming, and repeated application in the same method.

Instrument commands can be executed in two ways, namely: exclusively, i.e. no other command is processed until the current one is completed and nonexclusively, i.e. the method is allowed to continue executing successive steps of the command sequence while the current one is being completed.

The window "Execute" enables method initiation, pause, stop, as well as the control of the execution and continuously acquired data. Zoom, scale, and shift functions are also available.

Taking advantage of these features allows the creation and optimization of highly versatile applications. Based on 32-bit technology, AutoAnalysis allows multitasking operations such as simultaneous method execution and data processing, multiple loading, and working with other programs running at the same time.

6.3.2.3 Implementation of methods

Once a new method has been established or an existing one loaded, it can be implemented. A tool bar allows the implementation process and the settings of the graphics area (zoom, scales) to be controlled. The graphics area displays the analytical data as they are acquired and a status bar shows the current position in the command execution sequence.

Figure 6.3 shows the data acquisition window while the method with online detection is executed, e.g. spectrophotometric method. Upper window displays the fiagram, i.e. analytical signal vs time, while the lower window displays the spectrum at a preestablished wavelength.

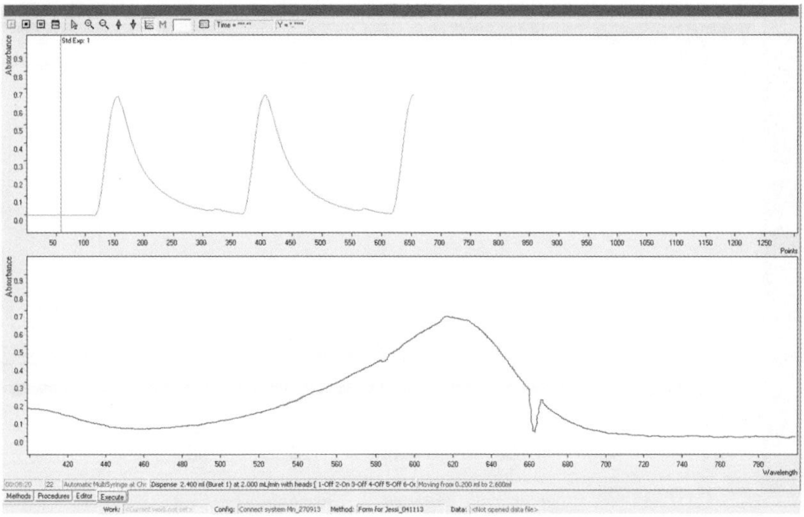

FIGURE 6.3

Data acquisition window during method execution. (For color version of this figure, the reader is referred to the online version of this book.)

6.3.2.4 Data processing

AutoAnalysis offers tools for data processing after method termination such as zoom functions, visualization of the original acquired data as well as of a smoothed function, baseline, and peak data. It allows peak height and peak area calibration, data export and saving, basic calculations, and overlay with prior acquired data (Figure 6.4). For peak maximum, start-, and end-point identification, thresholds for the numerical first and second derivative and a peak height minimum can be defined. Manual, mouse-drag peak correction, adding of nonidentified peaks or elimination of error-peaks is further possible. Online data processing is also possible by the use of conditionals.

6.3.3 Device types usable with AutoAnalysis

As noted earlier, using a given device in a system controlled via AutoAnalysis requires having a specific DLL for the device itself and another to manage it through an appropriate communication channel.

Devices most often controlled with AutoAnalysis can be classified in four types, namely: electronic boards, apparatuses, instruments and integrated modules. The following sections describe their desirable features and available DLLs. However, AutoAnalysis is in continuous evolution toward development of new DLLs [5].

6.3.3.1 Electronic boards

Electronic boards can be external and have their own communication interface, e.g. those from Ibercomp, or internal, for insertion into a PC slot. External boards have the advantage to be easily coupled

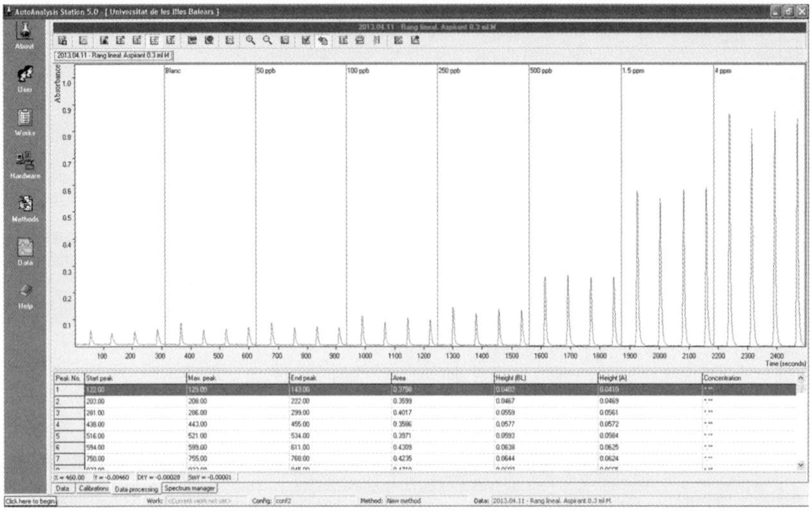

FIGURE 6.4

Fiagram and register table of analytical signals. (For color version of this figure, the reader is referred to the online version of this book.)

to instruments in order to facilitate operation as stand-alone systems. Internal boards are more common and do not have communication interface; however, they also require a DLL for communication and another for control.

Analog/digital converter boards can be very useful to incorporate instruments having no communication channel but allowing data to be obtained from their recorder outputs. This facilitates the use of obsolete laboratory instruments. Their greatest weakness is that communication is one-way (from the instrument to the computer, so the former cannot be controlled by the latter).

As noted in previous chapters, digital I/O boards are useful for governing solenoid valves, and also for controlling the power of Prolabo microwave ovens. Table 6.1 lists available DLLs for electronic boards.

6.3.3.2 Apparatuses

Apparatuses are devices which can perform mechanical operations but not make measurements. Many are essential elements of flow assemblies, e.g. syringe pumps, peristaltic pumps, selection valves, micropumps, external solenoid valves. Although commercially available in a wide variety of forms and models, many apparatuses lack an interface for connection to a computer. This should be an important consideration at the time of purchase. However, since they rarely need to feed back information to the computer, they can operate one-way (to receive commands from the computer). Table 6.2 lists available DLLs for apparatuses.

6.3.3.3 Instruments

Instruments are pieces of equipment capable of delivering measurable data. Therefore, they should be considered and treated as precision elements. Ideally, they should possess their own two-way

Table 6.1 Electronic Boards for which AutoAnalysis Dedicated DLLs are Available

Manufacturer and Model	Board Type and Features
Decision Super ADDA 14	A/D, 14 bits, 16 channels
Silmon Scope	A/D, 8 bits, 4 channels. Programmable gain (±80, 200, 400 and 800 mV; 2, 4, 8, 20 and 32 V)
Keithley/Metrabyte Dash8	A/D, 12 bits, 8 channels
Keithley/Metrabyte Dash801/CE	A/D, 12 bits, 8 channels. Programmable gain (±10, 50 and 500 mV; 5 and 10 V)
Ibercom	A/D 12 bits, 8 channels. Independent scale settings from 0 to 24 V. RS485 and RS232 interfaces
Decision 8255-8253	I/O, 48 lines, three 16-bit counters
Decision Industrial SmartLab	I/O 16 optocoupled inputs, 16 relay outputs
Flytech I/O Digital	I/O, 16 optocoupled inputs, 16 relay outputs
Ibercom	A/D, 10 bits. Independently adjustable gain for each of 8 channels. RS232C interface
LabJack U3-HV	A/D/A
EasyDAQ	8/16/24 pins relay board
Crison Pump and Valve	Relay board, pump and valve control card
Voltcraft Tester	Multimeter
Ibercom	Relay/input card

Table 6.2 Apparatuses for which AutoAnalysis Dedicated DLLs are Available

Manufacturer and Model	Features
Crison Pump 2060	One 8-port selection valve, one 6-port injection valve
Crison Pump 2060	Two 8-port selection valves
VICI	2-Position injection valve
VICI	8-Port selection valve
KNAUER	6-Port selection valve
Crison Microbur 2031	Single-syringe autoburette
Crison 738	Single-syringe autoburette
Crison 4S-D	Multisyringe autoburette with four 12 V outputs
Hamilton ML900	Multisyringe autoburette
Crison prototype	COMPACT stand-alone sampler
Ismatec Reglo-digital	Peristaltic pump with RS232C interface
Prolabo Maxidigest MX350	Focussed microwave oven
Micrux HV-Stat	Potentiostat for electrophoresis

interface for receiving commands intended to set their operating parameters and to transmit measured data and information about their working status. As with apparatuses, purchasing decisions on instruments should consider whether they possess an appropriate interface and are supplied with the required communication protocols and command set for computer control. Unfortunately, many instruments come with completely closed software, which prevents their coupling to other elements needed to set up a flow assembly. However, manufacturers can provide users with the information required to flexibly program their instruments. Table 6.3 lists available DLLs for instruments.

6.3.3.4 Integrated systems

As implied by their name, integrated systems use a combination of boards, apparatuses and/or instruments in order to facilitate specific tasks. They have the disadvantage with respect to modular systems that if one element breaks the whole system is rendered unusable.

Some manufacturers market integrated systems for the more classical flow techniques (SFA, FIA). Table 6.4 lists available DLLs for integrated systems.

Table 6.3 Instruments for which AutoAnalysis Dedicated DLLs are Available

Manufacturer and Model	Features
Crison 2001 and 2002	pH potentiometer plus temperature probe
Crison 501 and 517	pH potentiometer
Crison GLP 22	pH potentiometer
Crison 525	Conductimeter
Crison Digilab 517	Conductimeter
Crison Microsampler 2040	Conductimeter
Ocean Optics PC1000	Spectrophotometer
Ocean Optics PC2000 (master)	Spectrophotometer
Ocean Optics PC2000 (slave)	Spectrophotometer
Ocean Optics USB2000	Spectrophotometer
Ocean Optics SAS256	Spectrophotometer
Hewlett Packard HP 8452A	Spectrophotometer
Hewlett Packard HP 8453	Spectrophotometer
Perkin Elmer LS 5	Spectrofluorimeter
Perkin Elmer LS 50	Spectrofluorimeter
Sciware Systems RS232 commander	RS232 commander
PSA Analytica Escalibur, Merlin	Atomic fluorescence photometer
LDC SM4000	HPLC photometric detector
AMEL 533	Polarograph
Sciware 01	MSFIA/MPFS system

Table 6.4 Integrated Systems for which AutoAnalysis Dedicated DLLs are Available

Manufacturer and Model	Features
Crison COMPACT	SIA system with two burettes, one switching valve, two analog inputs, one stirrer and one sampler
Sciware MSFIA1	MSFIA system with four syringes, four 12-V outputs, one platform with four solenoid valves and one integrated CCD spectrophotometer
Amel 473	Electroanalytical system affording amperometry; dc and DPP polarographies; dc and DPP voltammetries; and linear stripping, square-wave and differential pulse voltammetries

References

[1] G.D. Marshall, J.F. Van Staden, Computer-aided flow-analysis for laboratory use and process analysis, Instrum. Sci. Technol. 20 (1992) 79–100.

[2] FIAlab: http://www.flowinjection.com.

[3] F. Mas, A. Cladera, J.M. Estela, V. Cerdà, New approach to sequential injection analysis: using the sample as carrier, Analyst 123 (1998) 1541–1546.

[4] E. Becerra, A. Cladera, V. Cerdà, Design of a very versatile software program for automating analytical methods, Lab. Rob. Autom. 58 (1999) 131–140.

[5] Sciware Systems: http://www.sciware-sl.com.

CHAPTER

Environmental Applications

7

CHAPTER OUTLINE

Flow methods have been consolidated as routine techniques of analysis, allowing a high degree of automation and leading to the concept of the total analytical approach, thanks to the ease they provide for integration of all steps of the analytical procedure, from sample collection to acquisition and processing of signals, on a single setup. The huge effort on research and development for monitoring of key indicators and surrogates has delivered a broad spectrum of operational and cost-effective methodologies based on flow analysis, some of which are nowadays commercially available and

Flow Analysis. http://dx.doi.org/10.1016/B978-0-444-59596-6.00007-3

have become standard methods, or are presently undergoing standardization, for the determination of many environmental parameters. Its recognized flexibility makes them suitable for the development of highly specialized systems as well as alternatives devoted to screening and multiparameter analysis.

This chapter gives an overview of online monitoring and environmental applications using flow-based methodologies. Their potential in this field is illustrated through representative examples covering a wide range of parameters and sample matrices, e.g. natural and wastewaters, gaseous compounds and atmospheric aerosols, soils, solid wastes and leachates, foodstuff, and plant extracts.

7.1 Flow analysis for environmental monitoring

Environmental studies are focused in the determination of amounts and distribution of chemical species in air, water, soil, sediments, sludge, biota and animal and human tissues, and fluids. These species may be elements or compounds, simple or complex in structure, organic or inorganic, natural or anthropogenic in origin, and occur at vastly different levels of concentration in the different environmental compartments.

There are two main reasons for monitoring in the environmental field:

1. To control pollutant releases, determine emission patterns and verify compliance with respect to defined standards and regulations;
2. To develop research programs and studies focused on the determination of the presence of target analytes, concentration levels, and distribution in particular areas or ecosystems.

There is also an increasing need for a real-time data collection linked to the development of risk assessment, environmental impact assessment, and toxicology studies.

Environmental analysis poses some difficulties due to:

- Sample matrices are typically complex—they may be solid, liquid or gaseous—and the source of a wide variety of potential interferences.
- The range of analyte types is very wide.
- Target analytes may have to be measured at trace levels with high precision, owing to the very low concentrations at which the harmful action of some pollutants is exerted and increasingly stringent national and international legislation.
- Speciation is of great concern, as the chemical form of the pollutant influences its mobility in the environment, bioavailability, physiological effects and properties, and consequently its toxicity and level of health hazard.
- Some analyses require in situ determination.
- Field sampling programs often generate a large number of samples to be processed.

Therefore, the ideal analytical method for measuring chemical species in environmental samples must offer:

- Quantitative analysis.
- Very low detection limits for trace and ultratrace determinations.
- Wide linear working ranges.
- Interference-free data.
- Reliability.

- Modular construction and simplicity of design.
- Hardware and software flexibilities.
- Potential to perform simultaneous multiparametric determinations and total or speciation analysis.
- Simple sample preparation.
- Possibility to accommodate and couple external devices, e.g. gas diffusion or dialysis units, packed columns, mixing chambers.
- High throughput and low cost.
- Automation capabilities through computer-compatible instruments and object-orientated software.

As explained in previous chapters, flow analysis techniques are well-established and powerful tools fitting most of these requirements, i.e. facilitating automation and allowing (1) increased sample throughputs, (2) high versatility, (3) new analytical improvements based on operating modes under controlled nonstationary conditions, (4) online sample pretreatment and analyte separation from interfering matrices, (5) decrease of sample handling, (6) more environmentally friendly procedures, and (7) use of alternative detection systems by improving selectivity and sensitivity and thus simplifying the operating conditions.

Procedures using flow schemes and commercially available instruments are more and more frequently employed in environmental analysis and introduced to the legal acts, which regulate environmental analysis. In particular, flow injection analysis (FIA) is up to know the preferred alternative in routine laboratories, with a wide range of environmental applications described in the literature.

Moreover, in situ measurements, in which the target analyte is determined in its own environment, are nowadays more demanded since they provide on-time results affording early warning about abnormal situations, while avoiding errors, delays, and costs related with the collection of samples for subsequent laboratory-based analysis. When intended to be used for in situ applications, in addition to the above specified analytical requirements, a field monitor for routine analysis must also afford the following:

- Automated introduction and handling of a variety of samples;
- Limited consume or no reagents;
- Reagent stability;
- Long-term autonomous operation;
- Robustness to continuously operate under nonideal environmental conditions;
- Resistance to biofouling, signal drift and blockages;
- Limited and easy on-site maintenance;
- Automatic zeroing and instrument calibration;
- Self-cleaning possibilities;
- Portability and miniaturization;
- Online data acquisition, processing, and transmission; and
- Full remote control, validating routines and internal diagnostics to detect malfunctions, with the possibility to incorporate computationally expert systems into the decision-making process.

However, online alternatives developed for environmental analysis are commonly direct adaptations of traditional analytical methods conducted in batch. Main limitations for most of these methods are sensitivity or precision problems due to the low concentrations of the individual species to be analyzed together with interferences and sample matrix effects. These are also not originally designed for field applications where they should operate under extreme and/or variable measurement conditions while providing accurate and reproducible results. In addition, analyzers often suffer from cross-contamination, blockages, biofouling, and signal-drift responses due to matrix influence particularly in long-term unattended operation. Consequently, these methods require frequent calibration and maintenance and often consume large quantities of chemical reagents that produce secondary forms of pollution that require safe disposal. The sampling step remains itself as a critical part of the overall analytical process. Design and construction of adequate inline sampling and sample conditioning systems remain as the weak points in implementing on-site and remote applications and becomes a crucial aspect for waters with high suspended solids that require filtration in order to prevent clogging of tubes and valves.

These factors have greatly limited the widespread use of online instrumentation, especially for large-scale monitoring. It is a real challenge to develop novel schemes capable to overcome these drawbacks and limitations and at the same time turn on the classical methodologies in more environmentally friendly ones. Recent advances in materials science, microelectronics, miniaturization, electro-optics, and computer capabilities have made that possible, enabling better instrumentation developments which, combined with statistical tools, provide emerging techniques and novel applications.

"Lab-on-valve", "Lab-on-chips", "Micro Total Analytical Systems (μ-TAS)" or "Micro Electro Mechanical Systems (MEMS)" that operate on micro- and nanometer scales are good examples of new trends and achievements in the area of instrumentation. These enable complex processes such as sampling, reagent addition, calibration, temperature control, and analysis, to be incorporated into compact, low powered devices.

In subsequent sections of this chapter, representative examples of online monitoring and environmental applications using flow techniques are given. Information is organized by analytes, according to the following categories: sulfur compounds, cyanides, carbon speciation, phosphorous compounds, nitrogen compounds, priority organic pollutants, and heavy metals.

As it is going to be illustrated through several examples, online wet chemical analyzers are extremely versatile and can be used to measure a wide range of components in environmental samples. The design of the flow system depends both on the chemical aspect of the determination and on the detector employed. Flow systems may include separation modules, where processes of diffusion, dialysis, and extraction are used, together with reactors where redox processes, derivatization, sorption, ion exchange, and enzymatic or immunochemical reactions take place. Flow techniques can also be applied to multicomponent determinations through appropriate arrangements and configurations or by taking advantage of the difference on kinetics of the reactions of individual analytes.

Certain analyses of environmental samples are difficult to achieve in a completely automated manner, i.e. soils, sediments, plant material, while aqueous samples are especially well suited to be analyzed by flow techniques. From a practical point of view, it is more attractive to focus on flexible modular systems easily adapted to specific casuistic.

Several international corporations, such as Technicon, Skalar, Tecator AB, Burkard Scientific Ltd., Timberline Instruments, Lachat Instruments, Bran and Luebbe, Alliance Instruments, Global FIA, Sciware Systems, Perkin Elmer, FIAlab Instruments Inc., have contributed to the development and commercialization of analyzers and assemblies based on the original Skeggs's idea followed by the various generations of flow systems.

7.2 Sulfur compounds

7.2.1 Sulfate ion

Turbidimetric flow methodologies are regarded as the simplest approaches for the determination of sulfate and have been successfully applied to environmental analysis. Apart from segmented flow analysis (SFA) and FIA modalities, SIA, FIA with multicommutation and binary sampling (FIA–MBS), FIA with sandwich sampling (FIASS), and gravity flow-based manifolds are described in the literature [1]. Exceptionally, a long path-length (100 cm) flow cell based on a liquid core waveguide [2] and electrode-separated piezoelectric quartz crystal sensor [3] have been used.

Flow schemes including online solid-phase extraction/enrichment of sulfate onto anion exchangers have been exploited for low-content sulfate samples ($\leq$0.1 mg/l) analysis. On the contrary, online sample dilution via either a large reaction coil or a dilution step as part of an SIA timing protocol allows the automated extension of the linear working range [4]. The manifold of the SIA system with the large reaction coil, shown in Figure 7.1(b), is more complex than the system including the dilution step

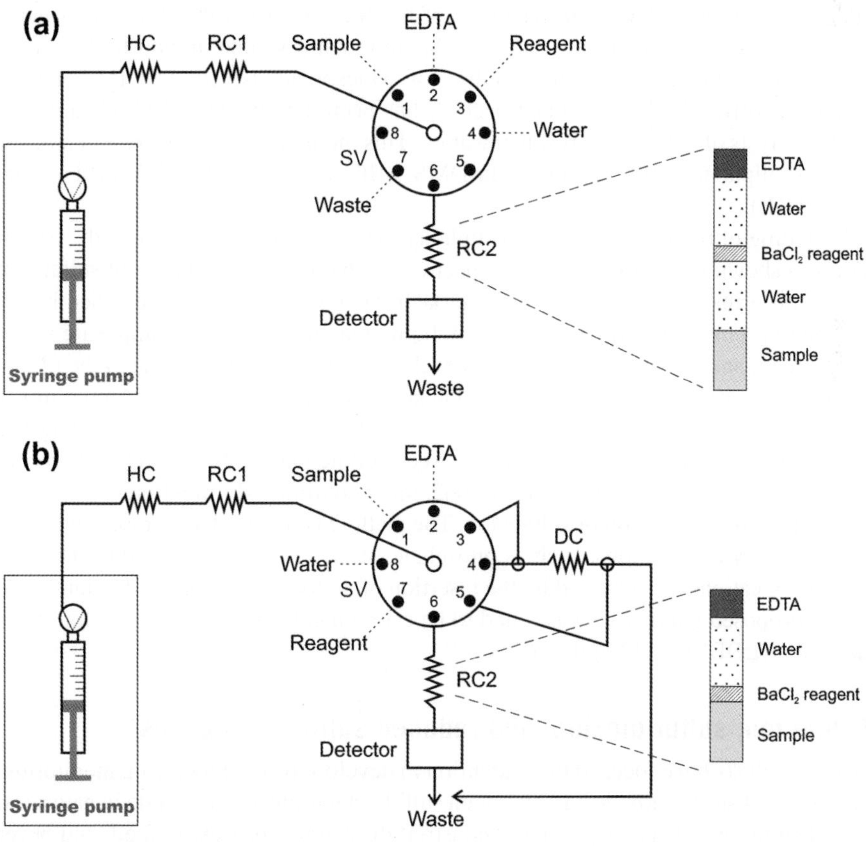

FIGURE 7.1

SIA system for sulfate determination. (a) Manifold including a dilution step. (b) Large reaction coil system. DC: dilution coil; HC: holding coil; RC: reaction coil; SV: selection valve.

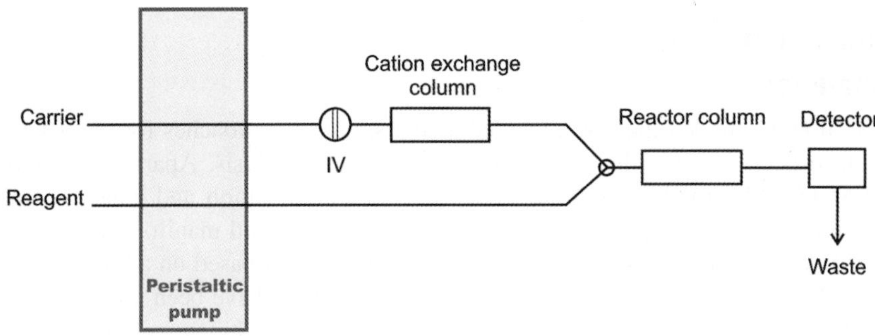

FIGURE 7.2

FIA system using a crushed $BaSO_4$ reactor column for the determination of sulfate. IV: injection valve.

(see Figure 7.1(a)) needing more complicated programming as well. Shorter analysis time was achieved with the dilution step, but the wide linear range of the SIA system with dilution coil makes it very suitable to be used as a process analyzer and suitable to monitor sulfate in industrial effluents.

Other alternative spectrophotometric method of widespread use for sulfate monitoring is the dimethylsulphonazo(III) (DMSA(III)) approach. It has been automated in FIA [5] and SIA [6] and applied to the analysis of natural and residual waters. This method involves the reaction of sulfate with barium-dimethylsulphonazo(III) complex, Ba-DMSA(III), displacing Ba^{2+} from the complex and forming DMSA(III).

The color fading of the barium-dimethylsulphonazo(III) complex by sulfate is the option adopted in the FIA setup shown in Figure 7.2 for the spectrophotometric determination of sulfate ions in tap, river, and mineral drinking waters [5]. This approach incorporates an online crushed barium sulfate reactor column to promote the efficiency of solid–liquid reaction. 200 μl of sample are injected into a water carrier stream and pumped through an ion exchange column, made of glass tube (2 mm internal diameter (i.d.) × 15 cm), packed with cation exchange resin (Na^+-type IR124T, Amberlite). Then, it is merged with the stream of reagent solution containing Ba-DMSA(III) before entering the crushed $BaSO_4$ reactor column. The exchange reaction between sulfate ion and Ba-DMSA(III) proceeds during flowing in this reactor column. Absorbance at 662 nm is continuously monitored. Signal decrease is used for the indirect determination of sulfate ion. The method is reported to be rapid and relatively free from interferences by other anions, such as phosphate, silicate, carbonate/bicarbonate, chloride, and nitrate. Interfering cations are removed by the insertion of an ion-exchange column into the analytical manifold. The proposed alternative was found to be economical and very effective for improving the limit of detection (LOD) (ca. 30 μg/l SO_4^{2-}).

7.2.2 Sulfite ion, sulfur dioxide, and reduced sulfur compounds

Environmental analysts have focused their attention in developing and modifying monitoring methods for routine control of sulfur dioxide and hydrogen sulfide in gas phase/air, sulfide in natural waters and sediments and sulfite in urban wastewaters, industrial discharges, processed food, and beverages.

Most flow methods developed for sulfur dioxide and hydrogen sulfide in atmospheric samples are based on the independent collection, concentration, and transfer of sulfur gaseous compounds into scrubbing solutions and final determination of sulfite or sulfide in an aqueous phase. These absorbers provide liquid samples that can be directly introduced into flow manifolds [7].

Few applications based on flow analysis can be found for the automated and direct determination of gaseous sulfur dioxide [8,9] or hydrogen sulfide [10] in atmospheric samples. This is mainly due to the difficulty encountered in the online sampling of gases, which has been partially overcome with the use of chromatomembrane cells or tubular and planar permeation denuders and diffusion cells for the collection of the analyte from air into an absorbing solution and its introduction into the flow manifold, followed by conductivity, potentiometric, piezoelectric, or spectrophotometric detection.

For instances, the injection of the donor phosphate buffer solution containing the gaseous SO_2 permeated through a gas diffusion membrane into a carrier stream of 5,5′-dithiobis(2,2′-dinitrobenzoic acid) (DTNB), a selective chromogenic reagent, allows the measurement of the formed colored compound at 410 nm by means of a miniaturized fiber optic spectrometer [8]. In order to overcome the lack of sensitivity required for atmospheric analysis, a preconcentration step has to be performed with the aid of the gas permeation device, by stopping the flow in the acceptor stream for 5–8 min. Detection limits rely on the enrichment time; e.g. for 5 and 8 min are 50 and 35 $\mu g/m^3$, respectively. The experimental setup provided with online sampling and preconcentration is shown in Figure 7.3. It is well suited to be applied in field.

Spectrophotometric detection based on the methylene blue (MB) method is the approach most commonly adopted for online derivatization and determination of sulfide in water due to its inherent selectivity. The so-called Fischer's reaction involves oxidative coupling of *N*,*N*-dimethyl-*p*-phenylenediamine (DMPD) with sulfide ions in the presence of an oxidizing reagent (namely, Fe(III)) in acidic medium, thus rendering a heterocyclic thiazine dye, the MB. Different flow techniques described in the literature, such as FIA, SIA [11,12], and multisyringe flow injection analysis (MSFIA) [13–17] have exploited this reaction, covering a broad spectrum of possibilities with respect to analytical features.

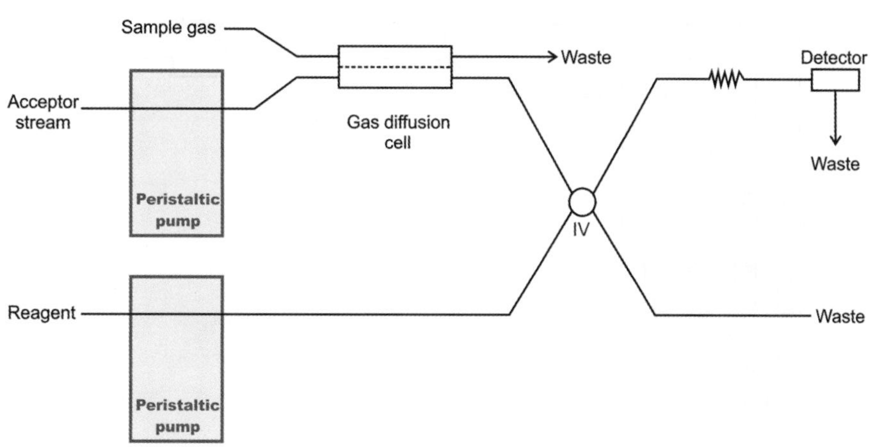

FIGURE 7.3

Schematic diagram of an FIA system for atmospheric SO_2 monitoring. IV: injection valve.

The potential of the MSFIA method based on the MB chemistry has been thoroughly investigated via the determination of trace concentrations of sulfide in water samples of different complexity, namely, mineral water, tap water, groundwater, seawater, and wastewater. Several setup configurations, combined as follows, are assessed:

Alternative 1. MSFIA sandwich-type and multiple injection modalities have been designed and successfully applied to the determination of sulfide in different spiked water matrices (Figure 7.4). Working ranges of 0.2–2.0 and 0.5–5 mg/l sulfide for sandwich and multiple injection techniques, and detection limits of 0.09 and 0.15 mg/l, were obtained, respectively [13].

Alternative 2. By hyphenating an optical fiber sensor for implementing disk-based solid-phase extraction and direct optosensing at the solid surface, there is no need for further elution to perform the optical detection in the eluate phase, as demanded in conventional sorbent-extraction protocols previously reported for sulfide, thus yielding improved enrichment factors [14] (Figure 7.5).

Alternative 3. A software-controlled flow system coupled to gas-diffusion (GD) separation for the spectrophotometric determination of sulfide in complex environmental samples, e.g. wastewaters, containing suspended solids, without requiring any previous batch treatment like flocculation or filtration (Figure 7.6). The method carried out in a multicommutated system is based on analyte release as hydrogen sulfide from the acidified donor channel of the GD-module into a stagnant recipient solution composed of DMPD and Fe(III). The MB dye generated inline via oxidative coupling is subsequently delivered downstream to a miniaturized flow-through light-emitting diode (LED)-based optical fiber plug-in spectrophotometer. High selectivity is obtained with respect to former options, where metal ions interfered significantly. Careful selection of chemical variables may also allow discrimination between sulfide, polysulfide, sulfite, thiosulfate, and sulfate [15].

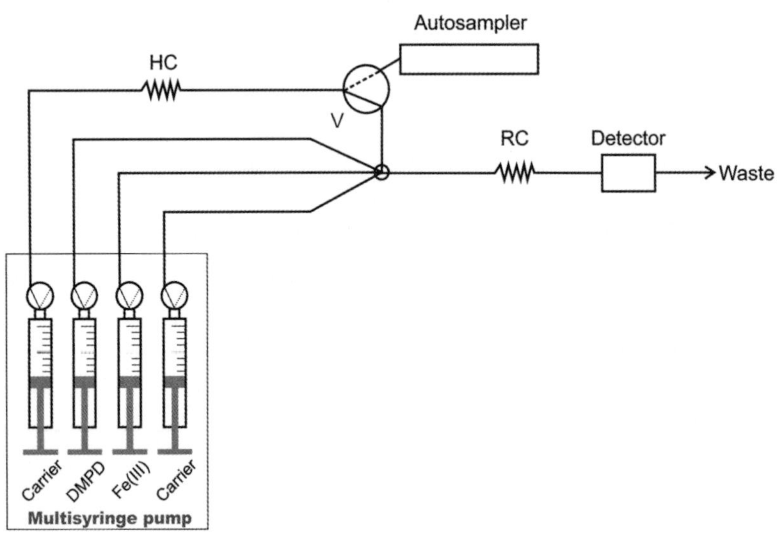

FIGURE 7.4

MSFIA manifold for the determination of sulfide. DMPD: *N,N*-dimethyl-*p*-phenylenediamine monohydrochloride; HC: holding coil; RC: reaction coil; V: solenoid valve.

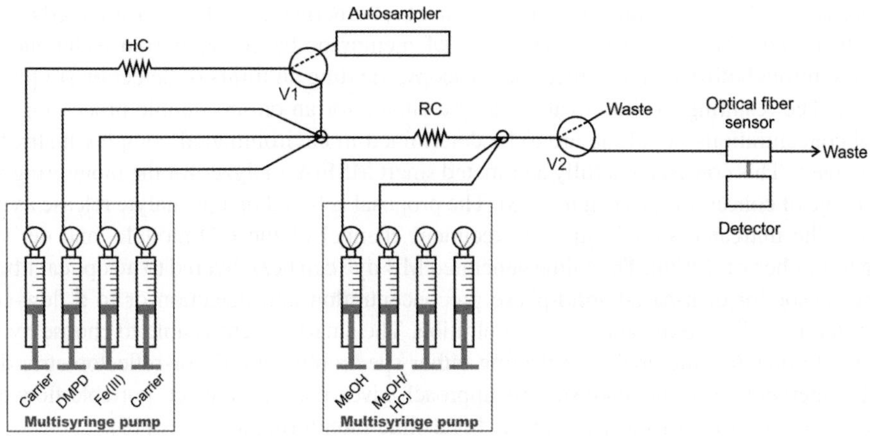

FIGURE 7.5

MSFIA disk-based extraction optosensing manifold for the determination of sulfide. DMPD: *N,N*-dimethyl-*p*-phenylenediamine monohydrochloride; HC: holding coil; RC: reaction coil; V: solenoid valve.

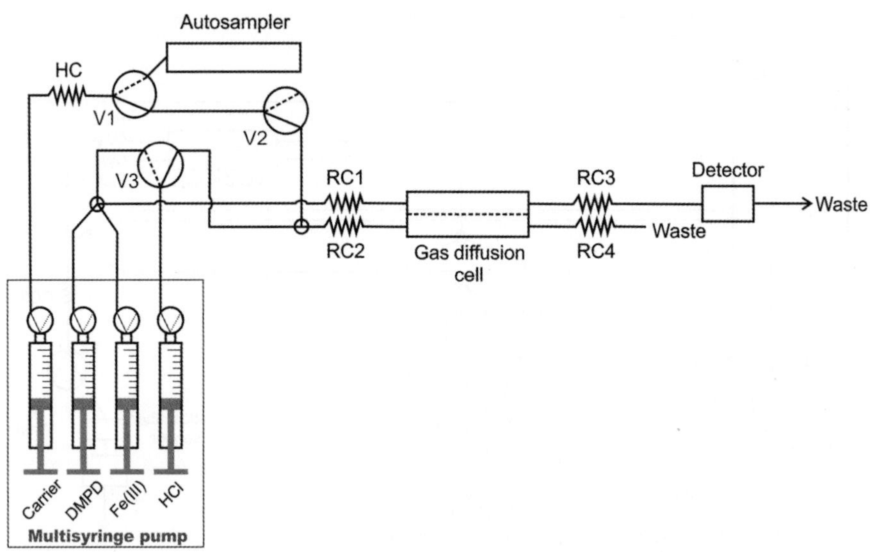

FIGURE 7.6

Gas-diffusion MSFIA manifold for the determination of sulfide. DMPD: *N,N*-dimethyl-*p*-phenylenediamine monohydrochloride; HC: holding coil; RC: reaction coil; V: solenoid valve.

Alternative 4. This is in fact a combination of alternative 2 and 3 (MSFIA–GD-optosensor coupling) that overcomes the loss of sensitivity of membrane-based separation techniques through sorptive flow-through diffuse reflectance spectroscopy, yielding an limits of detection (LOD) of 1 μg/l (Figure 7.7). The resulting method is an attractive choice for an online sample processing whenever trace-level concentrations of sulfide are to be determined in environmental complex matrices [16].

Alternative 5. This consists of a fully automated smart MSFIA analyzer for the monitoring of sulfide in a wide range of concentrations (Figure 7.8). The proposal is based on the analyte release as hydrogen sulfide from the untreated sample into the recipient channel of the GD-module, and derivatization according to Fischer's reaction. The inline generated MB dye can be delivered to an optical fiber diffuse reflectance sensor for disk-based solid-phase preconcentration and detection or to a flow-cell spectrophotometer according to the analyte concentration. The smart system is able to choose by itself the best strategy to quantify the analyte, selecting either spectrophotometric or reflectometric detection, without the intervention of the analyst. This approach gives rise to compact, portable flow analyzers readily adaptable to in-field measurements and real-time monitoring schemes [17].

While spectrophotometric detection of sulfide in natural/freshwaters is usually coupled to preconcentration in order to overcome lack of sensitivity, luminescence [18] and electrochemical [19] detection can be applied with no need for sample enrichment protocols.

For instance, quantification of ultratrace levels of sulfide is feasible with inexpensive flow analyzers relying on the catalytic action of the target analyte on the hydrogen peroxide–luminol chemiluminescent (CL) reaction. Furthermore, in combination with online gas diffusion processes, it is possible to suppress matrix effects, e.g. metals, ions, oxidants, and organic compounds, and improve selectivity as a result. The MSFIA–CL method depicted in Figure 7.9 features a linear working range

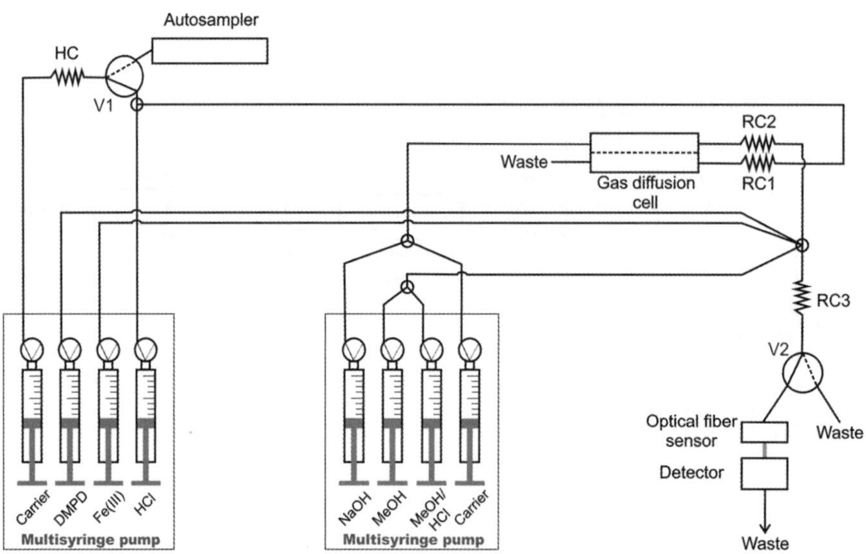

FIGURE 7.7

Gas-diffusion MSFIA manifold with integrated flow-through optical fiber diffuse reflectance sensor. DMPD: *N,N*-dimethyl-*p*-phenylenediamine monohydrochloride; HC: holding coil; RC: reaction coil; V: solenoid valve.

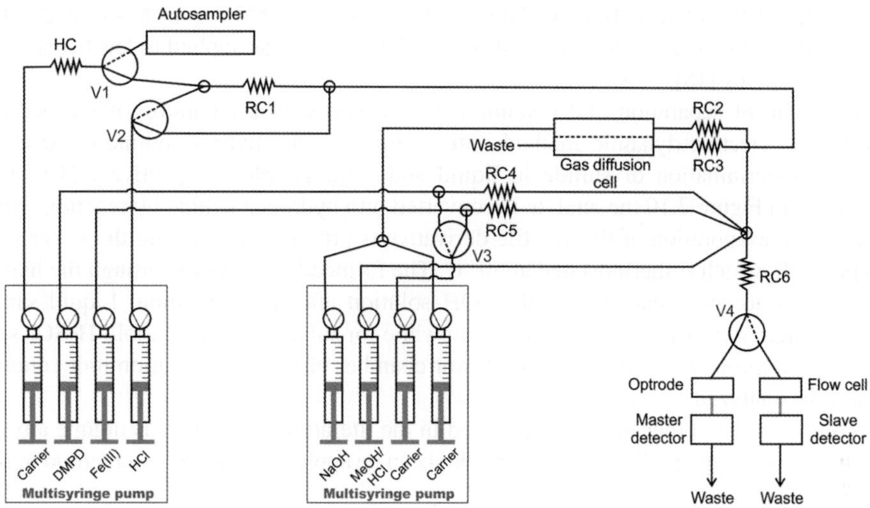

FIGURE 7.8

Smart MSFIA setup with integrated flow-through optical fiber diffuse reflectance sensor and flow-cell spectrophotometric sensor. DMPD: *N,N*-dimethyl-*p*-phenylenediamine monohydrochloride; HC: holding coil; RC: reaction coil; V: solenoid valve.

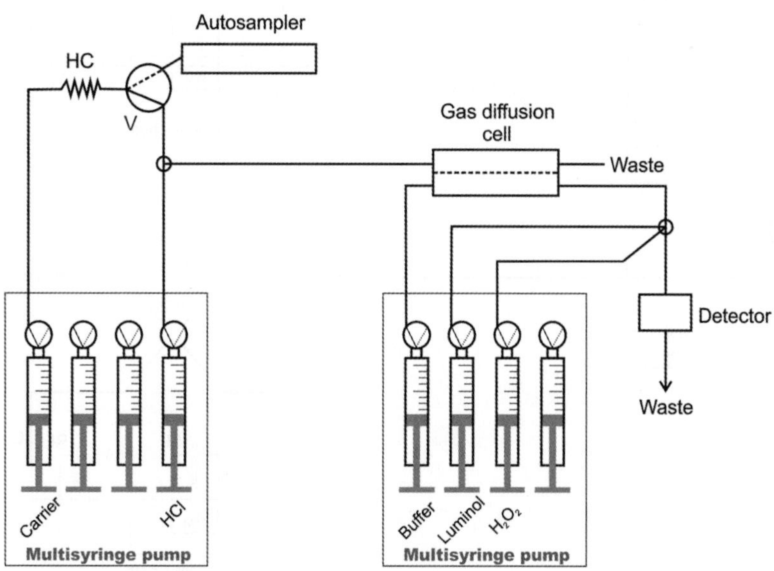

FIGURE 7.9

Gas-diffusion MSFIA–CL setup for the determination of sulfide. HC: holding coil; V: solenoid valve.

of 0.02–2 mg/l, a limit of detection of 3 μg/l, a throughput of 20 samples/h and a coefficient of variation of 2.4% for a 1 mg/l sulfide concentration. The method was applied to leachates and various types of water samples [18].

The integration of separation and potentiometric detection in a laboratory-made pervaporation module yields a selective dynamic method with continuous and hybrid continuous–discontinuous format for the determination of sulfide in liquid and solid samples, respectively [19]. Using the manifold shown in Figure 7.10 the analyte is converted into hydrogen sulfide by reaction with diluted sulfuric acid. The evaporation of the volatile derivative is promoted by heating the lower part of the pervaporation cell, which is thermostated at 50 °C. The formed H_2S diffuses through the hydrophobic membrane and is accepted into a diluted NaOH solution after pervaporation. Liquid samples are automatically injected into an acidic donor stream while for solid samples a 1 mol/l H_2SO_4 solution is manually injected through a septum into the lower chamber of the pervaporation unit containing the sieved soil to be analyzed.

When compared with other methods appeared in the literature, this offers as unique advantage its applicability to solid samples. The proposed method can be applied as a screening method to perform a

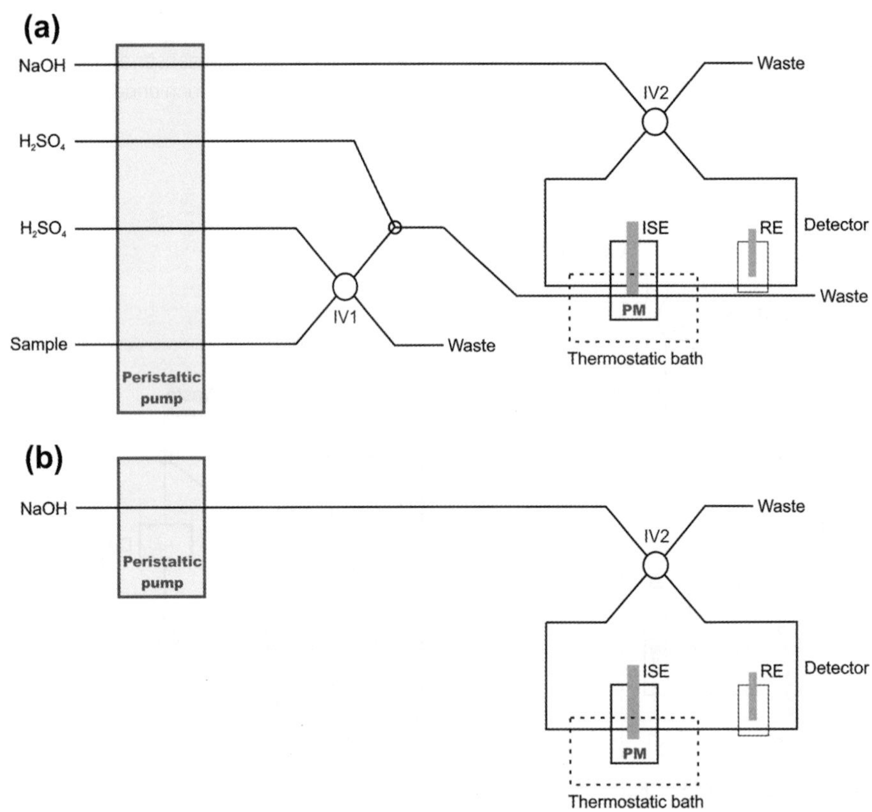

FIGURE 7.10

FIA manifolds for the determination of sulfide in (a) liquid and (b) solid environmental samples. ISE: ion-selective electrode; IV: injection valve; PM: pervaporation module; RE: reference electrode.

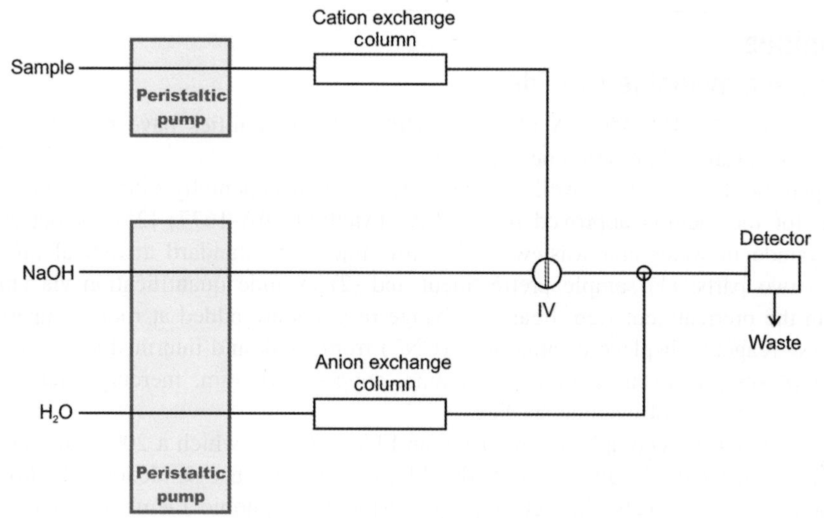

FIGURE 7.11

Schematic diagram of an FIA-CL system for the determination of sulfur dioxide in air. IV: injection valve; sample: gaseous sulfur dioxide absorbed from air samples into 1 mM NaOH solution.

fast discrimination between toxic and not toxic wastes in order to cut down the number of samples requiring time-consuming laboratory testing for USEPA reactivity.

A CL flow system with luminol immobilized on an anion-exchange column was developed for the indirect determination of sulfur dioxide in air (see Figure 7.11). Sulfite is detected by the CL reaction with the luminol bleeding from the column by hydrolysis [20].

Figure 7.12 depicts an FIA system based on the o-phthalaldehyde (OPA)–sulfite–NH_3 reaction in basic medium [21] for determining sulfite in fresh water and seawater.

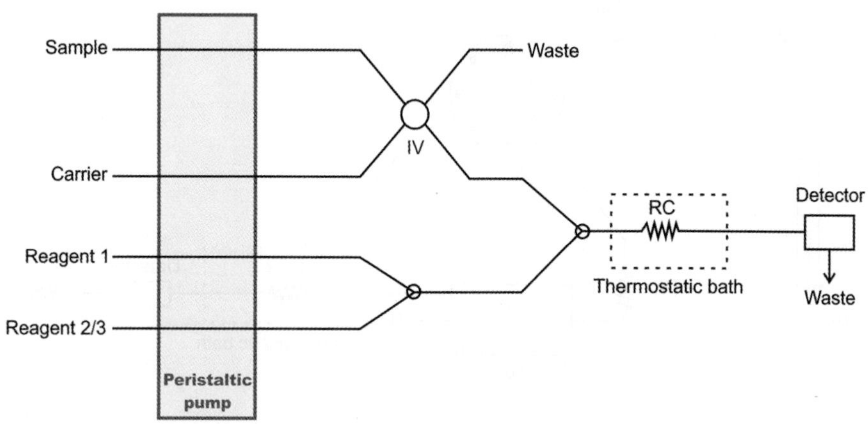

FIGURE 7.12

FIA-fluorescence manifold for the determination of sulfite in rainwater and seawater. IV: injection valve; RC: reaction coil.

7.3 Cyanides

7.3.1 Free and available cyanide

FIA [22–28], SIA [29–31], MCFIA [32], and MPFS [33] modalities have been reported for the determination of cyanide in environmental samples.

The amperometric detection is well recognized for its environmentally suitable working range and is the basis for the method approved by the EPA (Method OIA-1677) [25] for determination of available cyanide in water and wastewater by flow injection. Standard analytical procedures are divided into two parts: (1) sample pretreatment and (2) cyanide quantification via amperometric detection. In the pretreatment step, ligand-exchange reagents are added at room temperature to the sample. These reagents displace cyanide ions (CN^-) from weak and intermediate strength metallo-cyanide complexes, i.e. cyano-complexes of zinc, copper, cadmium, mercury, nickel, and silver, improving the recoveries of cyanide species.

Cyanide detection is accomplished by using an FIA system in which a 200 µl aliquot of the pre-treated sample is injected. The addition of hydrochloric acid converts cyanide ion to hydrogen cyanide (HCN) that passes selectively through a gas-diffusion hydrophobic membrane into an alkaline-receiving solution where it is converted back to cyanide ion. The latter is monitored amperometrically with a silver-working electrode, silver/silver chloride reference electrode, and platinum/stainless steel counter electrode, at an applied potential of zero volt. The current generated is proportional to the cyanide concentration present in the original sample.

The method detection limit is 0.5 µg/l and the minimum quantification level is 2 µg/l. The working range is approximately 2 µg/l–5 mg/l CN^-. Higher concentrations can be determined by dilution of the original sample or by reducing volume of the sample loop. Total analysis time is approximately 2 min. A slightly modified method for aquatic free cyanide is available. Free cyanide and metal-cyanide complexes dissociable at pH 6–8 are determined without using ligand-exchange sample pretreatment.

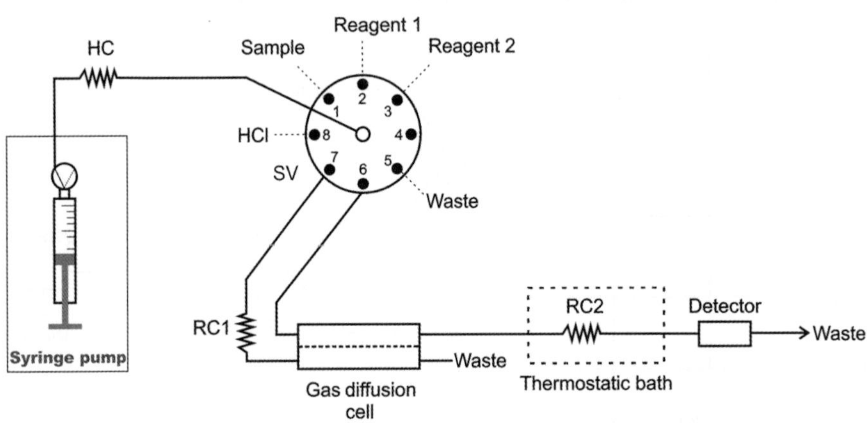

FIGURE 7.13

Schematic depiction of an SIA setup for the determination of free and available cyanide. Detector: spectrophotometer (λmax = 510 nm); HC: holding coil; RC: reaction coil; SV: selection valve.

The spectrophotometric detection of free and available cyanide can be achieved by the online reaction of the analyte with ninhydrin in carbonate medium to form a colored product (510 nm λmax). Cyanides are removed from sample matrix by acidification through a gas-diffusion step incorporated in the SIA manifold shown in Figure 7.13. The assay was validated in terms of linearity (up to 200 μg/l), limit of detection (2.5 μg/l), limit of quantitation (7.5 μg/l), precision (RSD $< 2.5\%$ at 100 μg/l), and selectivity. All interferences related to ionic species are removed because just gaseous molecules diffuse. High tolerance against critical species such as sulfides and thiocyanates was achieved. The method was applied to the quantitation of cyanide in tap and mineral waters [30].

Flow systems with gas diffusion separation and spectrofluorimetric detection have been proposed for the determination of cyanide anion in aqueous samples (Figure 7.14). The cyanide transferred

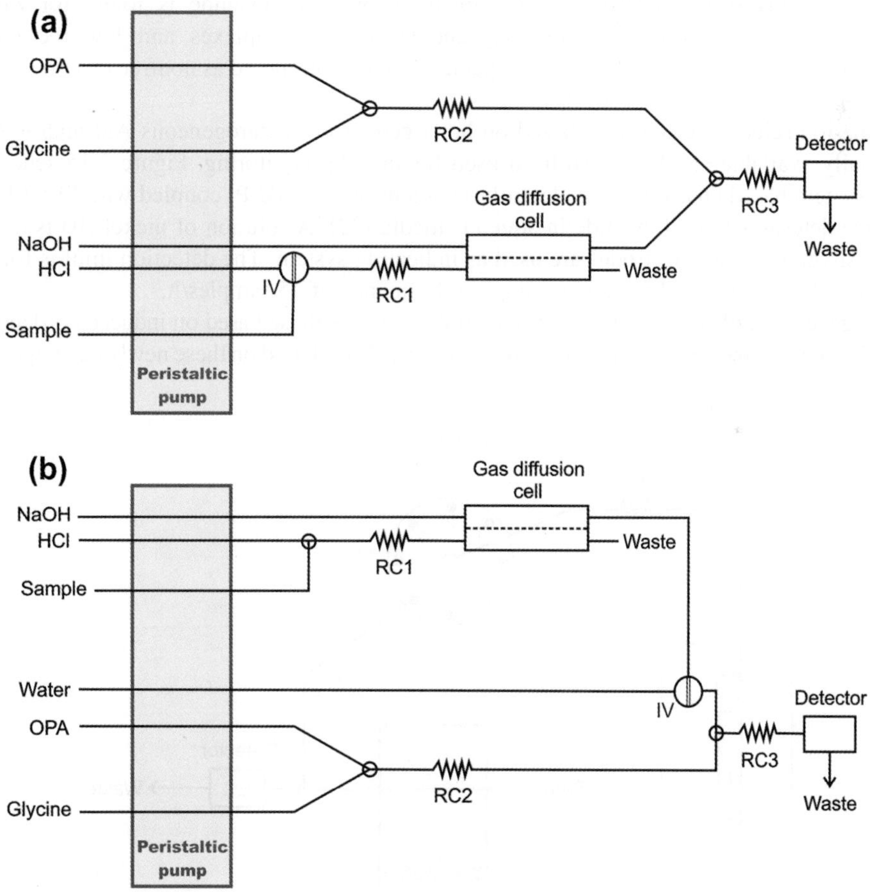

FIGURE 7.14

FIA-fluorescence manifolds for the determination of acid dissociable cyanide. (a) Conventional approach. (b) Introduction of an enrichment step. Detector: fluorimeter (λex/λem = 331/379 nm); IV: injection valve; OPA: *o*-phthalaldehyde; RC: reaction coil.

reacts with OPA in the presence of glycine to form a highly fluorescent isoindole derivative [24]. Thus, the cyanide concentration in the samples is proportional to the observed fluorescence intensity.

In conventional FIA manifolds (see Figure 7.14(a)), a volume of sample is injected into the donor stream; it diffuses through the membrane and is then collected in a flowing acceptor stream. Other manifolds improve the diffusion efficiency by means of an enrichment step. This can be achieved by continuously pumping the sample into the system and either working with a large sample/acceptor flow rate ratio using the manifold shown in Figure 7.14(b), or using a stationary solution in the acceptor side of the diffusion unit, which also serves as the injection loop. In the latter, the sample is pre-concentrated into a small volume of acceptor solution.

These methods are rapid, sensitive, selective, reproducible and easy to automate and can be used as simple routine tools for acid-dissociable cyanide monitoring. Extremely low levels of cyanide can be detected using this system ($\leq$0.5 μg/l). Complete recovery of cyanide is found for $Zn(CN)_4^{2-}$, $Cu(CN)_3^{2-}$, $Cd(CN)_4^{2-}$, $Hg(CN)_4^{2-}$, $Hg(CN)_2$ and $Ag(CN)_2^-$ complexes and low recovery from $Ni(CN)_4^{2-}$. No recovery is obtained from the species that are considered as nonfree cyanide producing, i.e. $Fe(CN)_6^{4-}$, $Fe(CN)_6^{3-}$ and $Co(CN)_6^{3-}$.

Cyanide ion-selective electrodes, based on homogeneous or heterogeneous AgI and/or Ag_2S are commercially available and have also been used for in-field monitoring. Figure 7.15 schematically depicts a system based on differential electrolytic potentiometry (DEP) coupled with SIA/FIA for the quantitative determination of cyanide in aqueous media [22]. A solution of nickel (II) is used as reagent and silver amalgam electrodes are used as indicating system. The detection limit is found to be 0.31 μg/ml with an RSD of 2.5% and a sampling frequency of 45 samples/h.

A new generation of solid-state cyanide electrodes, such as those based on induced co-deposition of metal chalcogenides, meet better online requirements [26]. Thus, based on these newly developed sensors,

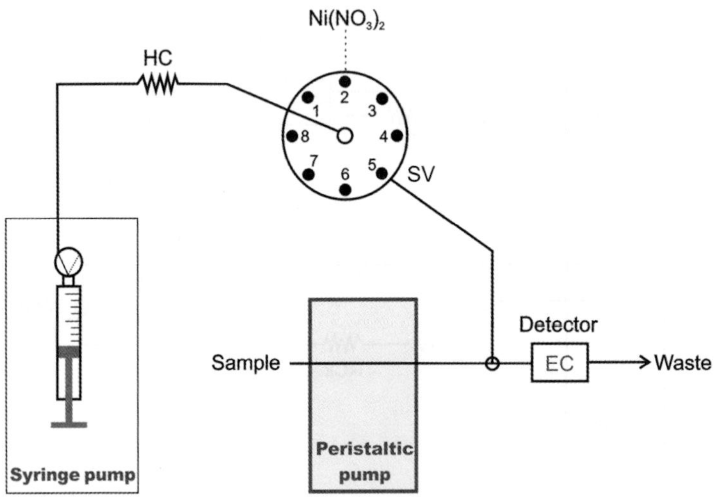

FIGURE 7.15

Flow setup used for the potentiometric detection of cyanides. EC: electrochemical cell containing two similar silver amalgam electrodes; HC: holding coil; SV: selection valve.

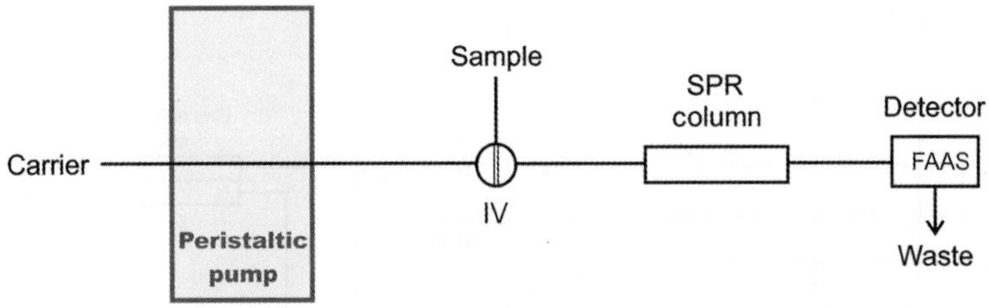

FIGURE 7.16

Single-line FIA system with solid-phase reagents for the indirect determination of cyanide by flame atomic absorption spectrometry. IV: injection valve; SPR: AgX (where X is Cl^-, Br^- and I^-) solid-phase reactor.

an FIA method for weak acid dissociable (WAD) cyanide determination without membrane separation was reported [27]. Cyanide speciation may be accomplished applying three consecutive measurements of: (1) nontreated, (2) ligand-exchange pretreated, and (3) electrochemically pretreated sample.

Obviously, cyanide cannot be directly measured by flame atomic absorption spectrometry (FAAS), but an indirect approach, as that schematically depicted in Figure 7.16, allows this possibility to be implemented, improving detection limits with regard to those reported previously for flow-based methods. The FIA manifold relies on the formation of soluble metal-cyanide complexes as the sample passes through a small column packed with solid-phase reagent (SPR). Different SPR have been tested for indirect determination of cyanide using FIA. In all cases the eluted complex is measured by FAAS. Detection limits close to 0.05 mg/l cyanide have been reported [28].

7.3.2 Total cyanide

Total cyanide determination requires appropriate destructive procedures to be implemented in order to decompose the sample matrix and liberate hydrogen cyanide. To this end, alternative approaches have been applied, e.g. microwave-assisted digestion [34] and photodissociation [35–37].

The standard method ASTM D 7511-09e2 for the determination of total cyanide in groundwaters, municipal wastewaters, industrial effluents, leachates and surface waters makes use of a narrow-band, low-watt UV irradiation source in combination with a gas-diffusion separation step and amperometric detection, the flow diagram is shown in Figure 7.17 [35].

The sample is injected into a carrier stream, segmented, and acidified. Under acidic conditions the WAD cyanide complexes convert to HCN and the strong metal-cyanide complexes are irradiated by ultraviolet light in the UV digestion module where they break down and release HCN. The HCN gas from all cyanide species present in the sample diffuses across a hydrophobic membrane into a basic acceptor solution, where it converts back to CN^- and is carried to a flow cell of an amperometric detector. Cyanide ions react with a silver electrode and generate a current proportional to the cyanide ion concentration.

Complete cyanide recoveries are typically attained for the most important metal-cyanide complexes found in environmental samples. Detection limits below 0.05 μg/l have been reported. The main drawback of this system is that the UV radiation promotes thiocyanate decomposition, what leads to sulfide and cyanide formation. To overcome this problem, it is recommended to use borosilicate glass instead of a quartz reactor, which cuts off radiation below 300 nm.

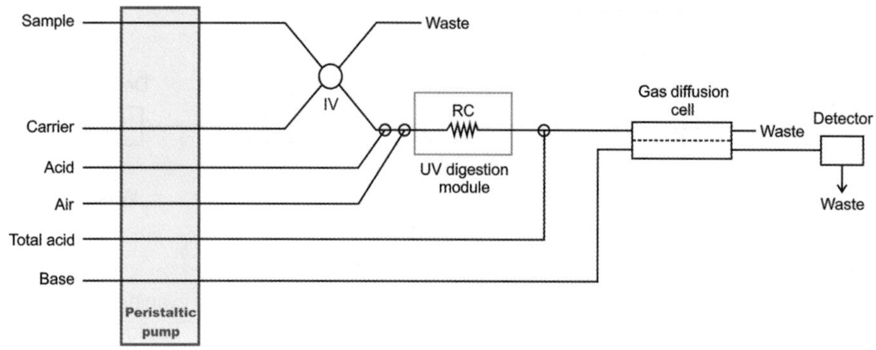

FIGURE 7.17

Flow system for total cyanide analysis by ASTM Method D 7511-09e2. Detector: silver working electrode, Ag/AgCl reference electrode, stainless steel counter electrode; IV: injection valve; RC: reaction coil.

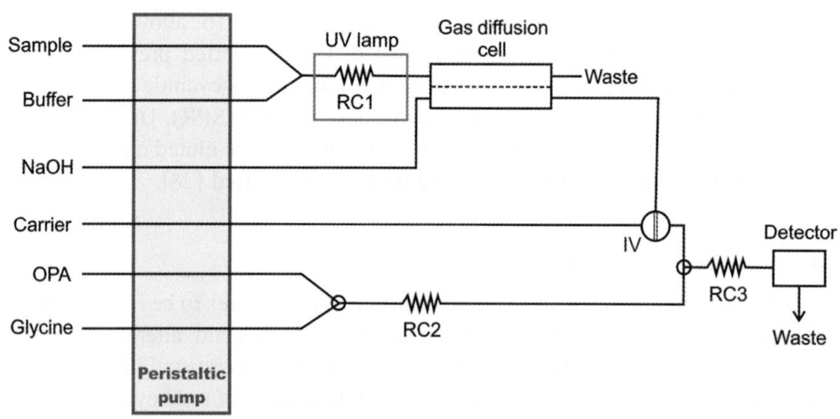

FIGURE 7.18

Gas-diffusion FIA-fluorescence manifold for the determination of total cyanide. Detector: fluorimeter; IV: injection valve; OPA: *o*-phthalaldehyde; RC: reaction coil.

The substitution of amperometric by fluorimetric detection based on the OPA-glycine method is possible by turning the manifold depicted in Figure 7.17 into the configuration shown in Figure 7.18. The sampling frequency for total cyanide is 4 samples/h and the detection limit is 0.4 μg/l. Recoveries of total cyanide from riverwater obtained with this method are about 90% of those obtained with APHA Method 4500-CN C for total cyanide [36].

7.3.3 Thiocyanate

Other compounds related to cyanide that may also require quantification are thiocyanates.

An SIA system has been proposed for the spectrophotometric determination of thiocyanate as $Fe(SCN)_2^+$. The system, shown in Figure 7.19, is fully computerized and able to monitor thiocyanate in

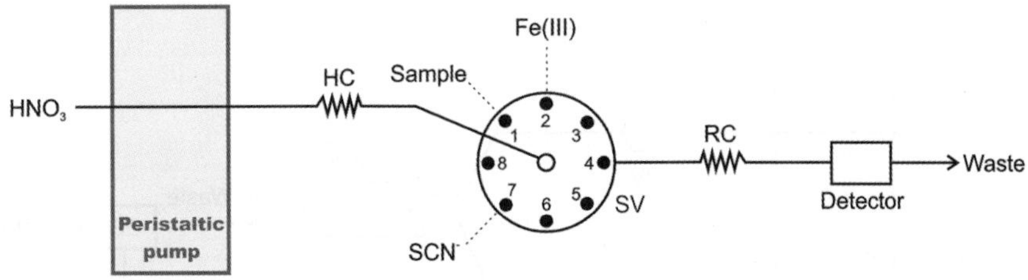

FIGURE 7.19

Schematic depiction of an SIA analyzer used for the determination of thiocyanate with Fe(III). Detector: spectrophotometer (λmax = 480 nm); HC: holding coil; RC: reaction coil; SV: selection valve.

samples at 24 samples per hour with a relative standard deviation of <1.2% for samples and standards with thiocyanate concentrations above 10 mg/l. The calibration graph is linear from 2 to 150 mg/l with a detection limit of 1.1 mg/l. This system has been applied to the determination of thiocyanate in process solutions and wastewaters [38].

7.4 Carbon speciation

7.4.1 Inorganic carbon

Numerous analytical procedures based on continuous systems have been developed for determining total inorganic carbon (TIC), most of them coupling gas permeation with FIA. It involves online CO_2 formation and diffusion through a semipermeable membrane from a donor toward an acceptor stream and the quantification of the trapped CO_2. In this technique, the sample is injected into, or merged with, an acid solution stream to form carbon dioxide, which diffuses across a polytetrafluoroethylene (PTFE) or silicone-rubber membrane with plane or tubular geometry into a recipient solution [39–49]. The trapped carbon dioxide has been determined either by electrochemical or spectrophotometric detection.

Figure 7.20 shows a simple FIA manifold coupling gas diffusion and electrochemical detection for inorganic carbon (TIC) determination in waters. The method is based on the diffusion of CO_2 across a gas permeable membrane from a donor stream containing 0.1 mol/l HCl to an acceptor stream of sodium acetate (10^{-5} mol/l and pH 10). The CO_2 trapped in the acceptor stream passes through an electrochemical flow cell that contains a tungsten oxide wire and a silver/silver chloride electrode, where it is detected [43]. The potential difference of the solution flowing through the flow cell is directly related to the composition of the acceptor solution and concentration of total CO_2 trapped in it. The electrode response is linear in a concentration range from 5 to 100 μg/ml CO_3^{2-}. Precision (RSD) is 1.4% for 20 μg/ml standard solution of CO_3^{2-} and the LOD is 0.20 μg/ml CO_3^{2-}. The sampling rate is 30 samples/h.

A large number of flow schemes for inorganic carbon analysis are based on conductometric detection. The inherent selectivity provided by the membrane separation process makes it possible. When NaOH is used as acceptor solution, an electrical conductivity decrease is recorded with increase in TIC concentration owing to the neutralization of the sodium hydroxide. On the contrary, a raise of the conductance signal is measured when either deionized water or weak bases such as tris(hydroxymethylamino)methane or sodium acetate are present in the receptor phase.

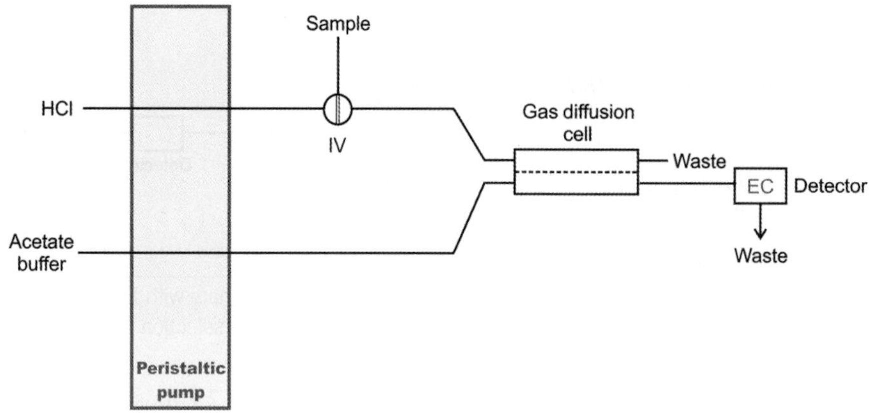

FIGURE 7.20

Schematic representation of an FIA manifold used for the determination of total inorganic carbonate. Detector: tungsten oxide electrode and Ag/AgCl reference electrode; EC: electrochemical cell; IV: injection valve.

The alternative approach is based on coupling gas diffusion and spectrophotometric detection, through monitoring the color change of acid–base indicators such as cresol red [47], 4-(2′,4′-dinitrophenylazo)-1-naphthol-5-sulfonic acid [48], bromothymol blue (BTB) [50], or the color fading of a phenolphthalein solution [45], all of them used as acceptor solution for the collection of the permeated CO_2. Several environmental applications are reported in the literature.

A reversed or reagent injection FIA (rFIA) system (Figure 7.21) was developed by coupling a GD unit with a portable FIA system. It has been used for long-term automatic and continuous in situ monitoring of CO_2 in natural waters, such as river- and seawaters [49].

In this system, only a 0.02 ml injection of acidic solution (diluted H_2SO_4) is sufficient for one measurement. The flow system incorporates two six-ports injection valves within the sample stream before the gas diffusion unit. One valve is used for injecting the acidic solution, the other is used for

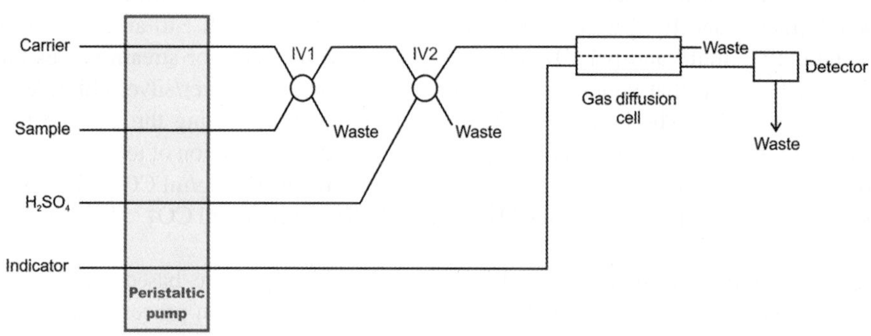

FIGURE 7.21

rFIA manifold used for the determination of total carbonate in water. Detector: spectrophotometer (λmax = 450 nm); IV: injection valve.

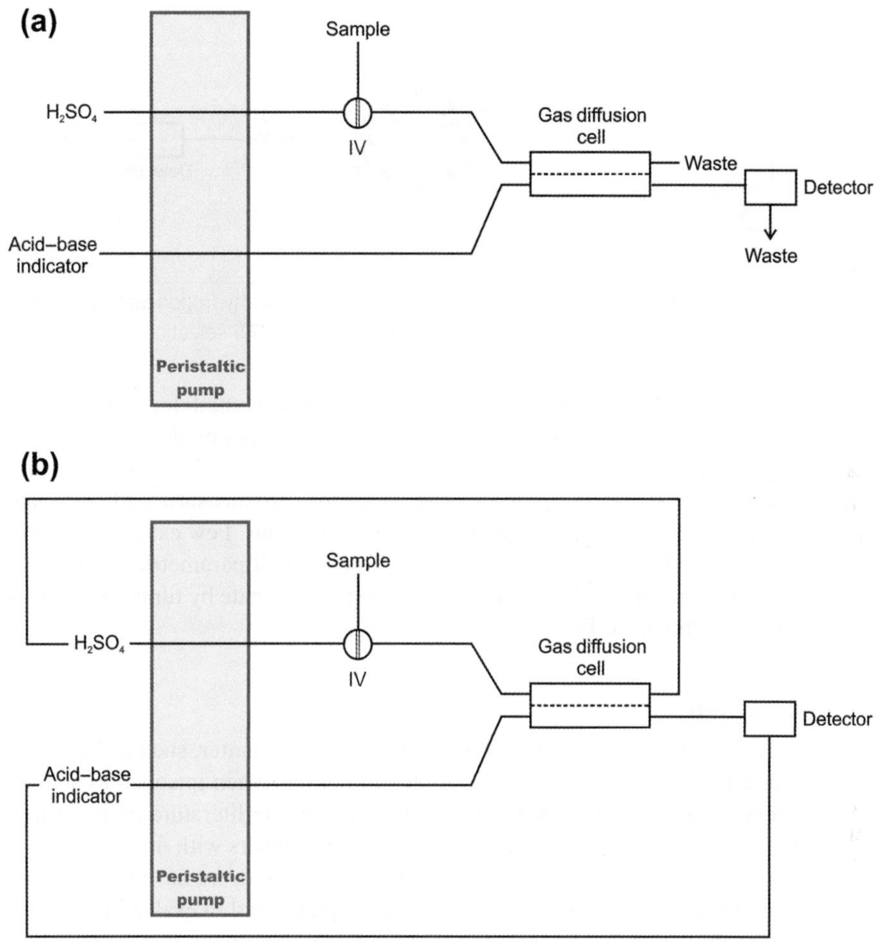

FIGURE 7.22

(a) Open and (b) closed gas diffusion-flow injection manifolds for the determination of carbon dioxide in gaseous samples. Detector: spectrophotometer (λmax = 554 nm); IV: injection valve.

injecting standard carbonate solutions for the preparation of calibration graphs. The detection limit of the method is 1.2×10^{-5} mol/l. The consumption of the acidic solution can be reduced to less than one hundredth, compared with a conventional FIA system.

Few examples for the direct determination of carbon dioxide in gaseous samples are reported in bibliography. The gas diffusion-flow injection systems shown in Figure 7.22 are some examples [44].

7.4.2 Total alkalinity

Carbon dioxide species in environmental samples have been conventionally measured through total alkalinity (TA). Alkalinity is, in fact, a measurement of the buffering capacity due to dissolved salts,

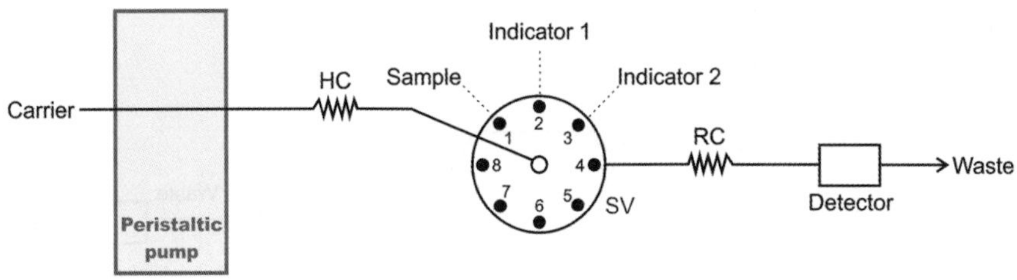

FIGURE 7.23

Sequential injection titration system for the determination of carbonate and hydrogencarbonate. Detector: spectrophotometer (λmax = 430 nm); HC: holding coil; RC: reaction coil; SV: selection valve.

predominantly carbonate and hydrogencarbonate. The standard method for the determination of alkalinity is by acid–base titration with a strong acid with the end points detected either with a pH electrode or using phenolphthalein and methyl orange or methyl red as indicators. These methods can be implemented in flow systems using the above indicators to measure carbonate and hydrogencarbonate colorimetrically resulting in much faster determinations. Few examples of flow titration are found in the literature [51–53], with some of them allowing multiparametric determinations [54].

The simultaneous determination of carbonate and hydrogencarbonate by titration can be performed using the SIA system illustrated in Figure 7.23.

7.4.3 Organic carbon

Traditional methods for obtaining information on the bulk organic matter, such as biological oxygen demand (BOD), are time consuming and suffer various limitations when intended to be implemented in flow-based systems. Therefore, few proposals are reported in the literature on this subject, being mainly restricted to studies on the capabilities of interfacing biosensors with flow schemes in order to carry out the BOD measurements in a kinetic mode. These BOD sensors have their widest environmental application in measuring waste loadings to treatment plants and in evaluating the efficiency of such systems. For example, online monitoring of biological treatment processes for waste and wastewaters has been carried out by means of a BOD sensor based on an immobilized mixed culture of microorganisms in combination with a dissolved oxygen electrode. Measurements were carried out using an FIA system, resulting in 25 s for one measurement followed by 4–8 min of recovery time. Good agreement was achieved between results of the sensor BOD measurement and those obtained from BOD_5 analysis of a wastewater sample from a food-processing factory [55].

The chemical oxygen demand (COD) test requires less time and results are more reproducible than that for BOD determination. It is a parameter established for the assessment of the amount of both organic and inorganic matter contained in a water body and measures the ultimate oxygen consumption of a sample, being widely employed in analytical laboratories for monitoring water quality.

More examples are found in the literature regarding the application of flow analysis to COD determination in which the digestion and detection steps are carried out online [56–67]. Spectrophotometric detection is commonly preferred as an alternative to the traditional titrimetric quantification. Flow analysis methods have been improved following strategies based on advanced sample

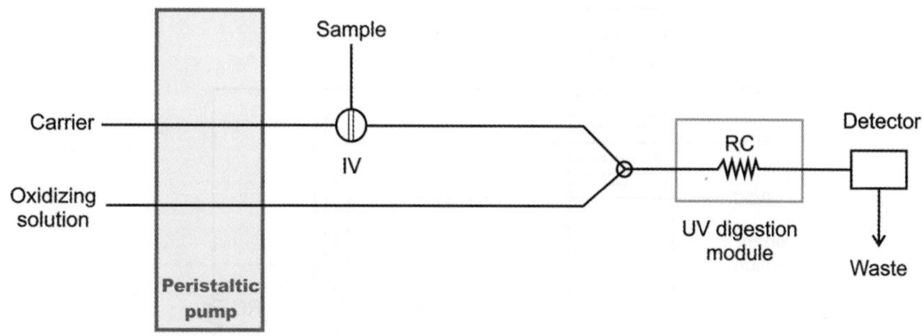

FIGURE 7.24

FIA manifold with on-line UV-photocatalytic oxidation used for the determination of chemical oxygen demand. Detector: spectrophotometer (λmax = 524 nm); IV: injection valve; RC: reaction coil.

digestion procedures instead of conventional heating, e.g. UV-photocatalytic oxidation [57,58], UV/TiO_2 solid-phase reactors [59–61], microwave [62,63] or ultrasound assisted digestion [64], or alternative detection techniques, such as flame atomic absorption spectrometry (FAAS) [62,64], electrochemical detection [59–61] or chemiluminescence by UV-ozonation [65], luminol-H_2O_2 [66] or luminol-H_2O_2-Cr^{3+} [67]-based reactions. By this, fast, sensitive, accurate, and cost-effective in situ COD determinations have been achieved, being in most cases more environmentally friendly options.

Three different options are analyzed in more detail below and their schematic diagrams outlined in Figures 7.24–7.26, respectively.

Option 1. Online UV-photocatalytic oxidation and spectrophotometric detection: its application to the determination of COD in fresh waters using FIA.

A 100 μl sample solution is injected into a 0.3 mol/l H_2SO_4 carrier stream containing 0.1 mol/l $(NH_4)_2SO_4$, merged with a permanganate solution (8×10^{-4} mol/l) and passed through a 250 cm FEP (fluoroethylene polymer) photoreaction coil wound around a 15 W UV lamp. The sample throughput

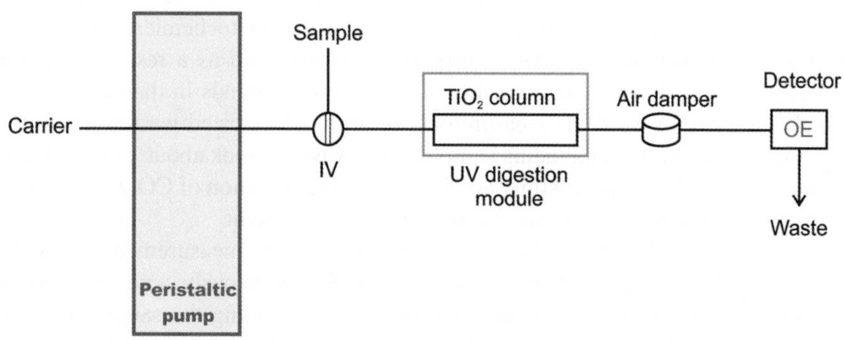

FIGURE 7.25

Schematic diagram of the photocatalytic FIA system for the determination of chemical oxygen demand. Detector: oxygen electrode (OE); IV: injection valve.

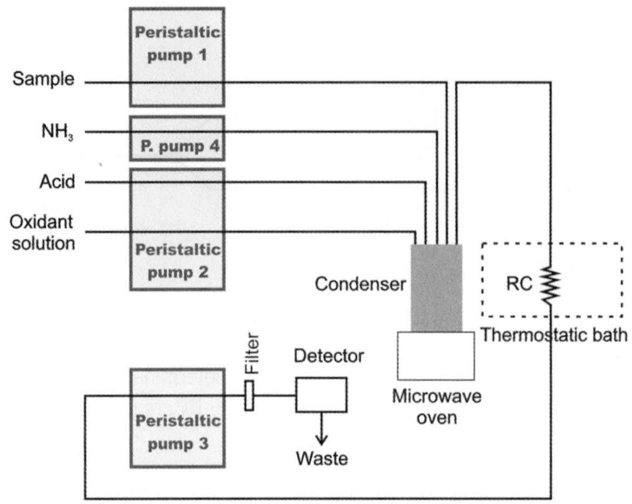

FIGURE 7.26

Scheme of COD automatic measurement device. Detector: spectrophotometer (λmax = 590 nm); RC: reaction coil.

is 30/h, with an LOD of 0.5 mg/l COD and a linear range of 0.5–50 mg/l COD. Results for a certified reference material (QC demand Quality Control Standard) were in good agreement with the certified COD value. Recoveries from fresh water samples for all tested COD standards were 83–111% and the COD values determined were in good agreement with those of a permanganate index reference method [57].

Option 2. Photocatalytic sensor for COD determination based on packed bead reactors and an oxygen electrode.

A photocatalytic sensor for the determination of COD with FIA based on the photocatalysis of organic compounds in the presence of titanium oxide beads in a photochemical column was described. The sensor is developed in conjunction with TiO_2 beads in the photochemical column and with an oxygen electrode as the sensing part. The sensor signal is observed as a result of the detection of dissolved oxygen changes due to photooxidation of organic compounds in the sample solution. This sensor responded linearly to the COD measurement of artificially treated wastewater in the range of 0.12–8 mg/l. A complete analysis, including sampling and washing, took about 10 min. The sensor was stable for over 15 days and was successfully applied to the determination of COD in lake samples [61].

Option 3. Fully automatic microwave-assisted COD measurement.

In the scheme shown in Figure 7.28 [63], sample goes into the measurement device through the pump 1, which leads the sample into the microwave oven. Simultaneously, acid and oxidant mixtures are added into the microwave oven through the pump 2. After adding the sample and reagents, the program starts automatically the reaction mixture irradiation with microwaves. Microwave radiation catalyzes oxidation of organic matter by potassium dichromate, so oxidation is completed just in 8 min. Then the reaction products go into the spectrophotometer by the aid of pump 3, where the absorbance of Cr(III) generated during sample oxidation is measured at 590 nm. Finally, the system is cleaned with concentrated ammonia added by pump 4. COD values are obtained in just 12 min. The application range extends from 0 to 15.000 mg/l O_2, and the Cl^- tolerance was improved up to 8.000 mg/l.

Speciation of dissolved inorganic (DIC) and organic carbon (DOC) has been automated through FIA and SIA [50,68] with online UV photooxidation.

In the SIA system depicted in Figure 7.27 total dissolved carbon and dissolved inorganic carbon are measured independently and the dissolved organic carbon is obtained as the difference between the two results. For inorganic carbon measurement, online sample acidification to pH <2 provides conversion of DIC to CO_2, which subsequently diffuses through a PTFE membrane into an alkaline, BTB acceptor stream. The absorbed CO_2 increases the concentration of the acidic form of the BTB indicator, with a resultant change in absorbance at 611 nm being directly proportional to DIC concentration. The total dissolved carbon (DIC + DOC) is determined after an online sample irradiation with a UV lamp coupled with acid peroxydisulfate digestion, with the subsequent detection of CO_2 as described above. DOC was determined by substraction of DIC from total dissolved carbon (TDC) (DIC + DOC). The method was applied to the analysis of bathing waters [50]. The developed SIA system allows the simultaneous determination of alkalinity with the same setup by introducing an online titration step based on the spectrophotometric measurement of bromocresol green color change after sample reaction with acetic acid.

Thus, the developed SIA method enabled the determination of DIC (0.24–3.5 mg/l C), CO_2 (1.0–10 mg/l C), TC (0.50–4.0 mg/l C), and alkalinity (1.2–4.7 mg/l C and 4.7–19 mg/l C) with limits of detection of: 9.5 μg/l C, 20 μg/l C, 0.21 mg/l C, 0.32 mg/l C, respectively.

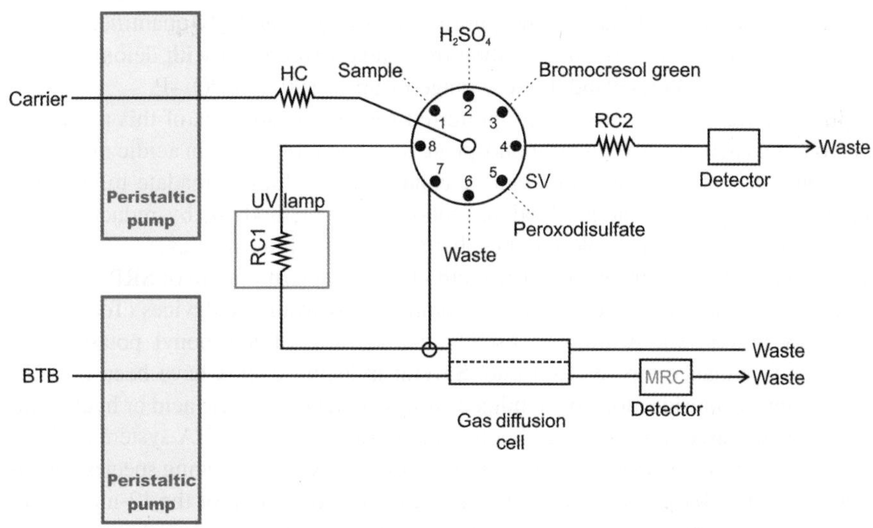

FIGURE 7.27

SIA manifold for carbon speciation in bathing waters. BTB: bromothymol blue indicator; Detector: spectrophotometer (λmax = 611 nm); Detector (MRC): multireflection cell coupled to light-emitting diode (λmax = 660 nm); HC: holding coil; RC: reaction coil; SV: selection valve.

7.5 Phosphorous compounds

The analysis of phosphorus has benefited from different flow techniques including gas-segmented continuous flow [69], FIA [70–84], rFIA [85–88], SIA [89–97], lab on valve (LOV) [98], MCFIA [99,100], the so-called all injection analysis (AIA) [101], the MSFIA [102–105] and MPFS [106].

7.5.1 Soluble reactive phosphorus

Flow-based arrangements have been extensively investigated and widely adopted for the determination of soluble reactive phosphorous (SRP) in environmental samples. Flow-based systems for SRP have been successfully used in shipboard determinations [69,73,76,94,95], in situ analysis of streams [107], in field monitoring and control of phosphorus removal in wastewater treatment plants [71,90,100,108,109] and also in other bioprocesses control and land-based laboratory applications.

Some of the proposed online methods for in situ determination integrate, or are connected to, sampling and sample filtration and pretreatment devices [69,71,73,76,95,109], which can deliver fresh drawn samples directly to the analytical system. Such devices avoid sample storage and all problems related to.

For in-field determinations of SRP in sea and estuarine waters methods must handle problems arising from the salinity gradient (schlieren or refractive index effect). Some rFIA manifolds using spectrophotometric detection solving these salinity interferences have been reported. Acidic phosphomolybdate reagent is injected into a carrier stream of sulfuric acid [85] or sodium chloride [86] of the same refractive index and sequentially merged with sample and reducing agent (either ascorbic acid [85] or tin chloride [86]). Reactive phosphorus has been successfully quantified in samples with salinities ranging from 0 to 30 psu using calibration standards prepared with deionized water, with a detection limit of 2–6 μg/l PO_4–P and a linear range of up to 100 μg/l PO_4–P.

Spectrophotometric methods are usually preferred for routine analysis of this parameter, most of them relying on the reaction between orthophosphate ions and molybdate in acidic medium in order to form a heteropoly acid. Color formation can be enhanced by adding vanadate to obtain the yellow vanadomolybdate complex (vanadomolybdophosphoric acid method) or by reducing the molybdophosphoric acid to yield strongly colored phosphomolybdenum blue species.

The molybdenum blue spectrophotometric method for the determination of SRP is universally used in routine analysis and has been implemented in commercially available devices (Tecator Ltd, Method Application ASN 60-03/82.). Antimony (usually in the form of antimonyl potassium tartrate) is frequently added to enhance the reaction rate. Several reducing agents have been reported for molybdenum blue: tin(II) chloride alone or stabilized with hydrazine, ascorbic acid or hydrazinium sulfate being the most widely used. As an interesting alternative, an automated SIA system analogous to that depicted in Figure 7.28 was proposed, not requiring unstable chemical reducing species but based on the online generation of the detectable species by electrochemical reduction of the 12-molybdophosphoric acid complex using a stainless steel tubular flow-through working electrode [91].

The established method is linear up to 20 mg/l P, with coefficients of variation ($n = 10$) of 2.4 and 1.8% for 2.0 and 10 mg/l P, respectively. The versatility of the SIA method to analyze samples containing high orthophosphate levels has been demonstrated by the implementation of a dilution chamber as well as flow-reversal techniques.

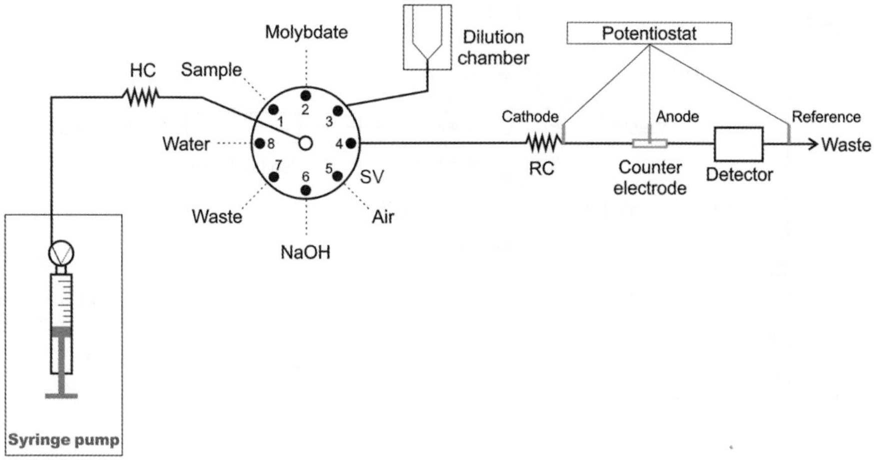

FIGURE 7.28

Schematic depiction of an SIA setup devised for the spectrophotometric determination of orthophosphate based on the electrochemical generation of molybdenum blue. HC: holding coil; RC: reaction coil; SV: selection valve.

Another example is a fully computerized MSFIA system for the spectrophotometric determination of available phosphorus in soil extracts. The molybdenum blue method is chosen for the colorimetric determination, using ascorbic acid as reducing agent, antimony to accelerate the reduction to the blue complex and applying the Egner–Riehm method to extract phosphorus from soil samples. It presents a linear calibration curve between 0.75 and 15 mg/l. A determination frequency of 15/h may be achieved, with good repeatability for 12 consecutive injections of soil extracts (RSD $<1.7\%$). Results obtained from 12 soil samples were statistically comparable to those attained by the usual batch method [102].

As an innovative alternative a sequential injection microcolumn extraction system (SI-MCE) may be used. It is based on the implementation of a soil-containing microcartridge as external reactor in a sequential injection network for dynamic partitioning of inorganic phosphorus in agricultural soils [92]. The extraction step is performed by means of the SIA arrangement illustrated in Figure 7.29, while orthophosphate levels are quantified using an FIA setup (see Figure 7.30).

In an attempt to integrate all the analytical process in a single device, a hyphenated MSFIA-microcolumn setup has been assembled for automated flow-through partitioning and accurate determination of the content of bioavailable forms of orthophosphate in soils and sediments utilizing the molybdenum blue method for extract processing [103].

Phosphate ion can also be determined by CL based on the oxidation of luminol by a heteropolyacid complex. Figure 7.31 shows a flow-through solid-phase-based CL optical sensor [104] for the determination of phosphate in different waters (mineral, ground, tap, and pond water and water–steam cycle of an incineration plant) exploiting the MSFIA concept with multicommutation.

The proposed system relies on the heteropolyacid complex formation and its transient immobilization in an *N*-vinylpyrrolidone/divinylbenzene copolymer packed in a spiral shaped flow through cell

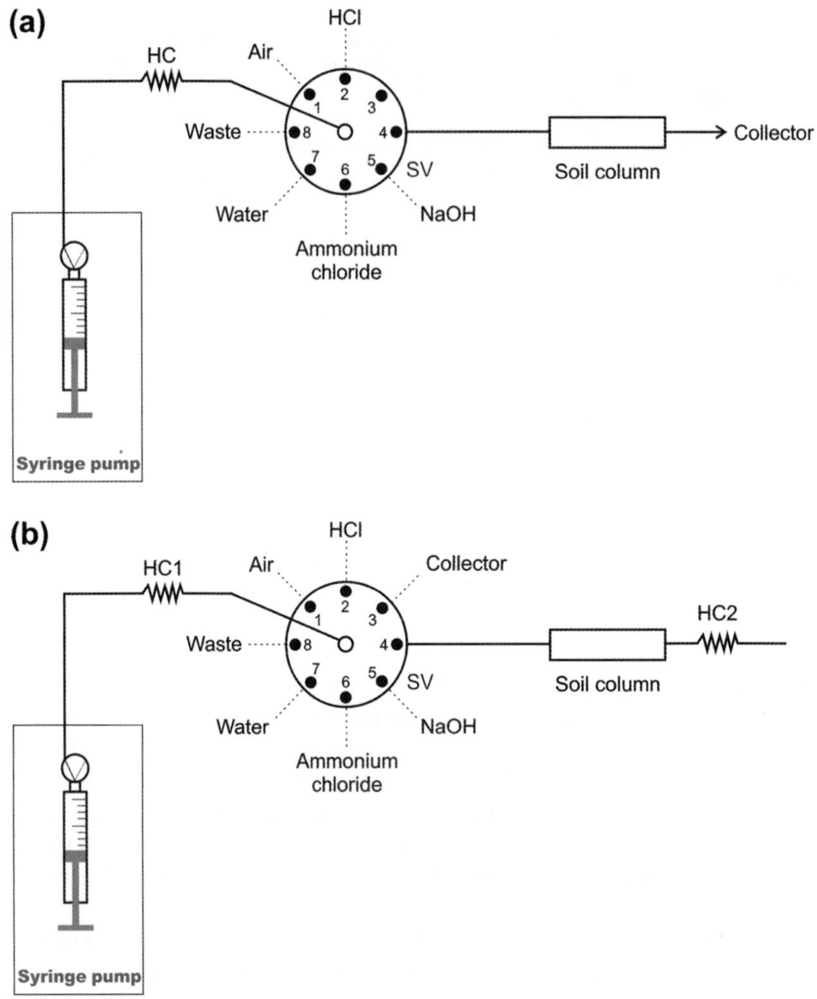

FIGURE 7.29

SIA system for dynamic fractionation of inorganic phosphorus in soils. (a) Unidirectional and stopped-flow configuration. (b) Bidirectional flow extraction. HC: holding coil; SV: selection valve; soil column: soil-containing microcolumn.

located in front of the window of a photomultiplier tube. Simultaneous injection of a well-defined plug of luminol in alkaline medium and methanol solution toward the packed reactor produces the light emission from the luminol oxidation by oxidant species retained on the sorbent material, which is readily detected. At the same time, the generated molybdenum blue compound is eluted by the minute amount of injected methanol, rendering the system ready for a new measuring cycle. The noteworthy features of the developed CL–MSFIA system are the feasibility to accommodate reactions with

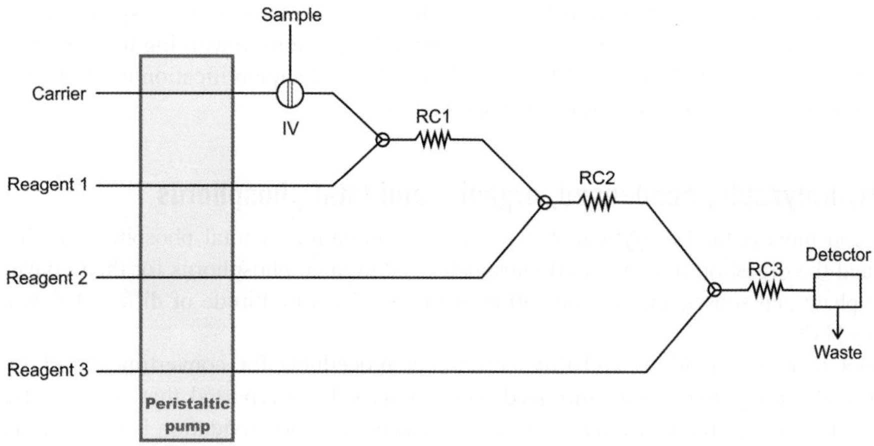

FIGURE 7.30

FIA manifold for automatic monitoring of orthophosphate in soil extracts. Detector: spectrophotometer (λmax = 690 nm); IV: injection valve; RC: reaction coil.

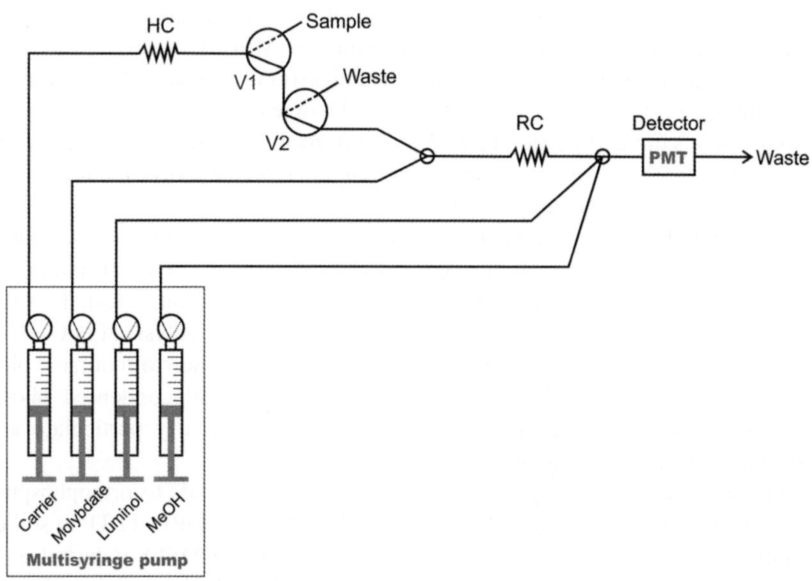

FIGURE 7.31

MSFIA–CL manifold for the determination of trace levels of orthophosphate in waters. Detector: chemiluminescence detector (flow-through solid-phase optical sensor + photosensor module); HC: holding coil; PMT: photomultiplier; RC: reaction coil; V: solenoid valve.

different pH requirements and the ability to determine trace levels of orthophosphate in high silicate content samples (Si/P ratios up to 500). Under optimized conditions, a working linear range from 5 to 50 μg P/l for a 1.8 ml sample, repeatability better than 3% and a quantification limit of 4 μg P/l were attained. The flowing stream system handles 11 analysis/h.

7.5.2 Hydrolyzable, condensed, organic, and total phosphorus

The first and most critical analytical step for the determination of total phosphorus is the acid hydrolysis and the digestion of condensed phosphates and organic phosphorus for their conversion into orthophosphate and subsequent determination of this ion by a multitude of different flow analytical techniques [110].

In order to avoid laborious and time-consuming procedures for converting the phosphorus to measurable phosphate ions some improved methods have been reported that may be effective for online monitoring, e.g. thermal induced digestion, ultraviolet photooxidation [110], and microwave-assisted digestion [81,97,105]. Whatever the choice, it is important to ensure that quantitative oxidation of relatively refractory organic phosphorus compounds, i.e. condensed phosphates, and further hydrolysis and release of P from polyphosphates occurs prior to the detection step.

Some FIA methods for TP/TDP have been reported involving the use of high temperatures, i.e. 140–170 °C, together with sulfuric acid and/or an oxidizing agent. Moreover, the mineralization step can be speeded up by inline implementation of a capillary digestor containing a platinum wire acting as a catalyst for the oxidation of organic species [110].

The use of online microwave heating coupled with flow methods allows the automation of the overall analytical procedure for TDP and TP measurements. An MSFIA system for spectrophotometric determination of total phosphorus (TP) involving inline digestion is illustrated in Figure 7.32. Sample and digestion solution are dispensed and directed toward a digestion vessel located inside a domestic microwave oven where sample digestion takes place. Afterward, the digested sample is merged with the necessary reagents for the colorimetric determination based on the molybdenum blue method. The system was applied to wastewater samples [105].

UV photooxidation alone is insufficient to successfully convert all organic and condensed phosphates into orthophosphates. As a consequence a two-stage process involving online UV-thermal-induced digestion system, using a combined oxidizing/hydrolyzing reagent, is usually performed [82]. The inclusion of both oxidative and hydrolytic processes allows the hydrolysis of P–O–P linkages, e.g. polyphosphates, and the oxidation of phosphoesters and C–P compounds to inorganic phosphate and guarantees the achievement of high recoveries of both organic and condensed phosphorus. An outstanding feature of the latter approach is the capability to quantify particulate and colloidal associated phosphates and, thus, making measurement of TP feasible.

The online photooxidative decomposition of phosphorus compounds to orthophosphate has also been applied to the determination of dissolved organic phosphorus (DOP) and TP in soil extracts and leachates, runoff waters, natural, and wastewaters. For instance, the MPFS depicted in Figure 7.33 allows the spectrophotometric determination of dissolved orthophosphate and DOP in wastewater samples [106]. The determination of orthophosphate is based on the vanadomolybdate method. Inline ultraviolet photooxidation is employed to mineralize organic phosphorus to orthophosphate prior to detection. A solenoid valve allows the deviation of the flow toward the UV lamp to carry out the determination of organic phosphorus.

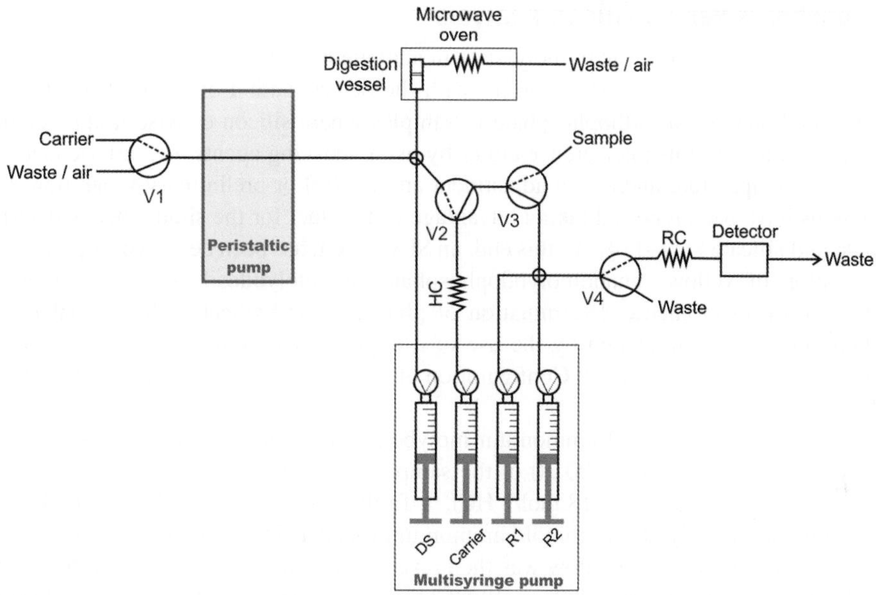

FIGURE 7.32

Schematic representation of an MSFIA manifold with inline microwave digestion used for the determination of total phosphorus. Detector: spectrophotometer (λmax = 880 nm); DS: digestion solution; HC: holding coil; R: reagent; RC: reaction coil; V: solenoid valve.

Calibration was found to be linear up to 20 mg/l P, with a detection limit of 0.08 mg/l P, an injection throughput of 75 injections/h and a repeatability (RSD) of 0.6% for the direct determination of orthophosphate. On the other hand, calibration graphs were linear up to 40 mg/l P, with a detection limit of 0.5 mg/l P, an injection throughput of 11 injections/h and a repeatability (RSD) lower than 2.3% for the procedures involving UV photooxidation.

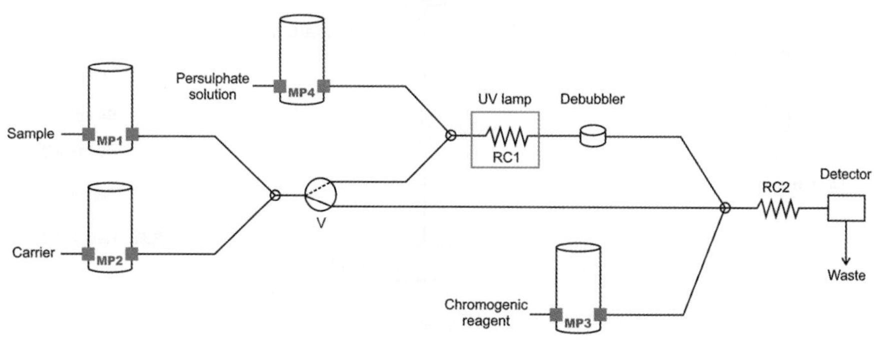

FIGURE 7.33

MPFS system for the spectrophotometric determination of dissolved orthophosphate and dissolved organic phosphorus in wastewater samples. MP: solenoid micropump; RC: reaction coil.

7.5.3 Phosphorus versus silicon analysis

Both phosphate and silicate react with molybdate in acid media to form the corresponding molybdate heteropoly acid. Thus, they seriously interfere with the determination of each other. Flow methods for environmental analysis of orthophosphate in samples where silicon coexist at appreciable levels should be adapted to avoid its interference either by using masking agents, i.e. tartaric acid-modified reagents, proper temperature and/or pH adjustment and control or preliminary separation. However, several methods have been reported taking advantage of this fact for the simultaneous determination of phosphate and silicate [111–114]. To this end, an SIA system has been designed (Figure 7.34) based on the formation of yellow vanadomolybdophosphate and molybdosilicate, respectively, for the simultaneous spectrophotometric determination of phosphate and silicate. The mutual interference between both analytes is eliminated by the use of a large sample volume, appropriate acidity, and sample segmentation with oxalic acid. Configuration of the system, which is totally automated, is very simple [111].

A portion (150 μl) of 0.1 mol/l ammonium molybdate in 0.4 mol/l HCl was aspirated into a reaction coil followed by a portion (800 μl) of the sample, after which the flow was stopped for 10 s. Next, 75 μl 5.6% oxalic acid in 0.18 mol/l HCl, a further portion (1200 μl) of sample and 75 μl 0.035 mol/l ammonium molybdate/3 mmol/l ammonium vanadate in 0.65 mol/l HCl were sequentially aspirated into the reaction coil. The flow was then reversed and the entire content of the reaction coil was propelled by a carrier stream of 0.05 mol/l HCl toward the detector. A flow rate of 4 ml/min was used for both aspiration and propelling purposes. The absorbance was measured at 400 nm for both phosphate and silicate determinations. Phosphate was determined from the first peak and silicate from the second. Calibration graphs were linear up to 12 and 36 mg/l, phosphate and silicate, respectively,

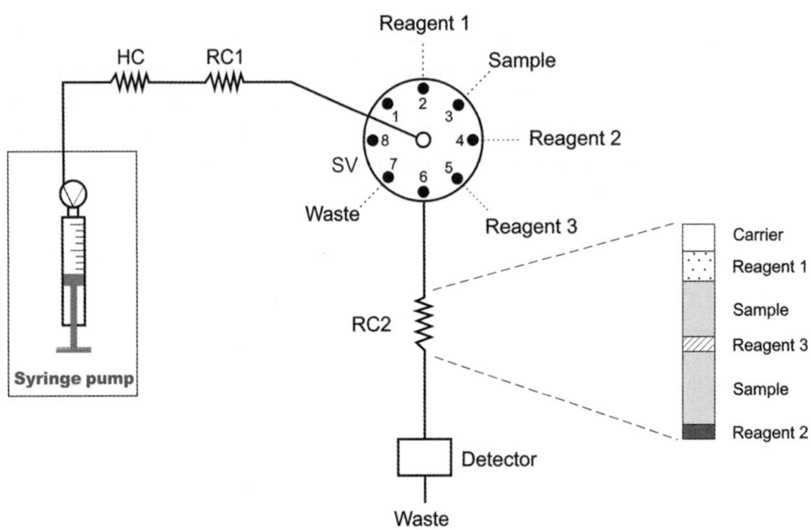

FIGURE 7.34

SIA system for the simultaneous spectrophotometric determination of phosphate and silicate. HC: holding coil; RC: reaction coil; SV: selection valve.

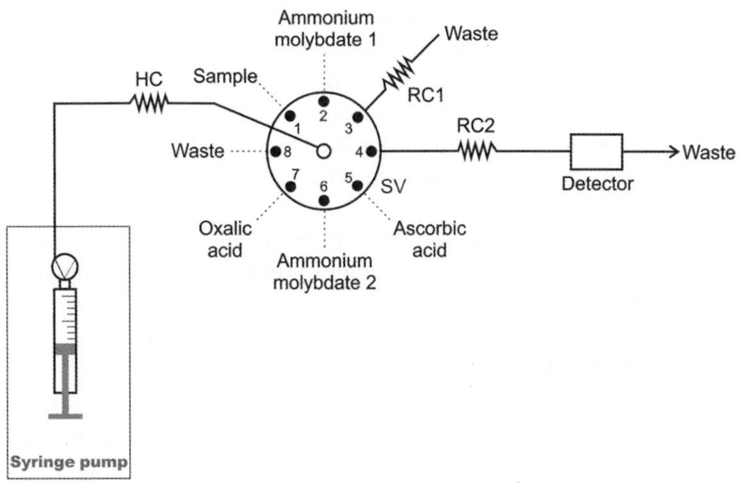

FIGURE 7.35

SIA system for the spectrophotometric determination of phosphate and silicate in environmental samples using the molybdenum blue reaction. HC: holding coil; RC: reaction coil; SV: selection valve.

and the corresponding detection limits were 0.2 and 0.9 mg/l, respectively. RSD ($n = 10$) were 1.4 and 3.9% phosphate and silicate, respectively. The injection throughput was 23 samples/h. The method was applied to wastewater.

The SIA manifold depicted in Figure 7.35 also allows the spectrophotometric determination of phosphate and silicate in environmental samples using the molybdenum blue reaction. The interference of silicate in the determination of phosphate is eliminated by using a reagent composed of 5 mmol/l ammonium molybdate in 0.2 mol/l nitric acid, containing 0.25% (w/v) oxalic acid to avoid the formation of molybdosilicic acid. The interference of phosphate in the determination of silicate is avoided adding a 10% (w/v) oxalic acid solution to the reaction zone where the molybdophosphoric and molybdosilicic acids are previously formed, in order to destroy the molybdophosphoric acid. To perform this task in the single-line SI system, obtaining a total sample and reagent zones penetration, a combination of sandwiching the sample zone between reagent zones and flow reversal through an auxiliary reaction coil was used [114]. The method has a phosphate sampling frequency of 75/h, with a linear working range between 0.2 and 7 mg/l and a detection limit of 0.1 mg/l PO_4^{3-}-P. For silicate, the sampling frequency is 40/h with a linear working range between 5 and 50 mg/l and a limit of detection of 1 mg/l SiO_3^{2-}-Si.

7.6 Nitrogen compounds

Inorganic nitrogen compounds can be determined by direct measurement of a physicochemical property, such as conductivity, optical absorption, the potential, or current. Although simple and rapid, these methods suffer from numerous interferences and therefore are usually accompanied separation techniques or replaced by more selective indirect methods.

The indirect determination of ammoniacal nitrogen, nitrite and nitrate is mainly based on the interconversion of one species into another, or conversion into other inorganic compounds such as hydroxylamine, nitrogen oxides, nitrosyl halides or molecular nitrogen. These transformations are performed by adding to the sample chemical oxidizing or reducing agents, either homogeneous or heterogeneous, or through photochemical or enzymatic reactions. Then, it is possible to use derivatization processes with organic reagents, accompanied by highly selective photometric, amperometric, or luminescent detection.

Organic nitrogen is also an important parameter in the field of environmental analysis. Routine monitoring is generally accomplished by conversion of organic nitrogen compounds into inorganic species through mineralization processes.

7.6.1 Ammoniacal nitrogen

Ammoniacal nitrogen is usually determined by flow methods involving derivatization reactions for spectrophotometric determination. Berthelot (or indophenol blue) [99,115–119] and Nessler methods [120] are the two most important examples, both adopted for the quantification of ammonium ion in matrices as diverse as natural waters, wastewaters, atmospheric samples, food, soil extracts, plant material, and biological fluids. Measuring the fluorescence intensity produced by the reaction of the ammonium ion with OPA is another widespread method for the analysis of this ion [121–125].

The reaction of ammonia with OPA in the presence of 2-mercaptoethanol is the basis of a simple, economic and transportable analyzer system for shipboard monitoring (Figure 7.36). The method makes use of a solenoid micropump system applying a reverse FIA concept, i.e. the reagent is injected into the sample flow, which is consequently also used as carrier. A specially made detection cell for fluorescence detection of the reaction product isoindol-1-sulfonat is made and combined with a commercial photomultiplier tube, a long-pass optical filter, and an UV-LED as excitation light source. An LOD of 13 nmol/l is achieved [121].

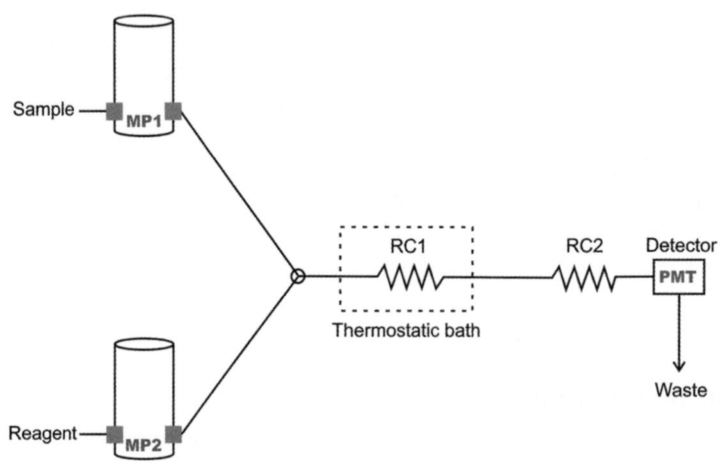

FIGURE 7.36

MPFS analyzer for ammonia determination. MP: solenoid micropump; PMT: photomultiplier; RC: reaction coil.

Although the fluorometric method has detection limits of nanomolar concentrations for ammonium, it often suffers from high background fluorescence and interference by methylamines. Primary amines and amino acids react in the same way as ammonia and thus constitute a major interference of the method. Interference is greatly reduced using flow systems, with alternative configurations—as those shown in Figure 7.37—where selectivity is gained, thanks to the introduction of diffusion cells with hydrophobic membranes [124]. This allows the application of the method to the determination of ammoniacal nitrogen without significant interference from other nitrogen-containing organic compounds, particularly amino acids and proteins, and the interference from volatile primary amines is minimized by kinetic discrimination or working at neutral pH.

Many flow-based methods for the determination of ammoniacal nitrogen incorporate gaseous diffusion processes, enabling the use of nonselective procedures relying on the measurement of changes in pH or conductivity, taking advantage of the selectivity conferred by the use of a microporous hydrophobic membrane. These methods are based on the diffusion of ammonia through the semipermeable membrane, the driving force for the transport being the difference between the partial pressures of the gas dissolved in the donor and acceptor solutions of the diffusion unit.

For ammonium ion detection, it is necessary a prior conversion into ammonia, which is done by adding an alkali solution to the sample. After diffusion, ammonia is ionized due to acid–base reactions in the acceptor solution, and causes a change of pH or conductivity, which may be related to the concentration of ammoniacal nitrogen present in the original sample.

When using conductometry the acceptor solution can be deionized water [126], a weak acid, e.g. boric acid [127] or a strong acid, e.g. hydrochloric acid [128], measuring an increase or decrease in conductivity accordingly.

In the SIA system schematically represented in Figure 7.38 [127], for the determination of ammonium with conductometric detection, the sample and alkaline solution are sequentially aspirated into a single tubular channel and mixed by flow reversal. The formed ammonia permeates through a gas permeable membrane and is collected on a static acceptor solution. The change in conductance of the acceptor solution is related to the concentration of ammonium ions in the sample. The manifold includes an additional valve of the type used for sample injection in FIA that is actuated when required to form a closed loop in the acceptor channel and prevent the liquid from leaving it when the sample is propelled through the donor channel. The detection limit can be lowered from 1.5 to 0.1 mg/l if preconcentration by successive sample flow reversal is used.

An MSFIA system exploiting a gas diffusion cell has been developed for the conductometric determination of ammonium in water samples of different matrices. These samples were coastal waters, pond waters, and compost aqueous extracts. The system based on the experimental setup shown in Figure 7.39 can be used for a wide concentration range of ammonia, from 0.075 to 360 mg/l, without sample dilution and with a precision better than 2% of RSD. The throughput of the method was 32 injections/h [128].

Photometric detection systems associated with gas diffusion processes are equivalent to those involved in the electrochemical detection of ammonia nitrogen and differ primarily in the detection mode and the acceptor solution used, which in this case consists of an acid–base indicator. Ammonia absorption released in this solution causes a pH change and a resulting change in the absorbance of the indicator, which can be measured photometrically [119–135]. These methods have been used mainly in FIA, but also in SIA, MCFIA, and MSFIA, by incorporating sandwich-type diffusion cells and performing conventional detection or measuring the change in absorbance of the indicator in the

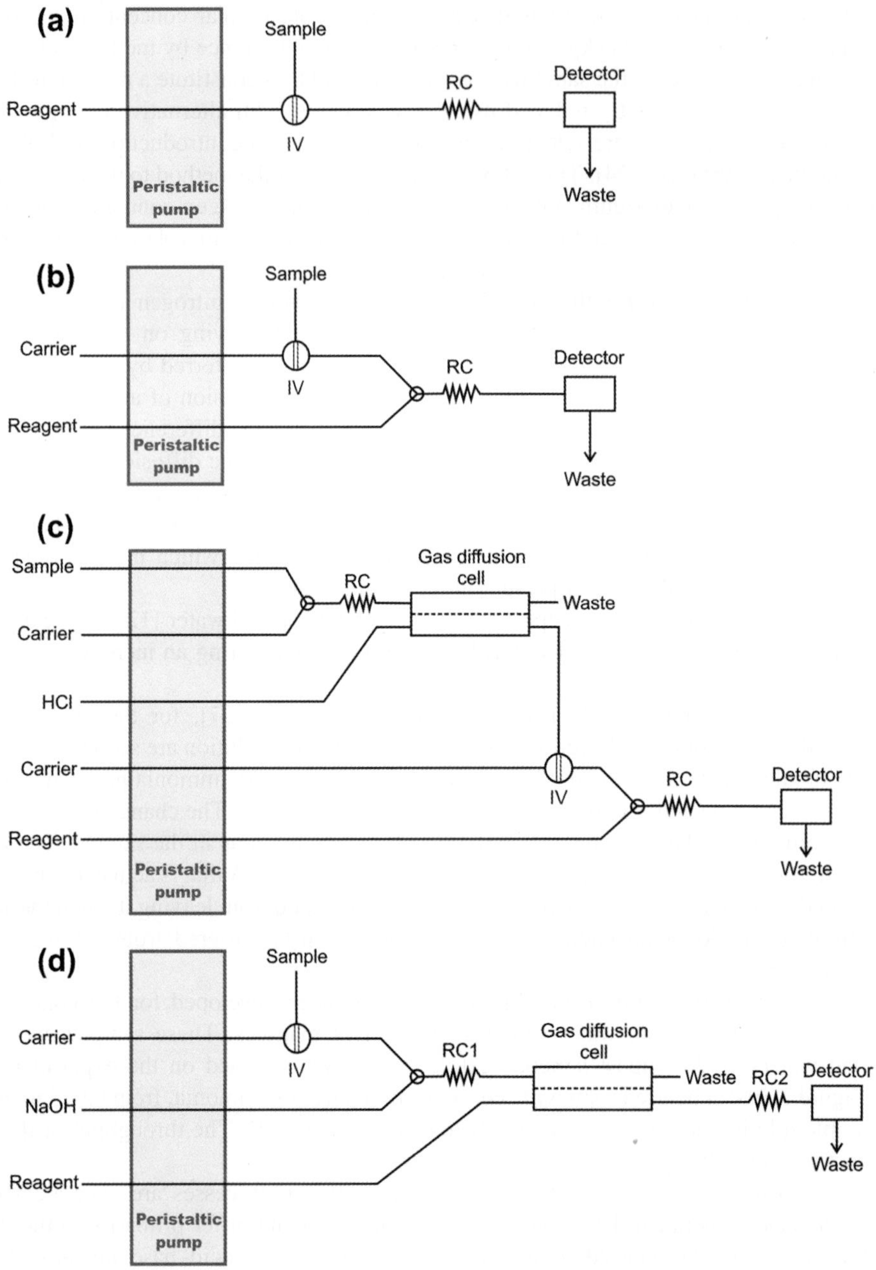

FIGURE 7.37

Different FIA setups for ammonium fluorimetric determination without (a, b) and with (c, d) gas diffusion units. Detector: fluorimeter (λex/λem = 415/485 nm); IV: injection valve.

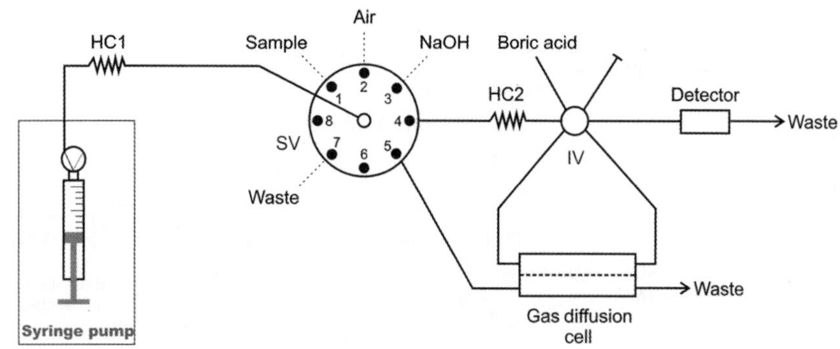

FIGURE 7.38

GD–SIA system used for ammonium quantification. Detector: two-electrode flow-through conductance cell; HC: holding coil; IV: injection valve; SV: selection valve.

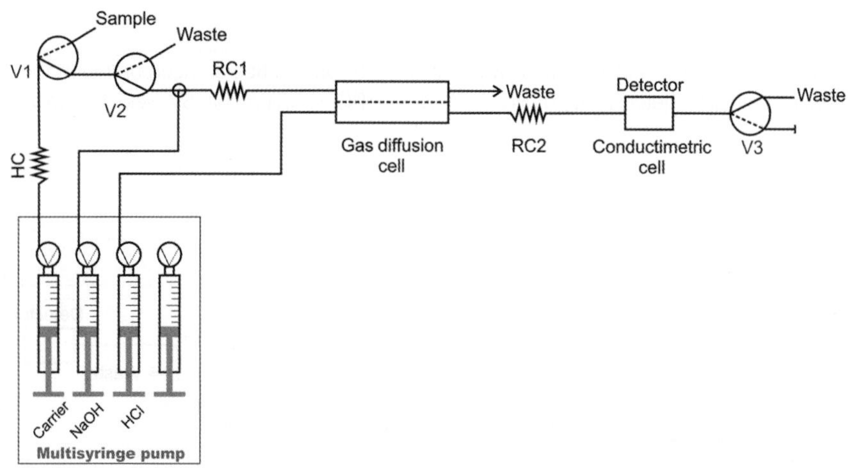

FIGURE 7.39

MSFIA–GD system for the conductometric determination of ammonium in water samples. Detector: homemade stainless steel conductometric cell connected to a conductometer; HC: holding coil; RC: reaction coil; V: solenoid valve.

acceptor channel of the diffusion cell, either by transmission or by reflection of the incident light. Integrating the processes of separation and detection is performed by optical fibers, incorporated for example in the diffusion cell, as shown in the SIA system illustrated in Figure 7.40 [130].

An MCFIA system exploiting a gas diffusion device has been developed for the spectrophotometric determination of ammonium nitrogen in surface and tap waters as well as sea and estuarine waters (Figure 7.41). The efficiency of complexing agents to prevent precipitation of metallic hydroxides, due

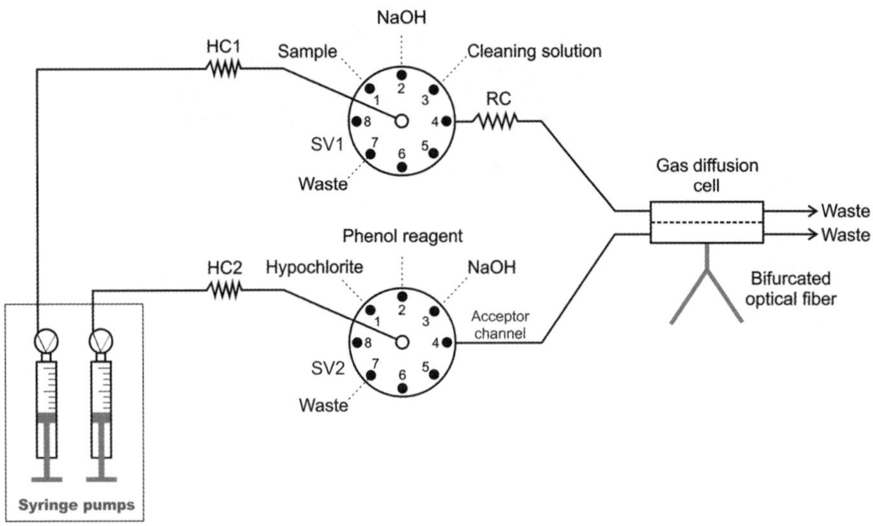

FIGURE 7.40

SIA system integrating processes of separation and detection by optical fibers, incorporated in the diffusion cell for ammoniacal nitrogen determination. HC: holding coil; RC: reaction coil; SV: selection valve.

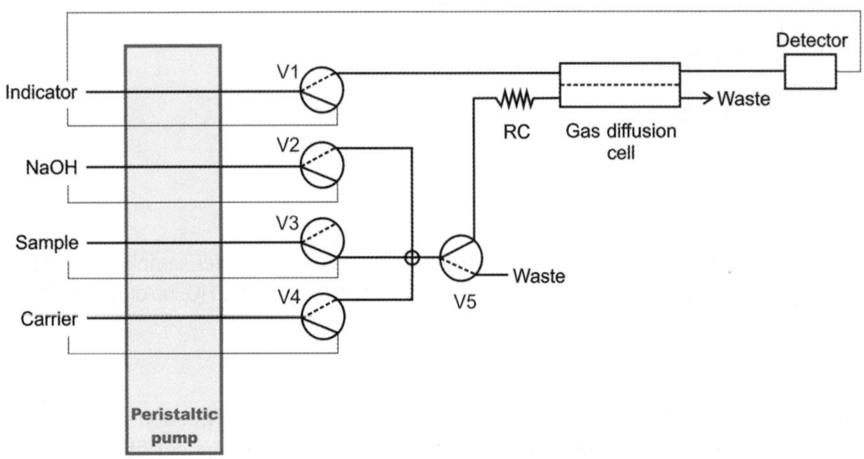

FIGURE 7.41

Schematic configuration of the multicommuted flow system for the spectrophotometric determination of ammonium in water samples. Detector: spectrophotometer (λmax = 620 nm); RC: reaction coil; V: solenoid valve.

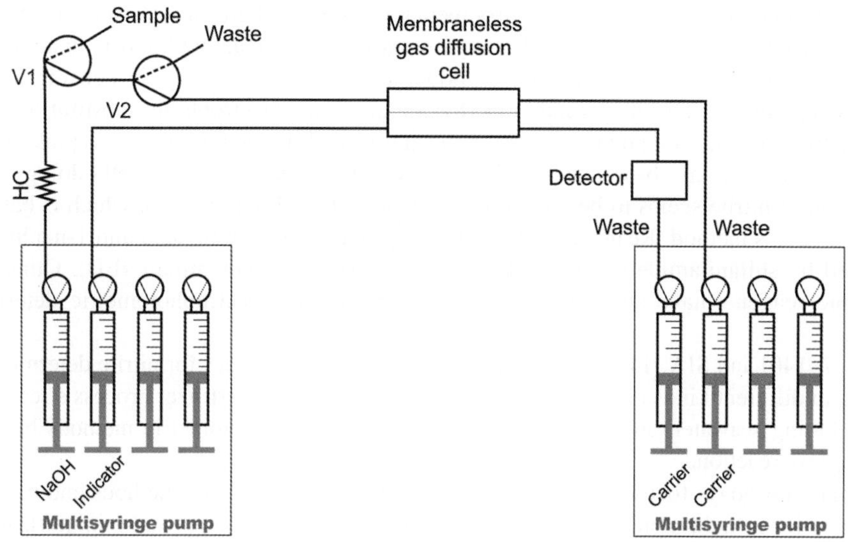

FIGURE 7.42

A membraneless gas diffusion (MGD) unit coupled to an MSFIA system for the spectrophotometric determination of ammonium. Detector: spectrophotometer (λmax = 635 nm); HC: holding coil; V: solenoid valve.

to the high pH value of the carrier solution, was studied. Under optimized conditions, no interference was observed [134,135]. The system allows both the recirculation of the acceptor solution and the use of a stopped flow approach. The proposed method provides the determination of NH_4^+ in concentrations ranging from 50 to 1000 μg/l, with detection and quantification limits of 18 and 35 μg/l, respectively. A determination rate of 20/h is achieved.

A membraneless gas diffusion (MGD) unit coupled to an MSFIA system may also be used for the spectrophotometric determination of ammonium using an acid–base indicator [136]. Ammonia diffuses through the headspace into the acceptor stream (BTB solution) of the flow manifold included in Figure 7.42. Several strategies concerning flow management were studied seeking to characterize and improve the analytical features of the methodology. Moreover, untreated environmental samples were analyzed without previous filtration. Stopped flow in acceptor channel with continuous flow in donor channel was chosen for wastewater and spiked riverwater samples. A linear concentration range between 10 and 50 mg/l NH_4^+, a LOD of 2.2 mg/l and a determination frequency of 11/h were obtained.

The above mentioned systems are very flexible and allow, when required, analyte preconcentration before measurement and therefore can be applied to the analysis of samples with a low content of ammonia nitrogen (0.1–1500 μg/l N), such as natural waters, including seawater and atmospheric samples or to industrial effluents with higher concentrations (1–100 mg/l N).

7.6.2 Oxidized inorganic nitrogen

Nitrite reactions with organic derivatives, mainly diazotization and coupling, followed by the photometric measurement of the resulting azo dye, are the basis of most existing flow methodologies for the determination of this ion in environmental samples. These are chosen due to the high selectivity and sensitivity they provide. Primary aromatic amines react with nitrite in acidic medium to form a diazonium salt through a process known as diazotization being, in general, a rapid and quantitative chemistry. The diazonium ion, by an aromatic electrophilic substitution (coupling reaction) with a phenolic compound or an aromatic amine leads to the formation of a pink-colored azo dye (~540 nm λmax), the absorbance of which is measured. The most-used methodology for online determination of nitrite seems to be a modified version of the Shinn reaction, which is based on the well-known Griess method. Reagents of the original proposal (sulfanilic acid and 1-naphthylamine) are replaced by sulfanilamide (SAM) and *N*-(1-naphthyl) ethylenediamine (NED). Other sensitive chromogenic reagents have also been tested as alternative in spectrophotometric determinations [137–148].

FIA [137–140] and SIA [141–146] are the most common techniques for nitrite determination and monitoring in different kinds of environmental samples (foodstuffs, fertilizer process streams, natural waters, including seawater, and wastewater effluents) by spectrophotometric methods based on the modified Griess reaction.

Efforts are nowadays focused on improving the sensitivity of the flow methods and making them more suitable for in-field monitoring, looking for greener flow procedures and miniaturization capabilities. To this end, the detection step using conventional spectrophotometry has been modified by introducing liquid core waveguide spectrophotometry [140], home-made flow-cell detectors, in combination with LEDs and photodiodes [147], novel paired emitter-detector diode (PEDD) [138], online preconcentration with hydrophilic-lipophilic-balanced (HLB) cartridge [145] and solid-phase enrichment with or without optosensing [144,146].

SIA (Figure 7.43) and FIA (see Figure 3.11) methods have been proposed for the isolation and preconcentration of the azo dye on a solid-phase extractant (C18). In the SIA manifold, the enrichment step is followed by elution with either methanol or ethanol and final spectrophotometric detection downstream [146].

The FIA system depicted in Chapter 3, Figure 3.11, constitutes a versatile and sensitive flow-through optical-fiber diffuse reflectance sensor, which implements disk-based solid-phase extraction. This methodology is based on inline nitrite derivatization with Shinn reagent and subsequent preconcentration of the azo dye onto octadecyl covalently bonded silica-gel particles tightly bound to an inert matrix (C18 disk), which is continuously monitored using a plug-in diode array spectrophotometer. After the analytical signal has been recorded, fast sensor regeneration is achieved with a methanolic eluent, rendering the system ready for the next extraction. An LOD of 0.1 ng/ml of nitrite, repeatability and reproducibility better than 3.2%, an analytical throughput of 11/h, and an enrichment factor of 140 are the figures of merit of the proposed optrode. The utility of the flow-through optosensing system was testified by the satisfactory results obtained in its application to tap, rain, ground, harbor, and aquarium water samples.

Potentiometric methods using ion-selective electrodes can be used for direct determination of nitrate in environmental samples [149–154]. However, nitrate ion-selective electrodes respond to other ions such as chloride, nitrite, hydrogencarbonate, sulfate, and phosphate according to the Hofmeister series. The presence of these interfering ions in environmental samples is virtually assured and much effort has been devoted to remove them. New trends are focused in signal processing with a multivariate data treatment. This is the case of the experimental setup depicted in Figure 7.44, which uses an artificial neural network applied to four potentiometric sensors featuring cross-term response for the

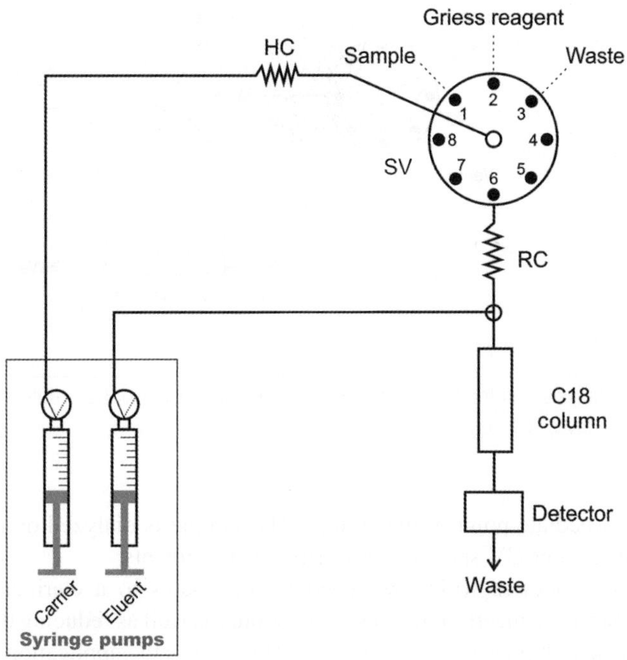

FIGURE 7.43

SIA system using an on-line solid-phase extraction and preconcentration method for nitrite determination. Detector: spectrophotometer ($\lambda max = 420$ nm); HC: holding coil; RC: reaction coil; SV: selection valve.

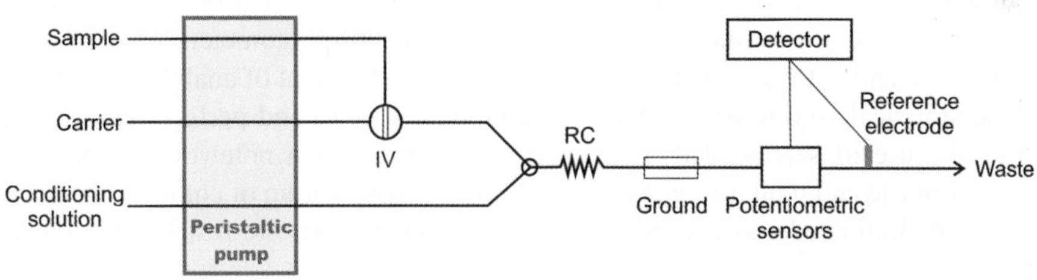

FIGURE 7.44

An FIA electronic tongue based on potentiometric sensors for the determination of nitrate in the presence of chloride. IV: injection valve; RC: reaction coil.

determination of nitrate in the presence of chloride. Nitrate ion could be quantified in concentrations ranging between 0.1 and 100 mg/l NO_3^- without the need to eliminate chloride interference [152].

Nitrate ion has two absorption bands in the ultraviolet region, a weak and a strong band, centered at 300 and 200 nm, respectively. Direct measurements in this region have been automated by flow-based arrangements [155]—as a way of example see the SIA setup depicted in Figure 7.45. It incorporates a

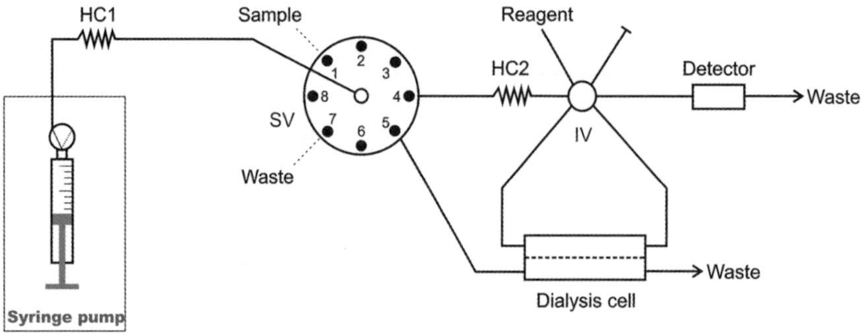

FIGURE 7.45

SIA system coupled to a dialysis unit for nitrate monitoring. Detector: spectrophotometer (210–270 nm); HC: holding coil; IV: injection valve; RC: reaction coil; SV: selection valve.

dialysis unit at an intermediate point of the system. The sample is dialyzed online in order to avoid interferences with subsequent UV spectrophotometric measurements.

Interfacing dialysis processes with flow schemes contributes to a clarification of the sample, removing suspended particles, macromolecules, and colloids as well as reducing the presence of humic substances, mainly those of higher molecular weight. This allows the application of the method to the analysis of very diverse waters such as riverwater, drinking water, or treated wastewater. As a result of the low efficiency of the process of mass transfer across the dialysis membrane (<5%) an automated sample dilution is in fact obtained, establishing a method with broad linear range (≤32 mg/l-N) but with a high LOD of 1.5 mg/l-N.

The similarity between the spectra of nitrate and nitrite in the ultraviolet region precludes the simultaneous determination of both ions by direct measurement in the region between 200 and 230 nm when both ions coexist in the sample. Introducing new spectrophotometers capable of measuring simultaneously at a range of wavelengths has allowed the development of analytical methods based on the use of mathematical algorithms to resolve overlapping spectra and perform multiparametric determinations. The use of second derivative spectrophotometry is a noteworthy possibility for the determination of nitrite and total nitrogen in natural waters. The overlap of chloride and nitrate spectra in seawater is not eliminated entirely and consequently the methods are restricted to the analysis of freshwaters [156].

The inline conversion of nitrate to nitrite and subsequent spectrophotometric determination after diazotization and coupling reactions is the basis of an important set of analytical methods for the simultaneous determination of both ions. Either homogeneous (titanium(III) chloride, hydrazinium sulfate, etc.) or heterogeneous (zinc, cadmium, amalgamated cadmium, copperized cadmium, etc.) reduction agents, photochemical or enzymatic reactions have been used [157]. The nitrite so produced is determined through spectrophotometric detection of the reddish purple azo dye formed as a result of the modified Griess chemistry.

Many flow injection systems for the monitoring of the sum of nitrate and nitrite have been reported with the use of copperized cadmium as reducing agent, applied to the quantification of oxidized nitrogen in natural waters, wastewaters, plant tissue, biological samples, hydrothermal vent biological communities, soils, food, etc. Detection limits of flow methods described in the literature range between 20 and 50 μg/l-N [158–164]. To increase the sensitivity, different proposals have been made, such as the use of stopped-flow mode or the incorporation of mixing chambers.

With many of these systems, it is not possible to differentiate between nitrite and nitrate ions and both are determined jointly. However, those configurations allowing speciation and individual quantification of both ions are preferable as they provide more information on the sample composition.

Nitrogen speciation has been facilitated with the introduction of SIA, which allows multi-parametric determinations with simpler instrumental setups and reduced reagents' consumption [165–169]. SIA systems using the Griess reaction have been exploited for determination of nitrite and nitrate in natural waters, wastewaters, vegetables, dairy products, and soils with incorporation of the solid-phase reactor either on one port of the selection valve or in the holding coil.

The SIA method depicted in Chapter 2, Figure 2.12, allows the separated determination of nitrate and nitrite with the modified Griess reaction. Both analytes are quantified by using a single sample injection and a sandwich technique. A copperized cadmium column is incorporated into the aspiration channel of the burette for the reduction step, its position being selected in order to ensure a partial contact between sample and reductor. Nitrate and nitrite are simultaneously determined after the addition of the Griess reagent to both ends of the sample zone. Nitrite is determined at the sample front by direct reaction with the Griess reagent while nitrate is similarly determined at the sample tail with previous reduction to nitrite. The same reaction product is formed at the two interfaces between sample and reagent and thus spectrophotometric measurements are carried out at a single wavelength (540 nm), which simplifies the detection process.

The implementation of downsized flow setups based on the multicommutation concept and of other low-power portable analyzers with solid state, flow-through spectrophotometric detectors has allowed to carry out unattended remote measurements of oxidized nitrogen by means of the online copper–cadmium reduction of nitrate to nitrite and derivatization chemistry. Many of these units include submersible pumping units for sampling and analysis at great depths and have been used in biological studies on marine ecosystems. Applications to nitrate and/or nitrite quantification in estuarine, coastal waters, and open ocean deep-sea environments have been reported, attaining consistent results from submersed deployments. "On-chip" micro-flow injection [161] and "on-valve" sequential injection [98] are good examples of the miniaturization flow capabilities attained for nutrient sensors when taking profit of the new achievements in the area of instrumentation.

Alternatively to heterogeneous reduction a flow method which incorporates hydrazinium sulfate as homogeneous reductant can be used [170,171] (Figure 7.46).

Determination of gaseous nitrogen oxides (NO_x) requires trapping of NO_x as NO_2^- and NO_3^-. Triethanolamine (TEA) is preferentially used as aqueous absorbing solution [172–175]. Online collection and preconcentration for the continuous determination of the atmospheric pollutant in air samples is realized by coupling either diffusion denuder [175] or a chromatomembrane cell [173,176] with a flow system. The amounts of oxydized nitrogen in the absorbing solution may be measured by spectrophotometric, fluorometric, chemiluminescent, and electrochemical detection and referred to the concentration of NO_x in the air sample.

Despite the importance of the speciation of inorganic nitrogen, the number of flow procedures for the joint determination of nitrate, nitrite, and ammonium without changing the manifold is scarce, most of them based on spectrophotometric detection [177,178]. The difficulty lies in the need for

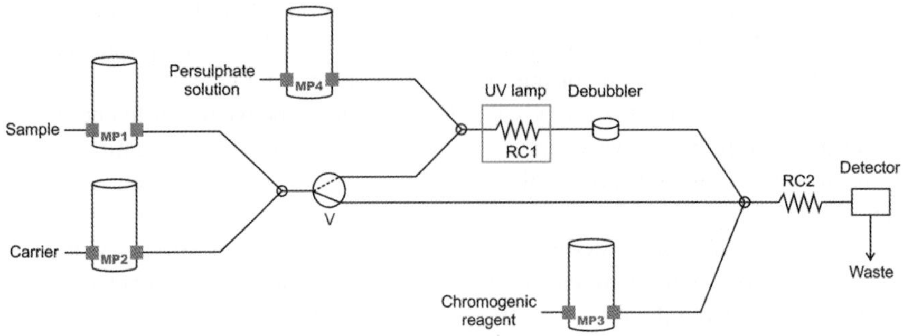

FIGURE 7.46

Experimental MPFS setup proposed for the determination of nitrite and nitrate. Detector: spectrophotometer (λmax = 540 nm); MP: solenoid micropump; RC: reaction coil; V: solenoid valve.

different chromogenic reagents and consequent requirement to detect the reaction product at different wavelengths. The first drawback can be overcome by using the multicommutation approach [178], employing discrete commutation devices to control the addition of each reagent and to perform all steps of the analytical procedure independently. Detection at various wavelengths may be performed by using fast scanning detectors (such as diode array spectrophotometers) or by an arrangement of flow-through units based on a combination of LEDs and photodetectors. These photometers are attractive to implement multidetection due to their favorable characteristics, such as low cost, high stability, fast response, and small size.

This has been implemented by making use of the MCFIA flow system shown in Figure 7.47 based on absorbance measurement of the azo dye obtained by the reaction of nitrite with *N*-(1-naphthyl) ethylenediamine dihydrochloride and sulfanilamide. Nitrate was reduced to nitrite in a copperized cadmium column while the modified Berthelot's reaction (indophenol blue reaction) with salicylate

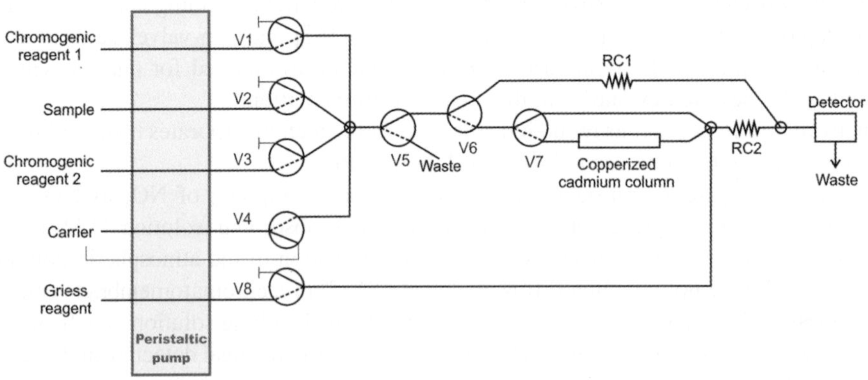

FIGURE 7.47

MCFIA system for speciation of inorganic nitrogen. Detector: dual-color LED homemade photometer; RC: reaction coil; V: solenoid valve.

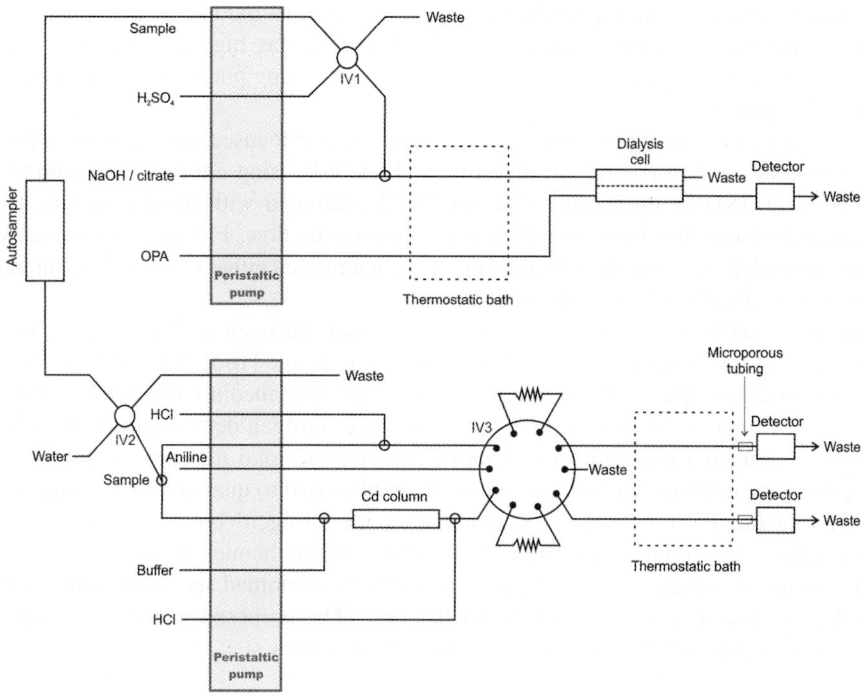

FIGURE 7.48

Sensor for the simultaneous determination of nanomolar concentrations of nitrite, nitrate, and ammonia in seawater by fluorescence detection. IV: injection valve.

was exploited for the determination of ammonium ions. Detection was carried out with a homemade photometer based on the combination of a dual-color (red/green) LED and a photodiode.

As an interesting alternative, a sensor package for the simultaneous determination of nanomolar concentrations of nitrite, nitrate, and ammonia in seawater by fluorescence detection is presented in Figure 7.48. Field experiments carried out demonstrated the capability of this sensor to determine horizontal nutrient gradients in oligotrophic surface waters [179].

7.6.3 Organic and total nitrogen

Routine monitoring of organic and/or total nitrogen is accomplished by conversion of organic nitrogen compounds into inorganic species through mineralization processes. The ions formed, usually ammonium or nitrate, are subsequently determined with any of the above mentioned techniques, which should be previously adapted to the pH conditions of the digestion mixture.

The oxidation of organic nitrogen compounds, nitrite and ammonium to nitrate for the determination of total nitrogen is the foundation of a broad set of analytical methods that has become very popular, since these are less polluting than the traditional Kjeldahl method, in addition, allowing a significant reduction in operation time and facilitating the automation of the digestion process.

Hydrogen peroxide or potassium persulfate are typically used as oxidizing agents and greater efficiency in the process of mineralization is obtained working at high pressure and temperature, accelerating the reactions by heating in a microwave oven, by using platinum as a catalyst or through photochemical reactions.

Recent developments in organic nitrogen determination are focused mainly in the automation of digestion procedures and in continuous digestion and analysis using either UV radiation for photo-oxidation [156,180,181] or microwave radiation [182] combined with the use of flow techniques. Online sample digestion has been described using continuous flow, FIA and SIA systems by integrating them in analyzers in which the formed nitrate is determined either by direct measurement in the ultraviolet region, or indirectly by reduction to nitrite.

An automatic method for the determination of total nitrogen in wastewater by SIA and mineralization with UV radiation is presented as a way of example [180]. The method is based on the mineralization of the samples with sodium persulfate in alkaline medium under UV radiation.

Figure 2.14 in chapter 2 depicts a flow system for total nitrogen determination providing a rapid method for the sequential determination of nitrite, nitrate, and total nitrogen. Nitrite is determined directly by using the modified Griess reaction, which is also used to quantify nitrate after reduction to nitrite with hydrazine. For total nitrogen determination, the nitrogen-containing compounds (organic substances, nitrite and ammonium ions) are oxidized photochemically using a UV lamp and converted into nitrate, which is then reduced to nitrite and determined spectrophotometrically [181]. The procedure is described in more detail in Chapter 2. The proposed method was applied to the determination of nitrate, nitrite, and total nitrogen in wastewaters.

7.7 Priority organic pollutants

In compliance with national and international directives, regular environmental monitoring is demanded, which requires appropriate analytical methods for fast and sensitive detection of relevant compounds such as pesticides (organophosphorus, carbamates, organochlorine, etc.), phenols, BTEX (benzene, toluene, ethylbenzene and total xylenes and *o*-, *m*- and *p*-xylene) and priority pollutants such as polychlorinated biphenyls (PCBs), polycyclic aromatic hydrocarbons (PAHs) and other endocrine disrupting compounds.

7.7.1 Phenolic compounds

The total content of phenols, or phenol index, is the quickest way for the obtainment of quantitative chemical information about phenol pollution. This index is commonly determined by oxidative coupling of phenolic compounds with 4-aminoantipyrine (4-AAP) in alkaline solution with $K_3[Fe(CN)_6]$ as oxidant and spectrophotometric detection.

Different procedures based on that chemistry and using FIA [183], SIA [166,184], MSFIA [185,186], and MCFIA [187] have been described. Most of them make use of online sample preparation, including micellar-assisted preconcentration (cloud point extraction (CPE)) or solid-phase extraction (SPE) [185,186,188]. To this end, either a microcolumn packed with Amberlite XAD-4 resin [185] or C18-modified silica material may be used.

One example of SPE coupled to flow-analysis techniques for the determination of the total phenolic index is the MSFIA system shown in Figure 7.49 [185]. The method employs online retention and

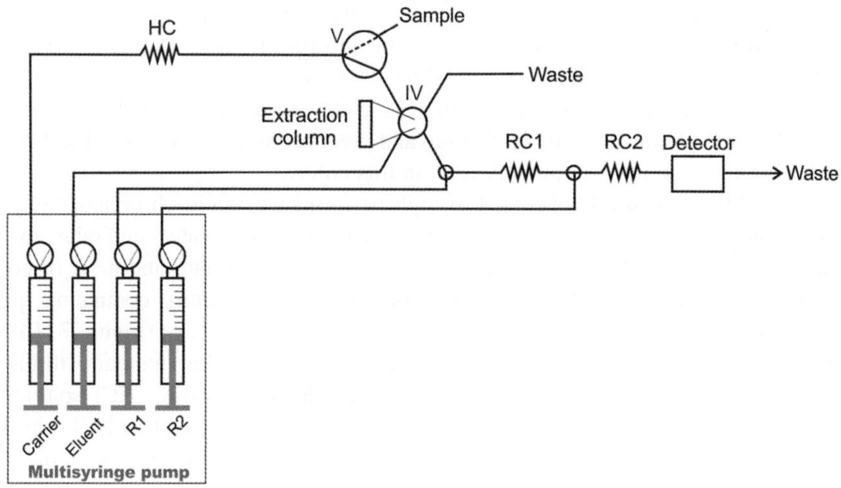

FIGURE 7.49

MSFIA exploiting solid-phase extraction for the determination of the total phenolic index. Detector: spectrophotometer; HC: holding coil; IV: injection valve; R: reagent; RC: reaction coil; V: solenoid valve.

preconcentration of the phenols in an acidified sample and subsequent elution from Amberlite XAD-4 resin into a flowing system with an alkaline solution, promoting the ionization of the retained phenolic compounds and their desorptions. The eluted analytes are then quantified spectrophotometrically as the products of reaction with 4-aminoantipyrine (4-AAP) and potassium ferricyanide on passing through the flow cell.

With the incorporation of C18-modified silica material to the flow scheme for the preconcentration of the imine product formed in the oxidative reaction between phenol and 4-AAP two detection variants have been used, namely (1) spectrophotometric measurement of the methanolic eluent containing the concentrated dye or (2) direct at-column optosensing of the retained reaction product followed by methanol elution. The LOD obtained for measurement of the absorbance in the eluent and on-column optosensing are 11 μg/l and 0.4 μg/l, respectively [185].

An MCFIA system based on the same chemistry was equipped with a 1 m long path-length liquid waveguide capillary cell (LWCC) for spectrophotometric measurements [187]. Sensitivity was 80-fold higher than that achieved with a 1 cm flow cell, making feasible the determination of phenols in the 10–100 μg/l range with a detection limit of 1 μg/l phenol. As the sensitivity is enhanced with the use of the LWCC, liquid–liquid extraction or SPE are not required. This MCFIA–LWCC system allows an analysis throughput of 90/h, with a coefficient of variation of 0.6% and a 200-fold reduction of reagent consumption in comparison with the reference batch method. Recoveries within 93 and 106% were achieved for determination of phenol in natural and wastewater samples.

Another more environmentally friendly alternative for the spectrophotometric determination of the phenolic index is based on the use of solenoid micropumps as microfluidic handling devices, using nitroprusside/hydroxylamine as reagents, instead of 4-AAP/ferricyanide [189].

Several alternative applications for the determination of this total index have also been proposed based on chemiluminescence or electrochemical detection [188]. Phenol determination is accomplished by online SPE (microcolumn filled with C18-modified silica) prior to luminol/ $K_3[Fe(CN)_6]$ chemiluminescent detection. By this an ultralow limit of detection of only 0.6 ng/l phenol is achieved with a working range from 4.7 to 470 ng/l.

Liquid chromatography in combination with C18 or C8 columns avoid the need to derivatize phenols and enables the online coupling of cleanup and preconcentration systems. The UV–visible, or diode array detection, is the most frequently used in this context.

An automatic SPE procedure hyphenated to high performance liquid chromatography-ultraviolet (HPLC-UV) was proposed [186] for screening of phenolic pollutants in waters and other environmental samples at ng/ml levels. The manifold consisted of a multisyringe burette and a set of commutation valves accommodating a lab-made extraction column (SPE column) containing polystyrene-divinylbenzene sorbent coupled to a chromatographic system injection valve (Figure 7.50).

The complete procedure included six steps: system cleaning (propelling methanol through the SPE column and chromatographic injection loop); sorbent conditioning (using HCl solution); sample loading toward the SPE column (flow rate of 8 ml/min); matrix removal (sending HCl solution through the SPE column); elution of the retained analytes (with methanol) and injection valve loop filling with eluent; before chromatographic run starts (eluate injection into the chromatographic system). High enrichment factors are attained for phenol and 10 of its derivatives, with LOD values lower than 1 ng/ml. The determination frequency is 4–10/h, providing an automatic, fast and reliable tool for water quality and environmental monitoring. The applicability of the proposed system has been corroborated by the recovery values attained (89–103%) when different water matrices were processed. Good results were also obtained for an aqueous extract from contaminated soil, indicating the suitability for determination in soil leachates.

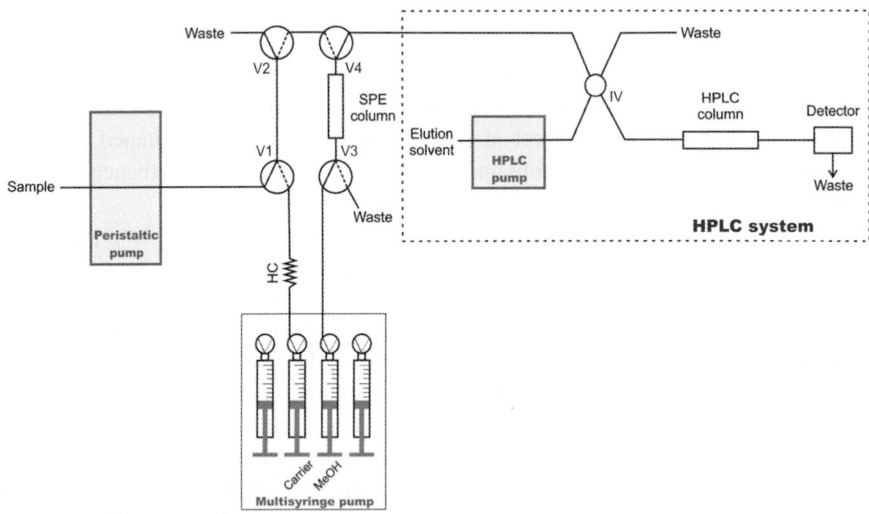

FIGURE 7.50

MSFIA–SPE–HPLC system for screening of phenolic pollutants. HC: holding coil; IV: injection valve; SPE column: monolithic column containing polystyrene-divinylbenzene sorbent; V: solenoid valve.

Other environmental applications are restricted to specific phenolic compounds included on the priority pollutant lists of the EU and USEPA, as is the case with the nitrophenolic isomers 2-nitrophenol, 4-nitrophenol, and the 2,4-dinitrophenol [190–192].

For instance, the hyphenation of MSFIA with capillary electrophoresis (MSFIA–CE) allowed the preconcentration, separation, and determination of mononitrophenols (NPs) in water samples following an analogous strategy to that previously explored for SIA-CE [191], but with increased versatility due to the incorporation of two additional syringes. The schematic representation of the manifold is given in Figure 7.51. The flow system accommodates a reverse-phase (C18) minicolumn for solid-phase preconcentration of phenolic compounds, followed by elution and injection in a lab-made CE system with a final visible wavelength spectrometry detection at 401 nm. The applied procedure comprises: conditioning of the SPE column (150 s), preconcentration (800 s), elution, and hydrodynamic injection (70 s). Three injections per hour are achieved by running the preconcentrating and the electrophoretic separation procedures in parallel. A preconcentration factor better than 115 and detection down to 0.11 mmol/l was achieved.

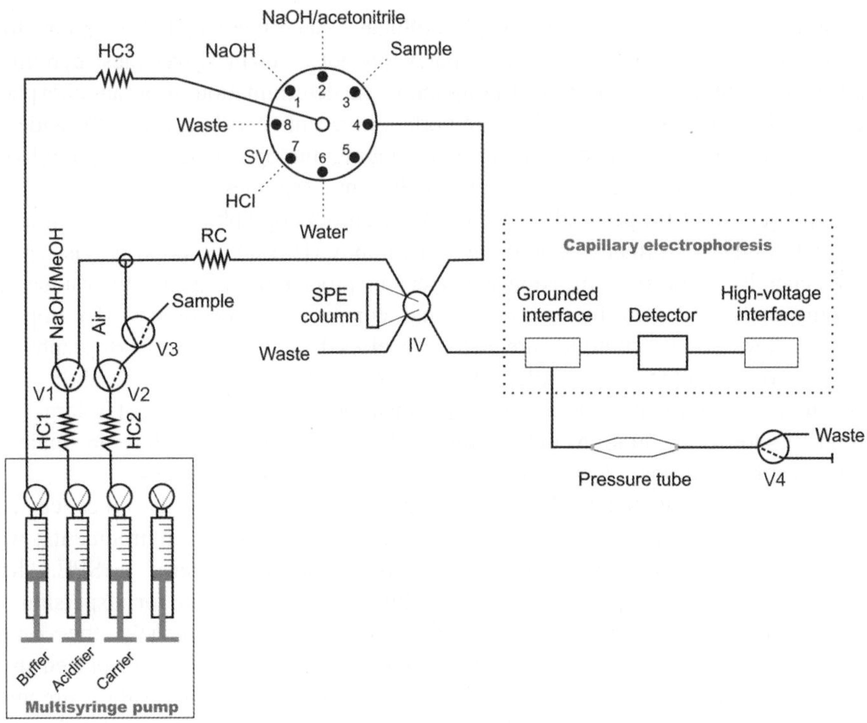

FIGURE 7.51

MSFIA–CE system applied to the preconcentration, separation, and determination of nitrophenols. Detector: photometric detector cell; HC: holding coil; IV: injection valve; RC: reaction coil; SV: selection valve; V: solenoid valve.

Other examples presented in the literature are those based on the use of copolymeric disk-based sorbent materials for the isolation and preconcentration of nitro-substituted phenol isomers and the concurrent removal of potentially interfering matrix components followed by online simultaneous determination of individual species by diode array spectrophotometry, via chemometric deconvolution of the overlapped spectra, without the need for chromatographic separation. In contrast to bead-extraction, no flow impedance is observed when using extraction disks while better enrichment factors are obtained because of the improved specific surface area. Compared with earlier methods for isolation/preconcentration of nitrosubstituted phenols based on liquid–liquid extraction, these systems should be regarded as environmentally friendly approaches because the use of harmful organic solvents is circumvented.

One of the proposals [192] combines flow-through diffuse reflectance optosensing with multivariate regression modeling for the simultaneous multiresidue monitoring of nitrophenol derivatives (2-, 4-, and 2,4-dinitrophenol) at trace level concentrations. Improved detectability is assured as a consequence of the direct optical detection onto the sorbent material.

7.7.2 Pesticides

Pesticides can be considered as priority organic pollutants since these include organochlorine substances, organophosphorous compounds, carbamates, phenols, triazines, triazoles, pyrethrins, pyrethroids, imidazoles, and mercaptoimides. Among them, the determination of organophosphorous and carbamate pesticides has become increasingly important in recent years because of the widespread use of these compounds, which is due to their wide-ranging biological activity and relatively low persistence compared to the actually banned organochlorine pesticides.

Earlier techniques used for pesticide detection were chromatographic methods such as gas chromatography (GC), high-performance liquid chromatography (HPLC) along with mass spectrometry (MS). They generally comprise extraction and cleanup procedures to separate residues from co-extracted compounds, followed by measuring and confirmation techniques. To improve these methods, newer flow-based techniques are being developed, including chromatographic and non-chromatographic ones, looking for simpler and more sensitive devices with the ability to perform on-site detection. They must provide reliable identification and quantitation of a large number of compounds at very low concentrations in foodstuff, soils, waters, and air and thus, no single analytical method is applicable to all possible residue combinations and sample matrices.

Figure 7.52 shows an automated SI–μSPE–LC hyphenated system for the preconcentration and determination of triazine residues with a broad spectrum of polarities in environmental waters and crude soil extracts, using carbon nanoparticles as reusable sorptive material in SPE. The method capitalizes on the design and exploitation of a flow-through stirred reactor for dispersion of carbon nanoparticles. Two stirred-flow sorptive microcolumn configurations, namely, the commercially available Mobicol cartridge (Mobicol M 1002, MoBiTec, Göttingen, Germany) placed in the horizontal position, and an in-house built cylindrical microcolumn in the upright position, are investigated as flow-through containers for handling of dispersed carbon nanotubes (CNTs) and carbon nanofibers (CNFs). The analytical procedure involves inline conditioning of the CNTs/CNFs, sample perfusion through the stirred-flow sorptive microcolumn, rinsing of carbon nanoparticles, sorbent drying, analyte elution, inline modulation of eluate composition, and online injection of eluate into an HPLC system for reversed-phase chromatographic separation [193].

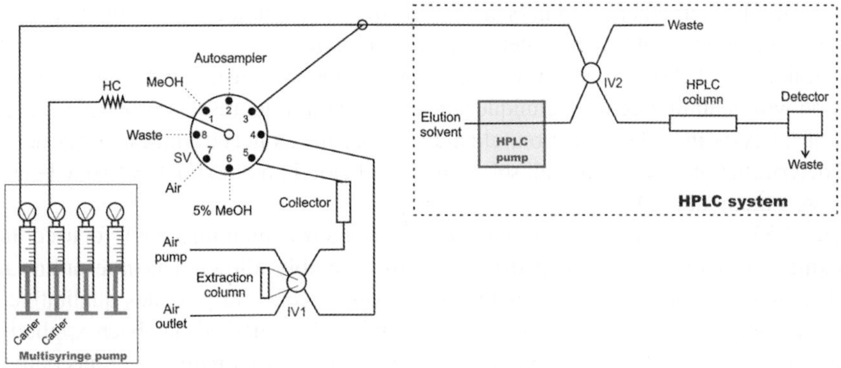

FIGURE 7.52

Schematic illustration of the SIA–LC hyphenated setup for automated CNF-based microextraction and determination of trace-level concentrations of triazine herbicides in environmental matrixes. HC: holding coil; IV: injection valve; SV: selection valve.

7.7.3 Polychlorinated biphenyls (PCBs)

Online sorptive preconcentration exploiting renewable solid surfaces, the so-called bead injection (BI), in the miniaturized LOV platform has been hyphenated to GC separation with electron-capture detection for automated determination of trace levels of PCBs in solid-waste leachates. To this end, the MSFIA system presented in Chapter 3, Figure 3.19, has been implemented. It comprises a multisyringe burette, an extra solenoid valve, an LOV, and an injection valve (IV) as well as reversed-phase copolymeric beads with hydroxylated surface (Bond Elut Plexa) included into the channels of a poly(ether imide) LOV microconduit, thus serving as a transient microcolumn-packed reactor for preconcentration of organic species [194].

The analytical procedure can be summarized in five steps: bead packing into LOV and conditioning; sample loading and bead drying; elution with 80 μl of ethyl acetate into a rotary injection valve; transportation of eluate by an air plug (to avoid dispersion) into the programmable-temperature vaporizer (PTV) injector of the GC (IV in the inject position); and bead discarding and system conditioning, then a new analysis cycle is initiated with fresh portion of beads, thus eliminating any possibility of cross-contamination between consecutive runs. By sampling 12 ml of leachate to which 50 vol% methanol was added to minimize sorption onto the components of the flow network, the automated analytical method features relative recovery percentages >81%, limits of quantification within the range of 0.5–6.1 ng/l, relative standard deviations better than 9% at the 50 ng/l level, and 25-fold decrease in cost of SPE consumables as compared with online robotic systems or dedicated setups.

7.7.4 Polycyclic aromatic hydrocarbons (PAHs)

Polycyclic aromatic hydrocarbons (PAHs) are ubiquitous environmental pollutants. Among the several 100 different PAHs already identified, 16 are considered as priority because they are supposed to be more harmful than the others; there is more information available about them and there is a greater possibility of people being exposed to them. These PAHs include the following: acenaphthene, acenaphthylene, anthracene, benzo(a)anthracene, benzo(a)pyrene, benzo(b)fluoranthene, benzo(g,h,i) perylene, benzo(k)fluoranthene, chrysene, dibenzo(a,h)anthracene, fluoranthene, fluorene, indeno(1,2,3-c,d)-pyrene, naphthalene, phenanthrene, and pyrene.

The quantification of PAHs on a routine basis is often conducted using liquid chromatography (LC) in conjunction with a luminescence technique since they exhibit fluorescent properties. In general, the algorithms of analysis include extraction, cleanup procedure, concentration, chromatographic separation, and determination. Liquid–liquid, solid-phase and cloud point extractions have been interfaced with flow-based methodologies.

In Figure 7.53 a flow-injection interface for fluorometric monitoring of focused microwave-assisted Soxhlet extraction is represented [195]. This assembly allows real-time online monitoring of the PAHs extracted from solid samples in each Soxhlet cycle and provides qualitative and semi-quantitative information from natural and spiked samples. The method has been applied to a certified reference material (CRM 524, BCR, industrial soil/organics) for quality assurance/validation. The proposed technique is as efficient as conventional Soxhlet to extract PAHs from soils but with a drastic reduction of both extraction time and organic solvent disposal.

Other alternative options for PAHs extraction coupled to flow injection are static, dynamic, and static–dynamic-pressurized liquid extraction. They have been applied to contaminated soils analysis, allowing the partial automation of the proposed approaches. Efficiencies close to 100% have been obtained with the three operational modes. However, the static-dynamic mode has proved as the most suitable alternative providing the shortest extraction time (25 min) versus the static (30 min) and the

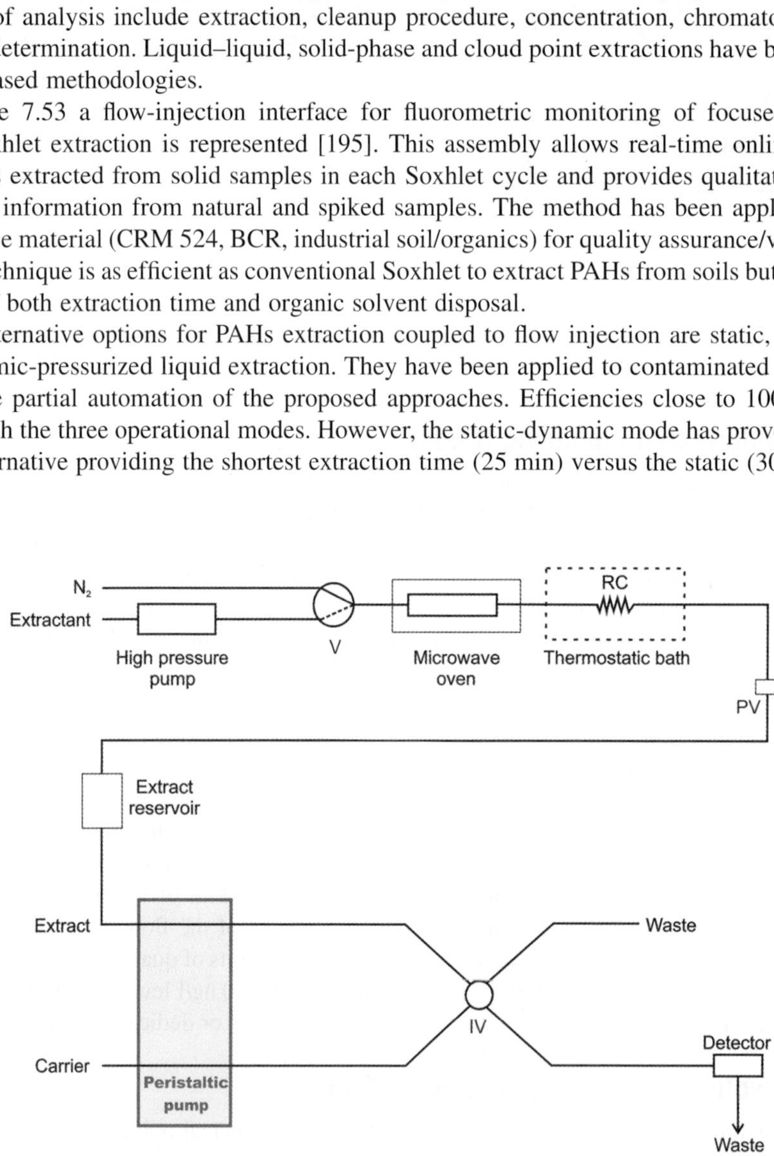

FIGURE 7.53

Flow injection interface for fluorometric monitoring of focused microwave-assisted Soxhlet extraction. IV: injection valve; PV: pressure valve; RC: reaction coil; V: solenoid valve.

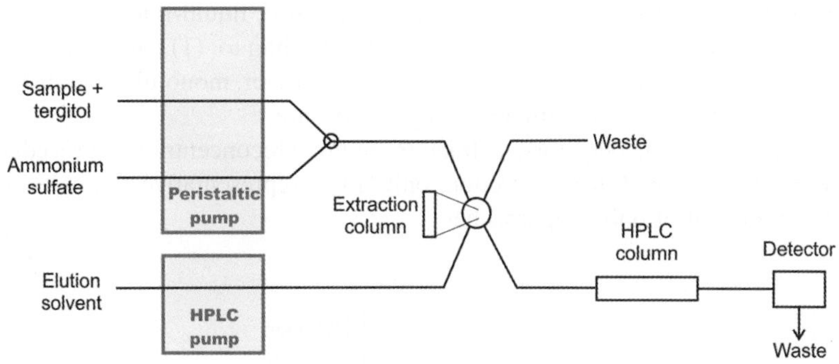

FIGURE 7.54

Flow injection-cloud point methodology (FI–CPE) coupled to HPLC and fluorescence detection.

dynamic (50–70 min) modes. The use of water as leaching agent has avoided the use of organic solvents providing an environmentally friendly method.

Flow injection-cloud point extraction methodology (FI–CPE) coupled to HPLC and fluorescence detection have also been used as analytical methods for PAHs quantification in soils [196].

Sample extracts obtained by both ultrasonic and microwave-assisted surfactant-mediated (Tergitol 15-S-7) extractions are preconcentrated and analyzed following the flow scheme depicted in Figure 7.54. Over 90% of recoveries for various PAHs in environmental soil samples are obtained under the optimal experimental conditions. The limit of detection of the FI–CPE–HPLC system ranges from 0.10 to 0.45 μg/l for the selected PAHs, i.e. phenanthrene, pyrene, chrysene, benzo(k)fluoranthene, and benzo(a)pyrene.

7.8 Metals and metalloids

Some metals, including heavy metals, affect negatively people's health, either individually or forming metal compounds, and therefore they must be controlled. Moreover, there are toxic semi-metallic elements the so-called metalloids, such as arsenic and selenium, which should also be monitored. Basically, this group of pollutants can be classified in: (1) metals and metalloids that are necessary to support life in very small amounts, but toxic in larger amounts, e.g. copper, zinc, chromium, arsenic, or selenium; (2) metals with unknown biological function which are extremely toxic even at trace levels, e.g. mercury, cadmium, or lead.

Metals and metalloids' anthropogenic impact is very elevated since they have been extensively used for thousands of years in a multitude of activities and applications, e.g. mining, industry, tannery, and electronics. So, the need to determine not only its presence, but also its speciation, bioavailability, and isotopic abundance leads to development of a great number of flow-based methodologies applied to several environmental matrices. Likewise, numerous kinds of detectors, from spectrophotometry to ICP–MS, have been used in order to determine these analytes online.

Mainly online developed methods for metals and metalloids determination include procedures to make the sample cleanup, eliminate interferences or preconcentrate the sample to achieve the required sensitivity. Among these procedures, it is worth mentioning SPE, liquid-liquid extraction (LLE), and hydride generation. Online speciation can be carried out according to: (1) the retention degree of each chemical species onto the adsorbent, e.g. membranes, resins, or monolith columns; and (2) their reactivity in redox reactions with or without UV-light exposure.

Numerous combinations of detectors, extraction and/or preconcentration procedures, and flow techniques have been developed. In this section, only a few representative examples are illustrated, leaving out a large amount of online approaches.

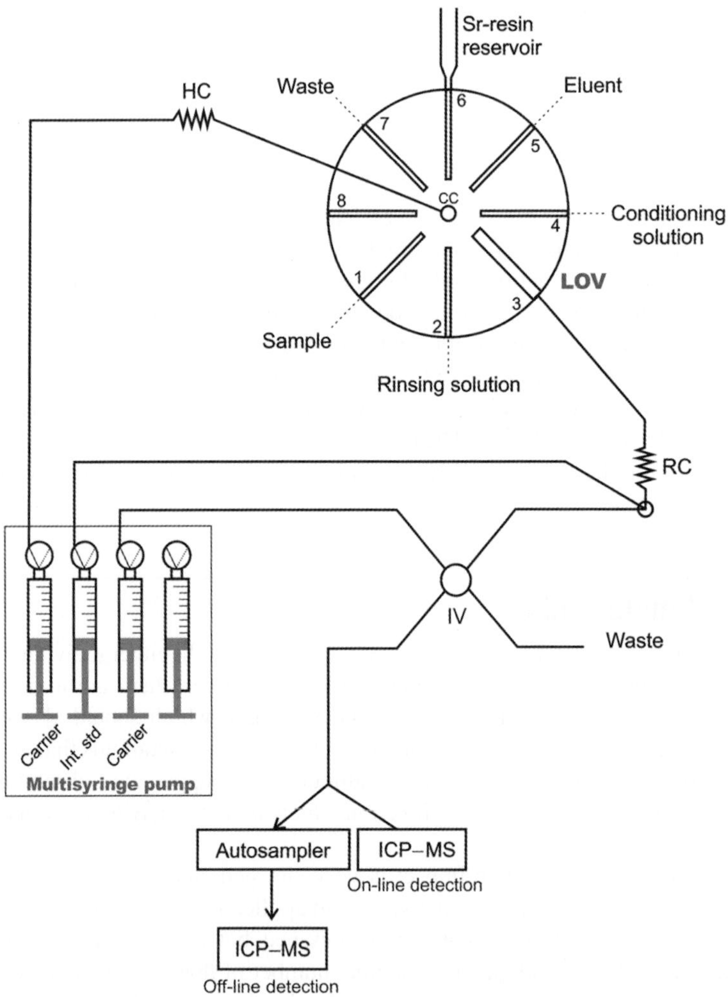

FIGURE 7.55

LOV–MSFIA–ICP–MS for simultaneous lead and strontium isotopic analysis applied to rainwater samples and airborne particulate filters (PM10). HC: holding coil, IV: injection valve; LOV: lab-on-valve; RC: reaction coil.

As a way of example, a LOV–MSFIA coupled to ICP–MS is described. This automated system was used for simultaneous lead and strontium isotopic analysis applied to rainwater samples and airborne particulate filters (PM10) (Figure 7.55) [197]. Pb and Sr are directly retained on the sorbent column of Sr-resin and eluted with a solution of 0.05 mol/l ammonium oxalate. LODs achieved were 0.04 ng for lead and 0.03 ng for strontium. Mass calibration curves were used since the proposed system allows the use of different sample volumes for preconcentration. Mass linear working ranges were between 0.1 and 50 ng for both elements. The repeatability of the method, expressed as RSD, was 2.1% and 2.7% for Pb and Sr, respectively. The injection frequency (including the sample treatment and the online detection) was between 5 and 3 injections/h varying the sample volume from 1 to 10 ml. Main features of this LOV–MSFIA–ICP–MS system are the capability to renew SPE at-will in a fully automated way, the remarkable stability of the column, which can be reused up to 160 times, and the potential to perform isotopic analysis. The synchronism of ICP–MS with the LOV–MSFIA system was performed through the digital output of the multisyringe burette. A relay interface (Sciware Systems) was used to trigger the events input of the ICP–MS. Nevertheless, because of the time needed to extract and preconcentrate the analytes and then measuring them using transient signal mode, the analyst should evaluate the offline detection, where the ICP–MS works in steady-signal mode, requiring less time of reading. In consequence, the consumption of argon gas decreases at the same time that the measurement frequency increases.

As an example of arsenic speciation, an automated system for on-line determination of dimethylarsinic and inorganic arsenic, As(III) and As(V), by hydride generation with atomic fluorescence spectrometry is presented [198] (Figure 7.56). An MSFIA module is coupled to hydride generation–atomic fluorescence spectrometry (HG–AFS) with UV photooxidation for dimethylarsinic (DMA), inorganic As and total As determination, depending on the pretreatment given to the sample

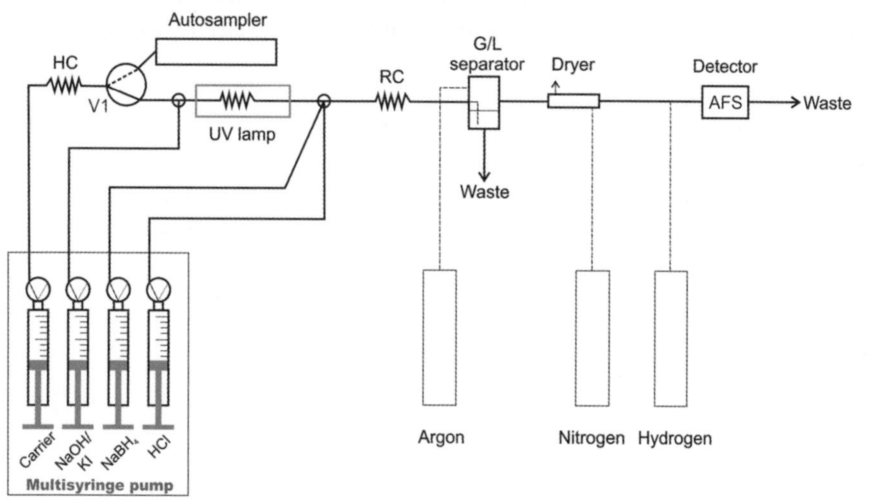

FIGURE 7.56

MSFIA system coupled to hydride generation UV–AFS for arsenic speciation. G/L separator: gas/liquid separator; HC: holding coil; RC: reaction coil; V: solenoid valve.

(extraction or digestion). The implementation of a UV lamp allows online photooxidation of DMA and the following arsenic detection, whereas a bypass leads the flow directly to the HG–AFS system, performing inorganic arsenic determination. DMA concentration is calculated by the difference of total inorganic arsenic and the measurement after the photooxidation step. The LOD for DMA and inorganic arsenic were 0.09 and 0.47 μg/l, respectively. Repeatability values accomplished were 2.4% and 1.8%. Injection frequencies were 24 and 28 injections/h for DMA and inorganic arsenic, respectively.

As an example of a system for mercury speciation, an MSFIA system for the inline preconcentration, cleanup and speciation of mercury by use of an anion-exchange membrane is described [199] (Figure 7.57). The speciation of mercury is based on retention of its tetrachloro complex onto the membrane while organic mercury flows freely through it. A multisyringe burette is used as liquid driver and a cold vapor atomic fluorescence detector is employed to ensure high sensitivity. Organic mercury is decomposed into inorganic mercury by using a UV lamp. The carrier and reductant streams consist of 1.5% (m/v) hydrochloric acid and 2% (m/v) tin chloride, respectively. The proposed system is more environmentally friendly than the classical liquid–liquid extraction procedure. Mercury recoveries from spiked samples and a reference material were all close to 100%. LODs of 14 and 16 ng/l were obtained for total and organic mercury, respectively, both with an RSD less than 1.3%.

Another representative example is an MSFIA system for automated selenium speciation using anion-chromatographic technique [200] (Figure 7.58). Selenite, selenomethionine, and selenate were analyzed. Selenomethionine was oxidized by UV-light to selenite, while selenate was reduced to selenite by UV-light under alkaline medium and halides. A 15 W lamp was employed as the UV-light source in the photoreactor. The photoreactor coil was constructed with classical Teflon tubing of 0.8 mm i.d. and 1.65 m length. The photoreactor was controlled by the PC through the additional 12 V output of the multisyringe, turning the lamp on only when the sample is flowing through the photoreactor by using a relay, minimizing the overheating, and reducing the electrical consumption. Hydride

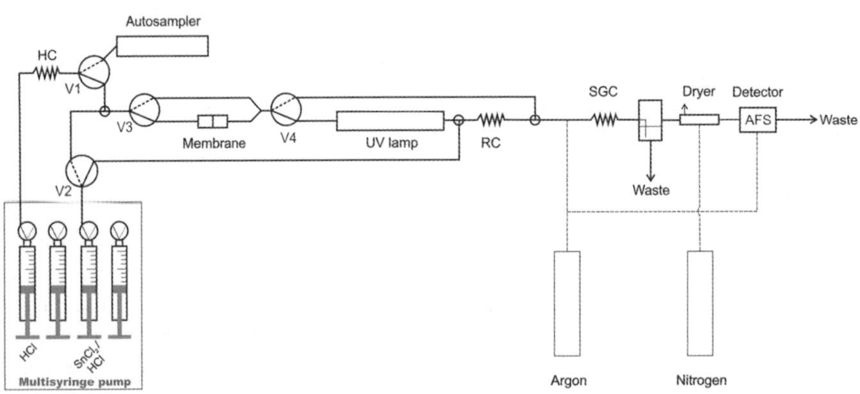

FIGURE 7.57

MSFIA system for the inline preconcentration, cleanup, and speciation of mercury by use of an anion-exchange membrane. HC: holding coil; RC: reaction coil; SGC: spray generator coil; V: solenoid valve.

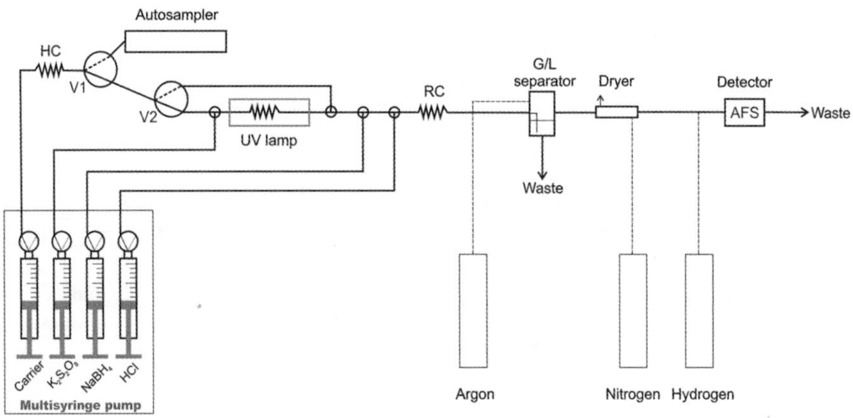

FIGURE 7.58

MSFIA system for selenium speciation by atomic fluorescence spectrometry. G/L separator: gas/liquid separator; HC: holding coil; RC: reaction coil; V: solenoid valve.

generation–atomic fluorescence spectrometry was selected as detection technique due to its sensitivity and selectivity. Detection limits of 0.11, 0.12, and 0.13 μg/l were obtained to selenite, selenomethionine, and selenate, respectively. RSDs obtained were always lower than 2.7%. Different environmental samples were analyzed with recoveries close to 100%. An injection throughput of 30 injections/h was accomplished.

7.9 Multiparametric monitoring

Apart from the huge amount of flow configurations reported in the literature for environmental speciation, as summarized in previous sections of this chapter, flow arrangements for multi-determinations devoted to monitoring quality of drinking water supplies, process control in wastewater treatment plants and aquaculture, or to gain an insight in nutrient content, dynamics and cycling (carbon, silicon, nitrogen, and phosphorous compounds) have devoted special interest.

The new generation of flow systems (SIA, MCFIA, etc.) has proved to possess exceptional capacities for operating in a multiparametric way in relation to segmented flow or FIA, giving rise to simpler manifolds and facilitating miniaturization.

Integration of various chemistries into this kind of flow manifolds for multiparametric determinations of sulfate together with chloride [201], Fe(II)/Fe(III), ammonia, phenolic compounds, nitrate and/or nitrite [166,167,202] have been proposed in the literature. Several researchers have reported the use of sequential injection for nitrogen speciation and aggregated additional reagents to the flow configuration in order to extend monitoring capabilities of the analyzers. Some examples are the determination of nitrite, nitrate, and sulfate in drinking water and wastewater [167], the determination of sulfate and phenolic compounds in wastewaters [166], phosphate [98], or iron speciation in surface waters [203]. Some of these examples are shown in Figures 7.59–7.61, corresponding to MPFS [201], SIA [167] and μSI–LOV [98] systems, respectively.

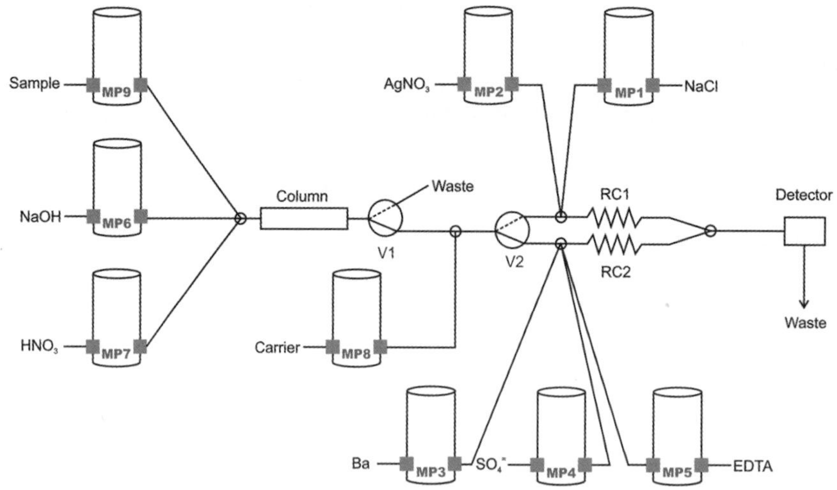

FIGURE 7.59

Expert MPFS system for turbidimetric determination of chloride and sulfate in natural waters. Detector: spectrophotometer (410/800 nm); MP: solenoid micropump; RC: reaction coil; V: solenoid valve.

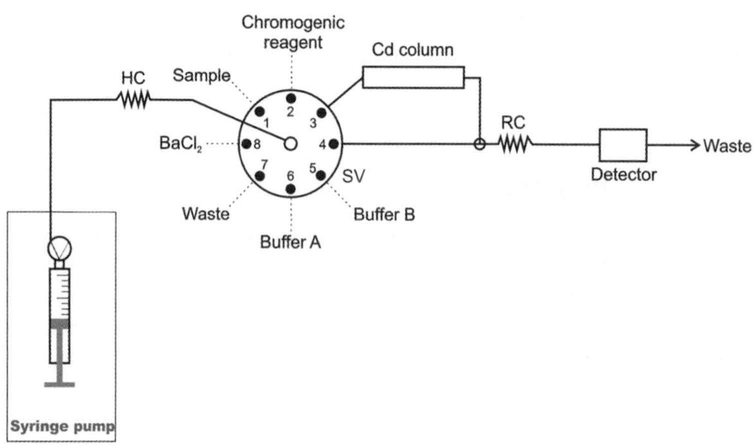

FIGURE 7.60

SIA monitoring system proposed for sulfate, nitrite, and nitrate. Detector: spectrophotometer ($\lambda max = 540$ nm); HC: holding coil; RC: reaction coil; SV: selection valve.

An important field of application of multicomponent determination is in the continuous monitoring of nitrogen and phosphorous in wastewater treatment plants. Figure 2.16 in Chapter 2 illustrates an SIA setup designed as an environmental wastewater monitor for the speciation of different forms of nitrogen (ammonium, nitrites, nitrates, and total nitrogen) and phosphorous (orthophosphates and total

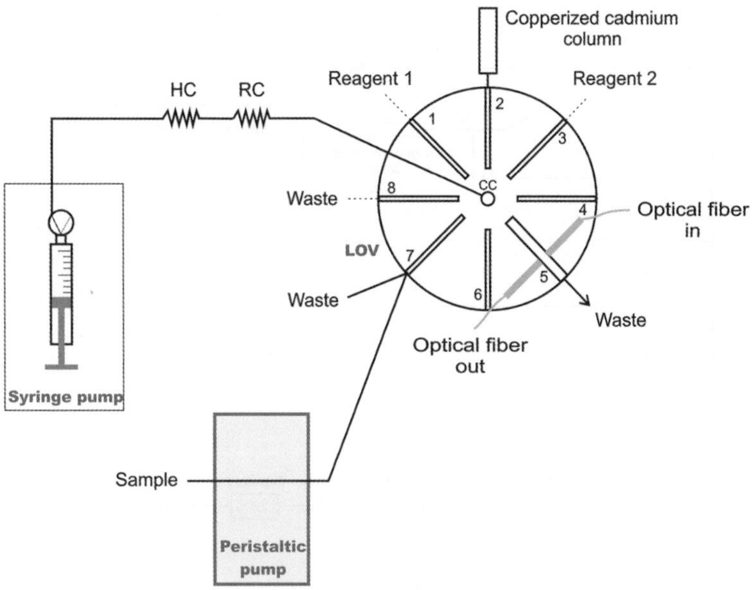

FIGURE 7.61

Micro-sequential injection: environmental monitoring of nitrogen and phosphate in water using lab-on-valve system furnished with a microcolumn. HC: holding coil; LOV: lab-on-valve; RC: reaction coil.

phosphorous). When a special diode array detector, produced by SECOMAM, is incorporated to the manifold, up to nine additional parameters may also be evaluated (BOD, COD, total suspended solids (TSS), surfactants, etc.). To this end, the authors proposed a system with two spectrophotometric detectors where all the parameters are measured in less than 15 min using either the direct exploitation of the UV spectrum of the sample, or a chemical reaction with one stable and simple reagent.

Other examples of "all-in-one" device systems that afford the in situ determination of nutrients are found in the bibliography. For instance, an FIA method intended for measuring nitrate, together with ammonium and organic carbon in soils, equipped system with a filtration probe that allows direct sampling [204]. Another representative example is a shipboard nutrient analyzer for simultaneously monitoring nitrate plus nitrite and phosphate at nanomolar concentrations in marine ecosystems by the coupling of an FIA system with two 2-m long path-length LWCCs as depicted in Figure 7.62 [205].

This analytical instrument was successfully used in several cruises for real-time monitoring of nutrient concentrations in seawater, incorporating a sampling unit for the in situ collection and online determination of phosphate, nitrate, and nitrite. The method is based on the initial determination of phosphate as its phosphoantimonylmolybdenum blue complex, which is then oxidized online by nitrite and the decrease in absorbance is monitored at 880 nm. Nitrate is determined as the difference between total and initial nitrite content in a separate flow after reduction to nitrite in a cadmium-reductive column.

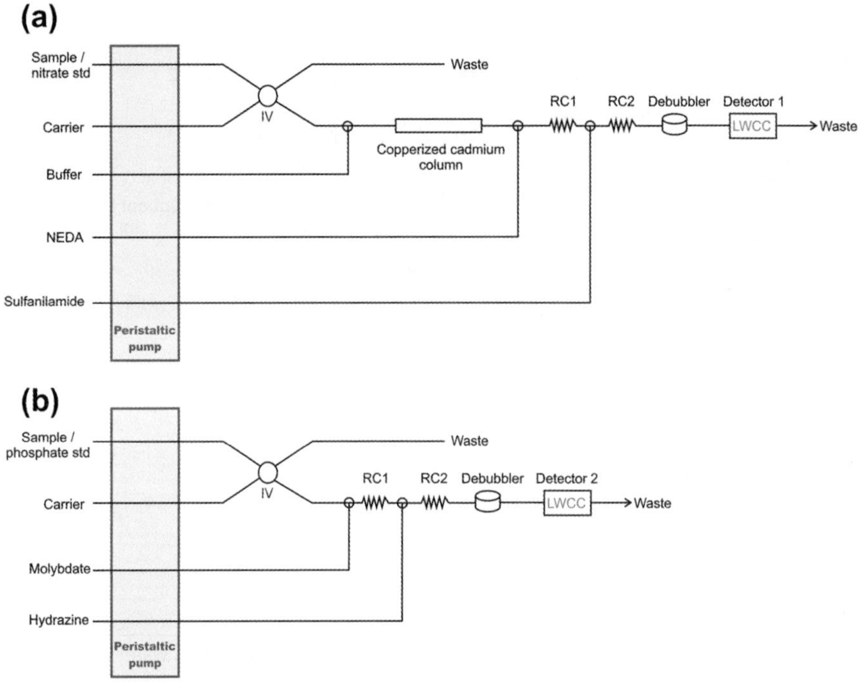

FIGURE 7.62

Underway monitoring of nanomolar nitrate plus nitrite and phosphate in oligotrophic seawater. (a) Nitrate and nitrite FIA manifold. (b) Phosphate FIA manifold. IV: injection valve; LWCC: liquid waveguide capillary cell; RC: reaction coil.

References

[1] J.A. Vieira, I.M. Raimundo, B.F. Reis, Turbidimetric determination of sulphate employing gravity flow-based systems, Anal. Chim. Acta 438 (2001) 75–81.

[2] W.R. Melchert, F.R.R. Rocha, An improved procedure for flow-based turbidimetric sulphate determination based on a liquid core waveguide and pulsed flow, Anal. Chim. Acta 616 (2008) 56–62.

[3] Y.S. Fung, C.C.W. Wong, J.T.S. Choy, K.L. Sze, Determination of sulphate in water by flow-injection analysis with electrode-separated piezoelectric quartz crystal sensor, Sens. Actuators 130 (2008) 551–560.

[4] J.F. van Staden, R.E. Taljaard, On-line dilution with sequential injection analysis: a system for monitoring sulphate in industrial effluents, Fresenius' J. Anal. Chem. 357 (1997) 577–581.

[5] R. Burakham, K. Higuchi, M. Oshima, K. Grudpan, S. Motomizu, Flow injection spectrophotometry coupled with a crushed barium sulfate reactor column for the determination of sulfate ion in water samples, Talanta 64 (2004) 1147–1150.

[6] V. del Río, M.S. Larrechi, M.P. Callao, Determination of sulphate in water and biodiesel samples by a sequential injection analysis-multivariate curve resolution method, Anal. Chim. Acta 676 (2010) 28–33.

[7] M. Kass, A. Ivaska, Spectrophotometric determination of sulphur dioxide and hydrogen sulphide in gas phase by sequential injection analysis technique, Anal. Chim. Acta 449 (2001) 189–197.

[8] Z.X. Guo, Y.Z. Li, X.X. Zhang, W.B. Chang, Y.X. Ci, Flow injection determination of gaseous sulfur dioxide with gas permeation denuder-based online sampling and preconcentration, Anal. Bioanal. Chem. 374 (2002) 1141–1146.

[9] P. Sritharathikhun, M. Oshima, Y. Wei, J. Simon, S. Motomizu, On-line collection/concentration and detection of sulfur dioxide in air by flow-injection spectrophotometry coupled with a chromatomembrane cell, Anal. Sci. 20 (2004) 113–118.

[10] R. Redondo, V.C. Machado, M. Baeza, J. Lafuente, D. Gabriel, On-line monitoring of gas-phase bioreactors for biogas treatment: hydrogen sulfide and sulfide analysis by automated flow systems, Anal. Bioanal. Chem. 391 (2008) 789–798.

[11] M.S.P. Silva, C.X. Galhardo, J.C. Masini, Application of sequential injection-monosegmented flow analysis (SI-MSFA) to spectrophotometric determination of sulfide in simulated waters samples, Talanta 60 (2003) 45–52.

[12] M.S.P. Silva, J.C. Masini, Coupling a gas diffusion cell to a sequential injection system for spectrophotometric determination of sulphide, Quim. Nova 27 (2004) 545–549.

[13] L. Ferrer, G. de Armas, M. Miró, J.M. Estela, V. Cerdà, A multisyringe flow injection method for the automated determination of sulfide in waters using a miniaturised optical fiber spectrophotometer, Talanta 64 (2004) 1119–1126.

[14] L. Ferrer, G. de Armas, M. Miró, J.M. Estela, V. Cerdà, Flow-through optical fiber sensor for automatic sulfide determination in waters by multisyringe flow injection analysis using solid-phase reflectometry, Analyst 130 (2005) 644–651.

[15] G. de Armas, L. Ferrer, M. Miró, J.M. Estela, V. Cerdà, In-line membrane separation method for sulfide monitoring in wastewaters exploiting multisyringe flow injection analysis, Anal. Chim. Acta 524 (2004) 89–96.

[16] L. Ferrer, G. de Armas, M. Miró, J.M. Estela, V. Cerdà, Interfacing in-line gas-diffusion separation with optrode sorptive preconcentration exploiting multisyringe flow injection analysis, Talanta 68 (2005) 343–350.

[17] L. Ferrer, J.M. Estela, V. Cerdà, A smart multisyringe flow injection system for analysis of sample batches with high variability in sulfide concentration, Anal. Chim. Acta 573–574 (2006) 391–398.

[18] F. Maya, J.M. Estela, V. Cerdà, Improving the chemiluminescence-based determination of sulphide in complex environmental samples by using a new, automated multi-syringe flow injection analysis system coupled to a gas diffusion unit, Anal. Chim. Acta 60 (2007) 87–94.

[19] B. Vallejo, P. Richter, I. Toral, C. Tapia, M.D. Luque de Castro, Determination of sulphide in liquid and solid samples by integrated pervaporation-potentiometric detection, Anal. Chim. Acta 436 (2001) 301–307.

[20] W. Qin, Z.J. Zhang, C.J. Zhang, Chemiluminescence flow system for the determination of sulphite, Fresenius' J. Anal. Chem. 361 (1998) 824–826.

[21] L.Q. Yin, D.X. Yuan, M. Zhang, Determination of sulfite in water samples by flow injection analysis with fluorescence detection, Chin. Chem. Lett. 21 (2010) 1457–1461.

[22] A.S. Taufiq, A.M.S. Abulkibash, Differential electrolytic potentiometric detector in flow injection analysis for cyanide determination, J. Flow Injection Anal. 24 (2007) 9–12.

[23] S.S.M. Hassan, M.S.A. Hamza, A.E.A. Kelany, A novel spectrophotometric method for batch and flow injection determination of cyanide in electroplating wastewater, Talanta 71 (2007) 1088–1095.

[24] E. Miralles, D.E. Prat, R. Compano, M. Granados, On-line gas-diffusion separation and fluorimetric detection for the determination of acid dissociable cyanide, Analyst 123 (1998) 217–220.

[25] U.S. EPA Method OIA-1677, Available cyanide by flow injection, Ligand Exchange and Amperometry, EPA-821-R-04–001, January 2004.

[26] A. Ipatov, M. Ivanov, S. Makarychev-Mikhailov, V. Kolodnikov, A. Legin, Y. Vlasov, Determination of cyanide using flow-injection multisensor system, Talanta 58 (2002) 1071–1076.

[27] A.R. Surleva, M.T. Neshkova, A new generation of cyanide ion-selective membranes for flow injection application: part III. A simple approach to the determination of toxic metal-cyanide complexes without preliminary separation, Talanta 76 (2008) 914–921.
[28] M. Noroozifar, M. Khorasani-Motlagh, A. Taheri, R. Zare-Dorabei, A comparative study of AgX (X = Cl^-, Br^-, I^- and N^-_3 solid-phase reactors for flow-injection determination of cyanide in electroplating wastewater, Anal. Sci. 24 (2008) 669–672.
[29] A.M. Abulkibash, S.M. Freihat, Sequential injection spectrophotometric determination of cyanide, J. Flow Injection Anal. 24 (2007) 17–21.
[30] D.G. Themelis, S.C. Karastogianni, P.D. Tzanavaras, Selective determination of cyanides by gas diffusion-stopped flow-sequential injection analysis and an on-line standard addition approach, Anal. Chim. Acta 632 (2009) 93–100.
[31] C. Zacharis, P. Tzanavaras, A. Voulgaropoulos, B. Karlberg, Amperometric determination of cyanides at the low ppb level by automated preconcentration based on gas diffusion coupled to sequential injection analysis, Talanta 77 (2009) 1620–1626.
[32] C.M. Infante, J.C. Masini, F.R.P. Rocha, A green flow-based procedure for fluorimetric determination of acid-dissociable cyanide in natural waters exploiting multicommutation, Anal. Bioanal. Chem. 391 (2008) 2931–2936.
[33] R.M. Frizzarin, F.R.P. Rocha, A multi-pumping flow-based procedure with improvised sensitivity for the spectrophotometric determination of acid-dissociable cyanide in natural waters, Anal. Chim. Acta 758 (2013) 108–113.
[34] M.C. Quaresma, M.F. de Carvalho, F.A. Meirelles, V.M. Santiago, R.E. Santelli, Application of hydrocyanic acid vapor generation via focused microwave radiation to the preparation of industrial effluent samples prior to free and total cyanide determinations by spectrophotometric flow injection analysis, Anal. Bioanal. Chem. 387 (2007) 1017–1025.
[35] ASTM D7511–09e2, Standard test method for total cyanide by segmented flow injection analysis, in-line ultraviolet digestion and amperometric detection, ASTM International.
[36] E. Miralles, R. Compano, M. Granados, M.D. Prat, Photodissociation/gas-diffusion separation and fluorimetric detection for the analysis of total and labile cyanide in a flow system, Fresenius' J. Anal. Chem. 365 (1999) 516–520.
[37] B. Vallejo-Pecharroman, M.D. Luque de Castro, Determination of cyanide by a pervaporation-UV photodissociation-potentiometric detection approach, Analyst 127 (2002) 267–270.
[38] J.F. van Staden, A. Botha, Spectrophotometric determination of thiocyanate by sequential injection analysis, Anal. Chim. Acta 403 (2000) 279–286.
[39] F.V. Almeida, J.R. Guimaraes, W.F. Jardim, Measuring the CO_2 flux at the air/water interface in lakes using flow injection analysis, J. Environ. Monit. 3 (2001) 317–321.
[40] A. Fonseca, J.C.B. Silva, A microflow analyzer with an integrated gas diffusion unit, J. Braz. Chem. Soc. 24 (2013) 5–11.
[41] V. Martinotti, M. Balordi, G. Ciceri, A flow injection analyser conductometric coupled system for the field analysis of free dissolved CO_2 and total dissolved inorganic carbon in natural waters, Anal. Bioanal. Chem. 403 (2012) 1083–1093.
[42] X.L. Su, L.H. Nie, S.Z. Yao, A novel gas-diffusion/flow-injection system coupled with a bulk acoustic wave impedance sensor for total inorganic carbonate and its application to determination of total inorganic and total organic carbon in waters, Anal. Chim. Acta 349 (1997) 143–151.
[43] L. Monser, N. Adhoum, S. Sadok, Gas diffusion-flow injection determination of total inorganic carbon in water using tungsten oxide electrode, Talanta 62 (2004) 389–394.
[44] S. Satienperakul, J.T. Cardwell, R.W. Cattrall, I.D. McKelvie, D.M. Taylor, S.D. Kolev, Determination of carbon dioxide in gaseous samples by gas diffusion-flow injection, Talanta 62 (2004) 631–636.

[45] P.C.C. de Oliveira, J.C. Masini, C.X. Galhardo, J.C.D. Lima, A.E.G. Santana, A.M.G. Vasconcelos, W.P. Nunes, O.L.C. Amaral, A new approach to construct diffusion/permeation cell for use in flow systems. Application in the spectrophotometric determination of bicarbonate ions, J. Braz. Chem. Soc. 17 (2006) 976–980.

[46] V. Grassi, M. Miyazawa, M.A. Pavan, M.Y. Kamogawa, Determination of residual carbonate in soil by flow analysis coupled pervaporation, Quim. Nova 25 (2002) 149–152.

[47] M. Oshima, P. Sritharathikhun, Y. Wei, S. Motomizu, Batchwise collection-flow injection method for the determination of carbon dioxide in ambient air, J. Flow Injection Anal. 20 (2003) 37–42.

[48] M. Oshima, Y. Wei, M. Yamamoto, H. Tanaka, T. Takayanagi, S. Motomizu, Highly sensitive determination method for total carbonate in water samples by flow injection analysis coupled with gas-diffusion separation, Anal. Sci. 17 (2001) 1285–1290.

[49] Y. Wei, M. Oshima, T. Takayanagi, S. Motomizu, Continuous in-situ analysis of total carbonate in water by reversed-flow injection method, J. Flow Injection Anal. 18 (2001) 156–161.

[50] I.C. Santos, R.B.R. Mesquita, A. Machado, A.A. Bordalo, A.O.S.S. Rangel, Sequential injection methodology for carbon speciation in bathing waters, Anal. Chim. Acta 778 (2013) 38–47.

[51] S.M. Gray, P.S. Ellis, M.R. Grace, I.D. McKelvie, Underway determination of alkalinity in estuarine waters by reagent- injection gas-diffusion flow analysis, Talanta 77 (2008) 533–540.

[52] P.J. Fletcher, J.F. van Staden, Determination of carbonate and hydrogencarbonate by titration using sequential injection analysis, Anal. Chim. Acta 485 (2003) 187–194.

[53] T.G. Jantsch, B. Mattiasson, A simple spectrophotometric method based on pH-indicators for monitoring partial and total alkalinity in anaerobic processes, Environ. Tech 24 (2003) 1061–1067.

[54] R.B.R. Mesquita, A.O.S.S. Rangel, A sequential injection system for the spectrophotometric determination of calcium, magnesium and alkalinity in water samples, Anal. Sci. 20 (2004) 1205–1210.

[55] J. Liu, L. Bjornsson, B. Mattiasson, Immobilised activated sludge based biosensor for biochemical oxygen demand measurement, Biosens. Bioelectron. 14 (2000) 883–893.

[56] M. Zenki, S. Fujiwara, T. Yokoyama, Repetitive determination of chemical oxygen demand by cyclic flow injection analysis using on-line regeneration of consumed permanganate, Anal. Sci. 22 (2006) 77–80.

[57] D. Dan, R.C. Sandford, P.J. Worsfold, Determination of chemical oxygen demand in fresh waters using flow injection with on-line UV-photocatalytic oxidation and spectrophotometric detection, Analyst 130 (2005) 227–232.

[58] R.T. Edwards, I.D. McKelvie, P.C. Ferrett, B.T. Hart, J.B. Bapat, K. Koshy, Sensitive flow-injection technique for the determination of dissolved organic carbon in natural and waste waters, Anal. Chim. Acta 261 (1992) 287–294.

[59] J. Chen, J. Zhang, Y. Xian, X. Ying, M. Liu, L. Jin, Preparation and application of TiO_2 photocatalytic sensor for chemical oxygen demand determination in water, Water Res. 39 (2005) 1340–1346.

[60] Y.C. Kim, S. Sasaki, K. Yano, K. Ikebukuro, K. Hashimoto, I. Karube, A flow method with photocatalytic oxidation of dissolved organic matter using a solid-phase (TiO_2) reactor followed by amperometric detection of consumed oxygen, Anal. Chem. 74 (2002) 3858–3864.

[61] Y.C. Kim, S. Sasaki, K. Yano, K. Ikebukuro, K. Hashimoto, I. Karube, Photocatalytic sensor for the determination of chemical oxygen demand using flow injection analysis, Anal. Chim. Acta 432 (2001) 59–66.

[62] A. Cuesta, J.L. Todoli, J. Mora, A. Canals, Rapid determination of chemical oxygen demand by a semi-automated method based on microwave sample digestion, chromium(vi) organic solvent extraction and flame atomic absorption spectrometry, Anal. Chim. Acta 372 (1998) 399–409.

[63] A.P. Beltrá, J. Iniesta, L. Gras, F. Gallud, V. Montiel, A. Aldaz, A. Canals, Development of a fully automated microwave assisted chemical oxygen demand (COD) measurement device, Instrum. Sci. Tech. 31 (2003) 249–259.

[64] Z.Q. Zhang, H.T. Yan, Y. Lin, Rapid determination of chemical oxygen demand by flame AAS based on flow injection on-line ultrasound-assisted digestion and manganese speciation separation, Atom. Spectrosc. 25 (2004) 191–196.
[65] B. Jin, Y. He, J. Shen, Z. Zhuang, X. Wang, F.S.C. Lee, Measurement of chemical oxygen demand (COD) in natural water samples by flow injection ozonation chemiluminescence (FI-CL) technique, J. Environ. Monit. 6 (2004) 673–678.
[66] T. Jinjun, H. Yonggang, Z. Jie, Chemiluminescence detection of permanganate index (CODMn) by a luminol-$KMnO_4$ based reaction, J. Environ. Sci. 20 (2008) 252–256.
[67] W. Liu, Z. Zhang, Y. Zhang, Chemiluminescence micro-flow system for rapid determination of chemical oxygen demand in water, Microchim. Acta 160 (2008) 141–146.
[68] O. Tue-Ngeun, R.C. Sandford, J. Jakmunee, K. Grudpan, I.D. McKelvie, P.J. Worsfold, Determination of dissolved inorganic carbon (DIC) and dissolved organic carbon (DOC) in freshwaters by sequential injection spectrophotometry with on-line UV photo-oxidation, Anal. Chim. Acta 554 (2005) 17–24.
[69] L.A. Zimmer, G.A. Cutter, High resolution determination of nanomolar concentrations of dissolved reactive phosphate in ocean surface waters using long path liquid waveguide capillary cells (LWCC) and spectrometric detection, Limnol. Oceanogr. Methods 10 (2012) 568–580.
[70] S. Motomizu, Z.H. Li, Trace and ultratrace analysis methods for the determination of phosphorus by flow-injection techniques, Talanta 66 (2005) 332–340.
[71] R.L. Benson, Y.B. Truong, I.D. McKelvie, B.T. Hart, G.W. Bryant, W.P. Hilkmann, Monitoring of dissolved reactive phosphorus in wastewaters by flow injection analysis. Part 2. One-line monitoring system, Water Res. 30 (1996) 1965–1971.
[72] J.Z. Zhang, J. Chi, Automated analysis of nanomolar concentrations of phosphate in natural waters with liquid waveguide, Environ. Sci. Technol. 36 (2002) 1048–1053.
[73] A.J. Lyddy-Meaney, P.S. Ellis, P.J. Worsfold, E.C.V. Butler, I.D. McKelvie, A compact flow injection analysis system for surface mapping of phosphate in marine waters, Talanta 58 (2002) 1043–1053.
[74] Y. Liang, D. Yuan, Q. Li, Q. Lin, Flow injection analysis of nanomolar level orthophosphate in seawater with solid phase enrichment and colorimetric detection, Mar. Chem. 103 (2007) 122–130.
[75] L.J. Gimbert, P.M. Haygarth, P.J. Worsfold, Determination of nanomolar concentrations of phosphate in natural waters using flow injection with a long path length liquid waveguide capillary cell and solid-state spectrophotometric detection, Talanta 71 (2007) 1624–1628.
[76] J. Ma, D. Yuan, Y. Liang, M. Dai, A modified analytical method for the shipboard determination of nanomolar concentrations of orthophosphate in seawater, J. Oceanogr. 64 (2008) 443–449.
[77] M.D. Hurtado, S. Carmona, A. Delgado, Automated modification of the molybdenum blue colorimetric method for phosphorus determination in soil extracts, Commun. Soil Sci. Plant Anal. 39 (2008) 2250–2257.
[78] S. Somnam, K. Grudpan, J.M. Jakmunee, Stopped-flow injection method for determination of phosphate in soils and fertilisers, Int. J. Sci. Tech. 2 (2008) 172–181.
[79] M.S.A.C. Neves, M.R.S. Souto, I.V. Toth, S.M.A. Victal, M.C. Drumond, A.O.S.S. Rangel, Spectrophotometric flow system using vanadomolybdophosphate detection chemistry and a liquid waveguide capillary cell for the determination of phosphate with improved sensitivity in surface and ground water samples, Talanta 77 (2008) 527–532.
[80] M. Yaqoob, A. Nabi, P.J. Worsfold, Determination of nanomolar concentrations of phosphate in freshwaters using flow injection with luminol chemiluminescence detection, Anal. Chim. Acta 510 (2004) 213–218.
[81] S. Hinkamp, G. Schwedt, Determination of total phosphorous in waters with amperometric detection by coupling of flow-injection analysis with continuous microwave oven digestion, Anal. Chim. Acta 236 (1990) 345–350.
[82] R.L. Benson, I.D. McKelvie, B.T. Hart, Y.B. Truong, I.C. Hamilton, Determination of total phosphorous in waters and waste waters by on-line UV/thermal induced digestion and flow-injection analysis, Anal. Chim. Acta 326 (1996) 29–39.

[83] D.M.W. Peat, I.D. McKelvie, G.P. Matthews, P.M. Haygarth, P.J. Worsfold, Rapid determination of dissolved organic phosphorus in soil leachates and runoff waters by flow injection analysis with on-line photo-oxidation, Talanta 45 (1997) 47–55.

[84] O. Tue-Ngeun, P. Ellis, I.D. McKelvie, P. Worsfold, J. Jakmunee, K. Grudpan, Determination of dissolved reactive phosphorus (DRP) and dissolved organic phosphorus (DOP) in natural waters by the use of rapid sequenced reagent injection flow analysis, Talanta 66 (2005) 453–460.

[85] S. Auflitsch, D.M.W. Peat, P.J. Worsfold, I.D. McKelvie, Determination of dissolved reactive phosphorus in estuarine waters using a reversed flow injection manifold, Analyst 122 (1997) 1477–1480.

[86] I.D. McKelvie, D.M.W. Peat, G.P. Matthews, P.J. Worsfold, Elimination of the Schlieren effect in the determination of reactive phosphorus in estuarine waters by flow-injection analysis, Anal. Chim. Acta 351 (1997) 265–271.

[87] J. Ma, D. Yuan, M. Zhang, Y. Liang, Reverse flow injection analysis of nanomolar soluble reactive phosphorus in seawater with a long path length liquid waveguide capillary cell and spectrophotometric detection, Talanta 78 (2009) 315–320.

[88] T. Perez-Ruiz, C. Martinez-Lozano, V. Tomas, J. Martin, Flow-injection spectrofluorimetric determination of dissolved inorganic and organic phosphorus in waters using on-line photo-oxidation, Anal. Chim. Acta 442 (2001) 147–153.

[89] A. Muñoz, F. Mas Torres, J.M. Estela, V. Cerdà, Evaluation of spectrophotometric methods for determination of orthophosphates by sequential injection analysis, Anal. Chim. Acta 350 (1997) 21–29.

[90] J.F. van Staden, R.E. Taljaard, Online monitoring of phosphate in natural water and effluent streams using sequential injection analysis, Microchim. Acta 128 (1998) 223–228.

[91] F. Mas-Torres, J.M. Estela, M. Miró, A. Cladera, V. Cerdà, Sequential injection spectrophotometric determination of orthophosphate in beverages, wastewaters and urine samples by electrogeneration of molybdenum blue using tubular flow-through electrodes, Anal. Chim. Acta 510 (2004) 61–68.

[92] J. Buanuam, M. Miró, E.H. Hansen, J. Shiowatana, On-line dynamic fractionation and automatic determination of inorganic phosphorus in environmental solid substrates exploiting sequential injection microcolumn extraction and flow injection analysis, Anal. Chim. Acta 570 (2006) 224–231.

[93] J. Ma, D. Yuan, Y. Liang, Sequential injection analysis of nanomolar soluble reactive phosphorus in seawater with HLB solid phase extraction, Mar. Chem. 111 (2008) 151–159.

[94] C. Frank, F. Schroeder, R. Ebinghaus, W. Ruck, Using sequential injection analysis for fast determination of phosphate in coastal waters, Talanta 70 (2006) 513–517.

[95] C. Frank, F. Schroeder, R. Ebinghaus, W. Ruck, A fast sequential injection analysis system for the simultaneous determination of ammonia and phosphate, Microchim. Acta 154 (2006) 31–38.

[96] C. Frank, F. Schroeder, Using sequential injection analysis to improve system and data reliability of online methods: determination of ammonium and phosphate in coastal waters, J. Autom. Methods Manag. Chem. (2007), 1/1–1/6.

[97] C.C. Oliveira, E.A.G. Zagatto, A.N. Araudjo, J.L.F.C. Lima, Sample preparation in sequential injection analysis – spectrophotometric determination of total phosphorus in food samples, Anal. Chim. Acta 371 (1998) 57–62.

[98] C.H. Wu, J. Ruzicka, Micro sequential injection: environmental monitoring of nitrogen and phosphate in water using lab-on-valve system furnished with a microcolumn, Analyst 126 (2001) 1947–1952.

[99] R.N. Fernandes, B.F. Reis, Flow system exploiting multicommutation to increase sample residence time for improved sensitivity. Simultaneous determination of ammonium and ortho-phosphate in natural water, Talanta 58 (2002) 729–737.

[100] M.E. Carrasco, J.A.G. Bautista, M.J.V. Garcia, Automated sequential monitoring of ammonium, phosphate and nitrite in wastewater by multi-commutated peristaltic and solenoid pumped flow system – a comparative study, Chem. Anal. 52 (2007) 757–770.

[101] N. Amornthammarong, P. Anujaravat, K. Sereenonchai, P. Chaisuwan, P. Sastranurak, P. Wilairat, D. Nacapricha, Method development based on all injection analysis for the determination of phosphorus in soil and sediment extracts, Talanta 68 (2005) 480–487.

[102] M.I.G.S. Almeida, M.A. Segundo, J.L.F.C. Lima, A.O.S.S. Rangel, Multi-syringe flow injection system for the determination of available phosphorus in soil samples, Int. J. Environ. Anal. Chem. 85 (2005) 51–62.

[103] J. Buanuam, M. Miró, E.H. Hansen, J. Shiowatana, J.M. Estela, V. Cerdà, A multisyringe flow-through sequential extraction system for on-line monitoring of orthophosphate in soils and sediments, Talanta 71 (2007) 1710–1719.

[104] I.P.A. Morais, M. Miró, M. Manera, J.M. Estela, V. Cerdà, M.R.S. Souto, A.O.S.S. Rangel, Flow-through solid-phase based optical sensor for the multisyringe flow injection trace determination of orthophosphate in waters with chemiluminescence detection, Anal. Chim. Acta 506 (2004) 17–24.

[105] M.I.G.S. Almeida, M.A. Segundo, J.L.F.C. Lima, A.O.S.S. Rangel, Multi-syringe flow injection system with in-line microwave digestion for the determination of phosphorus, Talanta 64 (2004) 1283–1289.

[106] C. Pons, I.V. Toth, A.O.S.S. Rangel, R. Forteza, V. Cerdà, Multi-pumping flow system for the determination of dissolved orthophosphate and dissolved organic phosphorus in wastewater samples, Anal. Chim. Acta 572 (2006) 148–154.

[107] D. Halliwell, J. Coventry, D. Nash, Inorganic monophosphate determination in overland flow from irrigated grazing systems, Int. J. Environ. Anal. Chem. 76 (2000) 77–87.

[108] I.A. Tsoulfanidis, G.Z. Tsogas, D.L. Giokas, A.G. Vlessidis, Design of a field flow system for the on-line spectrophotometric determination of phosphate, nitrite and nitrate in natural water and wastewater, Microchim. Acta 160 (2008) 461–469.

[109] K.M. Pedersen, M. Kuemmel, H. Soeeberg, Monitoring and control of biological removal of phosphorus and nitrogen by flow-injection analysers in a municipal pilot-scale waste-water treatment plant, Anal. Chim. Acta 238 (1990) 191–199.

[110] A.G. Vlessidis, M.E. Kotti, N.P.A. Evmiridis, Study for the validation of spectrophotometric methods for detection and of digestion methods using a flow injection manifold, for the determination of total phosphorus in wastewaters, J. Anal. Chem. 59 (2004) 77–85.

[111] F. Mas-Torres, A. Muñoz, J.M. Estela, V. Cerdà, Simultaneous determination of phosphate and silicate in waste water by sequential injection analysis, Analyst 122 (1997) 1033–1038.

[112] F. Mas-Torres, A. Muñoz, J.M. Estela, V. Cerdà, Simultaneous spectrophotometric determination of phosphate and silicate by a stopped-flow sequential injection method, Int. J. Environ. Anal. Chem. 77 (2000) 185–202.

[113] K. Grudpan, P. Ampan, Y. Udnan, S. Jayasvati, S. Lapanantnoppakhun, J. Jakmunee, G.D. Christian, J. Ruzicka, Stopped-flow injection simultaneous determination of phosphate and silicate using molybdenum blue, Talanta 58 (2002) 1319–1326.

[114] C.X. Galhardo, J.C. Masini, Spectrophotometric determination of phosphate and silicate by sequential injection using molybdenum blue chemistry, Anal. Chim. Acta 417 (2000) 191–200.

[115] J.F. van Staden, R.F. Taljaard, Determination of ammonia in water and industrial effluent streams with the indophenol blue method using sequential injection analysis, Anal. Chim. Acta 344 (1997) 281–289.

[116] T. Tsuboi, Y. Hirano, Y. Shibata, S. Motomizu, Sensitivity improvement of ammonia determination based on flow-injection indophenol spectrophotometry with manganese(II) ion as a catalyst and analysis of exhaust gas of thermal power plant, Anal. Sci. 18 (2002) 1141–1144.

[117] P.L. Qian, Z. Jia-Zhong, J.M. Frank, A.H. Dennis, Continuous colorimetric determination of trace ammonium in seawater with a long-path liquid waveguide capillary cell, Mar. Chem. 96 (2005) 73–85.

[118] B. Bucur, M. Catala, J. Martinez, Spectrophotometric determination of ammonium by an rFIA assembly, Rev. Roum. Chim. 51 (2006) 101–108.

[119] A. Cerdà, M.T. Oms, R. Forteza, V. Cerdà, Evaluation of flow injection methods for ammonium determination in wastewater samples, Anal. Chim. Acta 311 (1995) 165–173.

[120] A.C. Ariza, P. Linares, M.D. Luque de Castro, M. Valcarcel, Unsegmented flow approach for online monitoring of pH, conductivity, dissolved nitrogen and determination of nitrite and ammonium in aquaculture, J. Autom. Chem. 16 (1994) 59–62.

[121] B. Horstkotte, C.M. Duarte, V. Cerdà, A miniature and field-applicable multipumping flow analyzer for ammonium monitoring in seawater with fluorescence detection, Talanta 85 (2011) 380–385.

[122] A. Aminot, R. Kerouel, D. Birot, A flow injection-fluorometric method for the determination of ammonium in fresh and saline waters with a view to in situ analyses, Water Res. 35 (2001) 1777–1785.

[123] R.J. Watson, E.C.V. Butler, L.A. Clementson, K.M. Berry, Flow-injection analysis with fluorescence detection for the determination of trace levels of ammonium in seawater, J. Environ. Monit. 7 (2005) 37–42.

[124] H. Mana, U. Spohn, Rapid and selective determination of ammonium by fluorimetric flow injection analysis, Fresenius' J. Anal. Chem. 366 (2000) 825–829.

[125] N. Amornthammarong, J. Jakmunee, J. Li, P.K. Dasgupta, Hybrid fluorometric flow analyzer for ammonia, Anal. Chem. 78 (2006) 1890–1896.

[126] B.F. Reis, J.A. Vieira, F.J. Krug, M.F. Gine, Development of a flow injection system with two analytical paths for ammonium determination in soil extracts by conductometry, J. Braz. Chem. Soc. 8 (1997) 523–528.

[127] M.T. Oms, A. Cerdà, V. Cerdà, Preconcentration by flow reversal in conductometric sequential injection analysis of ammonium, Electroanalysis 8 (1996) 387–390.

[128] C. Henríquez, B. Horstkotte, V. Cerdà, Conductometric determination of ammonium by a multisyringe flow injection system applying gas diffusion, Int. J. Environ. Anal. Chem. (2012) 1–17.

[129] M.T. Oms, A. Cerdà, A. Cladera, V. Cerdà, R. Forteza, Gas diffusion techniques coupled sequential injection analysis for selective determination of ammonium, Anal. Chim. Acta 318 (1996) 251–260.

[130] I. Lukkari, J. Ruzicka, G.D. Christian, Determination of total ammonium-nitrogen and free ammonia in a fermentation medium by sequential injection analysis, Fresenius' J. Anal. Chem. 346 (1993) 813–818.

[131] K.N. Andrew, P.J. Worsfold, M. Comber, On-line flow injection monitoring of ammonia in industrial liquid effluents, Anal. Chim. Acta 314 (1995) 33–43.

[132] M.J. Bloxham, M.H. Depledge, P.J. Worsfold, In situ flow-injection monitoring of ammonia in landfill leachate, Lab. Rob. Autom. 9 (1997) 175–183.

[133] S.M. Gray, P.S. Ellis, M.R. Grace, I.D. McKelvie, Spectrophotometric determination of ammonia in estuarine waters by hybrid reagent-injection gas-diffusion flow analysis, Spectrosc. Lett. 39 (2006) 737–753.

[134] S.M. Oliveira, T.I.M.S. Lopes, I.V. Toth, A.O.S.S. Rangel, A multi-commuted flow injection system with a multi-channel propulsion unit placed before detection: spectrophotometric determination of ammonium, Anal. Chim. Acta 600 (2007) 29–34.

[135] S.M. Oliveira, T.I.M.S. Lopes, I.V. Toth, A.O.S.S. Rangel, Determination of ammonium in marine waters using a gas diffusion multicommuted flow injection system with in-line prevention of metal hydroxides precipitation, J. Environ. Monit. 11 (2008) 228–234.

[136] M.I.G.S. Almeida, J.M. Estela, M.A. Segundo, V. Cerdà, A membraneless gas diffusion unit – multisyringe flow injection spectrophotometric method for ammonium determination in untreated environmental samples, Talanta 84 (2011) 1244–1252.

[137] V.V. Kuznetsov, S.V. Zemyatova, Flow-injection spectrophotometry of nitrites based on the diazotization reactions of azine dyes, J. Anal. Chem. 62 (2007) 637–644.

[138] M. O'Toole, R. Shepherd, K.T. Lau, D. Diamond, Detection of nitrite by flow injection analysis using a novel paired emitter-detector diode (PEDD) as a photometric detector, Proc. SPIE Int. Soc. Opt. Eng. (2007) 6755.

[139] W.R. Melchert, C.M.C. Infante, F.R.P. Rocha, Development and critical comparison of greener flow procedures for nitrite determination in natural waters, Microchem. J. 85 (2007) 209–213.

[140] H. Takiguchi, A. Tsubata, M. Miyata, T. Odake, H. Hotta, T. Umemura, K. Tsunoda, Liquid core waveguide spectrophotometry for the sensitive determination of nitrite in river water samples, Anal. Sci. 22 (2006) 1017–1019.

[141] J.F. van Staden, T.A. van der Merwe, Online monitoring of nitrite in fertilizer process streams, natural and wastewater effluents with sequential injection analysis, Microchim. Acta 129 (1998) 33–39.

[142] M. Baeza, J. Bartroli, J. Alonso, Autoadaptative sequential injection system for nitrite determination in wastewaters, Talanta 68 (2005) 245–252.

[143] Y.Q. Zhao, Y.Z. He, W.E. Gan, L. Yang, Determination of nitrite by sequential injection analysis of electrokinetic flow analysis system, Talanta 56 (2002) 619–625.

[144] G. Chen, D. Yuan, Y. Huang, M. Zhang, M. Bergman, In-field determination of nanomolar nitrite in seawater using a sequential injection technique combined with solid phase enrichment and colorimetric detection, Anal. Chim. Acta 620 (2008) 82–88.

[145] M. Zhang, D. Yuan, Y. Huang, G. Chen, Z. Zhang, Sequential injection spectrophotometric determination of nanomolar nitrite in seawater by on-line preconcentration with HLB cartridge, Acta Oceanol. Sin. 29 (2010) 100–107.

[146] M. Miró, A. Cladera, J.M. Estela, V. Cerdà, Sequential injection spectrophotometric analysis of nitrite in natural waters using an on-line solid-phase extraction and preconcentration method, Analyst 125 (2000) 943–948.

[147] M.M. Baeza, N. Ibañez-Garcia, J. Baucells, J. Bartroli, J. Alonso, Microflow injection system based on a multicommutation technique for nitrite determination in wastewaters, Analyst 131 (2006) 1109–1115.

[148] G.M. Greenway, S.J. Haswell, P.H. Petsul, Characterisation of a micro-total analytical system for the determination of nitrite with spectrophotometric detection, Anal. Chim. Acta 387 (1999) 1–10.

[149] J.L.F.C. Lima, A.O.S.S. Rangel, M.R. Souto, Determination of nitrate levels in water using a flow-injection system and a tubular potentiometric detector, Rev. Port. Farm 46 (1996) 82–88.

[150] E. Santos, M.C.B.S.M. Montenegro, C. Couto, A.N. Araujo, M.F. Pimentel, V. Lins da Silva, Sequential injection analysis of chloride and nitrate in waters with improved accuracy using potentiometric detection, Talanta 63 (2004) 721–727.

[151] S.J. Birrell, J.W. Hummel, Membrane selection and ISFET configuration evaluation for soil nitrate sensing, Trans. ASAE 43 (2000) 197–206.

[152] J. Gallardo, S. Alegret, M. del Valle, A flow-injection electronic tongue based on potentiometric sensors for the determination of nitrate in the presence of chloride, Sens. Actuators B Chem. 101 (2004) 72–80.

[153] S.S.M. Hassan, H.E.M. Sayour, S.S. Al-Mehrezi, A novel planar miniaturized potentiometric sensor for flow injection analysis of nitrates in wastewaters, fertilizers and pharmaceuticals, Anal. Chim. Acta 581 (2007) 13–18.

[154] G.A. Alvarez-Romero, M.E. Palomar-Pardave, M.T. Ramirez-Silva, Development of a novel nitrate-selective composite sensor based on doped polypyrrole, Anal. Bioanal. Chem. 387 (2007) 1533–1541.

[155] W.R. Melchert, F.R.P. Rocha, A green analytical procedure for flow-injection determination of nitrate in natural waters, Talanta 65 (2005) 461–465.

[156] B.S. Gentle, P.S. Ellis, M.R. Grace, I.D. McKelvie, Flow analysis methods for the direct ultra-violet spectrophotometric measurement of nitrate and total nitrogen in freshwaters, Anal. Chim. Acta 704 (2011) 116–122.

[157] A. Cerdà, M.T. Oms, V. Cerdà, R. Forteza, Comparison of flow injection methods for nitrate determination in aqueous environmental samples, Anal. Methods Instrum. 2 (1995) 330–336.

[158] C. Gal, W. Frenzel, J. Moeller, Re-examination of the cadmium reduction method and optimisation of conditions for the determination of nitrate by flow injection analysis, Microchim. Acta 146 (2004) 155–164.

[159] S. Hirata, S. Karthikeyan, T. Kajiya, Portable flow-injection analyzer for nitrite and nitrate in natural water, Anal. Sci. 20 (2004) 567–569.

[160] A.R.J. David, T. McCormack, P.J. Worsfold, A submersible battery-powered flow injection (FI) sensor for the determination of nitrate in estuarine and coastal waters, J. Autom. Methods Manag. Chem. 21 (1999) 1–9.
[161] P.H. Petsul, G.M. Greenway, S.J. Haswell, The development of an on-chip micro-flow injection analysis of nitrate with a cadmium reactor, Anal. Chim. Acta 428 (2001) 155–161.
[162] D. Gabriel, J. Baeza, F. Valero, J. Lafuente, A novel FIA configuration for the simultaneous determination of nitrate and nitrite and its use for monitoring an urban waste water treatment plant based on N/D criteria, Anal. Chim. Acta 359 (1998) 173–183.
[163] P.C.F.C. Gardolinski, A.R.J. David, P.J. Worsfold, Miniature flow injection analyser for laboratory, shipboard and in situ monitoring of nitrate in estuarine and coastal waters, Talanta 58 (2002) 1015–1027.
[164] A. Kazemzadeh, A.A. Ensafi, Sequential flow injection spectrophotometric determination of nitrite and nitrate in various samples, Anal. Chim. Acta 442 (2001) 319–326.
[165] R.A.S. Lapa, J.L.F.C. Lima, I.V.O.S. Pinto, Sequential injection analysis-based system for on-line monitoring of nitrite and nitrate in wastewaters, Anal. Sci. 16 (2000) 1157–1160.
[166] R.A.S. Lapa, J.L.F.C. Lima, I.V.O.S. Pinto, Simultaneous determination of nitrite, nitrate, sulphate and phenolic compounds, by sequential injection analysis, in wastewaters, Analusis 28 (2000) 295–301.
[167] A. Ayala, L.O. Leal, L. Ferrer, V. Cerdà, Multiparametric automated system for sulfate, nitrite and nitrate monitoring in drinking water and wastewater based on sequential injection analysis, Microchem. J. 100 (2012) 55–60.
[168] S.M. Oliveira, T.I.M.S. Lopes, A.O.S.S. Rangel, Sequential injection determination of nitrate in vegetables by spectrophotometry with inline cadmium reduction, Commun. Soil Sci. Plant Anal. 38 (2007) 533–544.
[169] M.J.L. Reis, S.M.V. Fernandes, A.O.S.S. Rangel, Determination of nitrate and nitrite in dairy samples by sequential injection using an in-line cadmium-reducing column, Int. Dairy J. 16 (2006) 1442–1447.
[170] M.T. Oms, A. Cerdà, V. Cerdà, Sequential injection analysis of nitrites and nitrates, Anal. Chim. Acta 315 (1995) 321–330.
[171] C. Pons, J.L.M. Santos, J.L.F.C. Lima, R. Forteza, V. Cerdà, Multi-pumping flow system for the determination of nitrite and nitrate in water samples, Microchim. Acta 161 (2008) 73–79.
[172] Y. Wang, S.H. Fan, S.L. Wang, Chemiluminescence determination of nitrogen oxide in air with a sequential injection method, Anal. Chim. Acta 541 (2005) 129–134.
[173] Y.L. Wei, M. Oshima, J. Simon, S. Motomizu, The application of the chromatomembrane cell for the absorptive sampling of nitrogen dioxide followed by continuous determination of nitrite using a micro-flow injection system, Talanta 57 (2002) 355–364.
[174] K. Toda, Y. Hato, K. Mori, S.I. Ohira, T. Namihira, Sequential multiple analyses of atmospheric nitrous acid and nitrogen oxides, Talanta 71 (2007) 1652–1660.
[175] P. Mikuska, K. Motyka, Z. Vecera, Determination of nitrous acid in air using wet effluent diffusion denuder-FIA technique, Talanta 77 (2008) 635–641.
[176] Y. Wei, M. Oshima, J. Simon, L.N. Moskvin, S. Motomizu, Absorption, concentration and determination of trace amounts of air pollutants by flow injection method coupled with a chromatomembrane cell system: application to nitrogen dioxide determination, Talanta 58 (2002) 1343–1355.
[177] A. Tovar, C. Moreno, M.P. Manuel-Vez, M. Garcia-Vargas, A simple automated method for the speciation of dissolved inorganic nitrogen in seawater, Anal. Chim. Acta 469 (2002) 235–242.
[178] F.R.P. Rocha, B.F. Reis, A flow system exploiting multicommutation for speciation of inorganic nitrogen in waters, Anal. Chim. Acta 409 (2000) 227–235.
[179] R.T. Masserini, K.A. Fanning, A sensor package for the simultaneous determination of nanomolar concentrations of nitrite, nitrate, and ammonia in seawater by fluorescence detection, Mar. Chem. 68 (2000) 323–333.
[180] M.T. Oms, A. Cerdà, V. Cerdà, Sequential injection system for on-line analysis of total nitrogen with UV-mineralization, Talanta 59 (2003) 319–326.

[181] A. Cerdà, M.T. Oms, R. Forteza, V. Cerdà, Speciation of nitrogen in wastewater by flow injection, Analyst 121 (1996) 13–17.

[182] A. Cerdà, M.T. Oms, R. Forteza, V. Cerdà, Total nitrogen determination by flow injection using online microwave-assisted digestion, Anal. Chim. Acta 351 (1997) 273–279.

[183] M.C.B. Quaresma, R.J. Cassella, M.F.B. Carvalho, R.E. Santelli, Focussed microwave-assisted sample preparation: total phenol determination in petroleum refinery effluents by flow injection spectrophotometry, Microchem. J. 78 (2004) 35–40.

[184] R.A.S. Lapa, J.L.F.C. Lima, I.V.O.S. Pinto, Determination of phenolic compounds in waste waters by sequential injection analysis and spectrophotometry, Int. J. Environ. Anal. Chem. 76 (2000) 69–76.

[185] H.M. Oliveira, M.A. Segundo, S. Reis, J.L.F.C. Lima, Multi-syringe flow injection system with in-line pre-concentration for the determination of total phenolic compounds, Microchim. Acta 150 (2005) 187–196.

[186] H.M. Oliveira, M.A. Segundo, J.L.F.C. Lima, V. Cerdà, Multisyringe flow injection system for solid-phase extraction coupled to liquid chromatography using monolithic column for screening of phenolic pollutants, Talanta 77 (2009) 1466–1472.

[187] K.O. Lupetti, F.R.P. Rocha, O. Fatibello-Filho, An improved flow system for phenols determination exploiting multicommutation and long pathlength spectrophotometry, Talanta 62 (2004) 463–467.

[188] H. Qi, J. Lv, B. Li, Determination of phenol at ngL-1 level by flow-injection chemiluminescence combined with on-line solid-phase extraction, Spectrochim. Acta A Mol. Biomol. Spectrosc. 66 (2007) 874–878.

[189] E. Rodenas-Torralba, A. Morales-Rubio, M. de la Guardia, Determination of phenols in waters using micro-pumped multicommutation and spectrophotometric detection: an automated alternative to the standard procedure, Anal. Bioanal. Chem. 383 (2005) 138–144.

[190] B. Horstkotte, O. Elsholz, V. Cerdà, Development of a capillary electrophoresis system coupled to sequential injection analysis and evaluation by the analysis of nitrophenols, Int. J. Environ. Anal. Chem. 87 (2007) 797–811.

[191] B. Horstkotte, O. Elsholz, V. Cerdà, Multisyringe flow injection analysis coupled to capillary electrophoresis (MSFIA-CE) as a novel analytical tool applied to the preconcentration, separation and determination of nitrophenols, Talanta 76 (2008) 72–79.

[192] M. Manera, M. Miró, J.M. Estela, V. Cerdà, M.A. Segundo, J.L.F.C. Lima, Flow-through solid-phase reflectometric method for simultaneous multiresidue determination of nitrophenol derivatives, Anal. Chim. Acta 600 (2007) 155–163.

[193] W. Boonjob, M. Miró, M.A. Segundo, V. Cerdà, Flow-through dispersed carbon nanofiber-based micro-solid-phase extraction coupled to liquid chromatography for automatic determination of trace levels of priority environmental pollutants, Anal. Chem. 83 (2011) 5237–5244.

[194] J.B. Quintana, W. Boonjob, M. Miró, V. Cerdà, On line coupling of bead injection lab-on-valve analysis to gas chromatography: application to the determination of trace levels of polychlorinated biphenyls in solid waste leachates, Anal. Chem. 81 (2009) 4822–4830.

[195] S. Morales-Muñoz, J.L. Luque-Garcia, M.D. Luque de Castro, Static extraction with modified pressurized liquid and on-line fluorescence monitoring independent matrix approach for the removal of polycyclic aromatic hydrocarbons from environmental solid samples, J. Chromatogr. A 978 (2002) 49–57.

[196] C.F. Li, J.W.C. Wong, C.W. Huie, M.M.F. Choi, On-line flow injection-cloud point preconcentration of polycyclic aromatic hydrocarbons coupled with high-performance liquid chromatography, J. Chromatogr. A 1214 (2008) 11–16.

[197] B. Beltrán, J. Avivar, M. Mola, L. Ferrer, V. Cerdà, L.O. Leal, Automated method for simultaneous lead and strontium isotopic analysis applied to rainwater samples and airborne particulate filters (PM10), Environ. Sci. Technol. 47 (2013) 9850–9857.

[198] L. Chaparro, L. Ferrer, V. Cerdà, L.O. Leal, Automated system for on-line determination of dimethylarsinic and inorganic arsenic by hydridegeneration–atomic fluorescence spectrometry, Anal. Bioanal. Chem. 404 (2012) 1589–1595.

[199] A.M. Serra, J.M. Estela, V. Cerdà, An MSFIA system for mercury speciation base on an anion-exchange membrane, Talanta 78 (2009) 790–794.

[200] A.M. Serra, J.M. Estela, V. Cerdà, A MSFIA system for selenium speciation by atomic fluorescence spectrometry, J. Anal. At. Spectrom. 27 (2012) 1858–1862.

[201] P.R. Fortes, M.A. Feres, E.A.E.G. Zagatto, An expert flow system involving in-line prior assay for turbidimetric determination of chloride and sulphate in natural waters, Talanta 77 (2008) 571–575.

[202] J.S. Cosano, M.D. Luque de Castro, M. Valcarcel, Flow-injection analysis of water. I. Automatic preconcentration determination of sulfate, ammonia and iron(II)/iron(III), J. Autom. Chem. 15 (1993) 141–146.

[203] M.A. Feres, B.F. Reis, A downsized flow set up based on multicommutation for the sequential photometric determination of iron(II)/iron(III) and nitrite/nitrate in surface water, Talanta 68 (2005) 422–428.

[204] E. Ballesteros, A. Rios, M. Valcarcel, Integrated automatic determination of nitrate, ammonium and organic carbon in soil samples, Analyst 122 (1997) 309–313.

[205] Q.P. Li, D.A. Hansell, J.Z. Zhang, Underway monitoring of nanomolar nitrate plus nitrite and phosphate in oligotrophic seawater, Limnol. Oceanogr. Methods 6 (2008) 319–326.

CHAPTER

Automating Radiochemical Analysis

8

CHAPTER OUTLINE

Radiochemical analyses are increasingly demanded in routine laboratories for monitoring radionuclides from environmental samples, nuclear power plants, water treatment plants, nuclear medicine services and industrial uses, among others. Thus, benefits of automation are of great interest in the radiochemical field. Features such as the minimization of sample and standards handling together with the decrease of waste generation are crucial, since the safety of the analyst is ensured and more environmentally friendly methods are developed. In this chapter, different strategies used for automating radiochemical analyses are discussed, encompassing fully and partially automated flow methodologies. Thus, automated analytical protocols of extraction and preconcentration of radionuclides are presented, together with their coupling to radiometric detectors or inductively coupled plasma mass spectrometers (ICP-MS). Besides, sensors for radionuclide determination performed in a flow system approach and screening methods for radioactive elements determination are also described as alternatives to classical radiochemical analysis.

8.1 Strategies used for automating radiochemical analysis

Automation of radiochemical analyses has gained ground in last years because of the increase of both radionuclide uses and the number of samples to be analyzed together with the increasing environmental concern. The use of radionuclides has dramatically increased in energy production, medical diagnosis and therapy, and industrial applications, resulting mandatory the monitoring of radionuclides in the environment. Therefore, as with any other analyte of interest that requires the analysis of a large number of samples in a short period of time, automated methods based on flow analysis techniques have been developed. Among the advantages provided by automation exploiting flow techniques, the minimizing of sample and standards handling and the decrease of waste generation volume should be highlighted when dealing with radioactive samples. Since these features ensure the safety of the

Flow Analysis. http://dx.doi.org/10.1016/B978-0-444-59596-6.00008-5

analyst and care about the environment, respectively. Automated flow systems are especially interesting to be implemented in hot chambers, e.g. from power plants, since these are able to work autonomously reducing the analyst exposition dose.

Analytical methods based on flow analysis for determination of radionuclides encompass pretreatment, extraction and preconcentration steps, and sometimes also include the detection online. The main impediment to achieve full automation of radiometric methods when dealing with environmental samples is the low activity generally found in this kind of samples. Radionuclides are found in the environment at trace or ultra-trace levels, requiring long counting-times, usually close to 1 day by radiometric detection. Partly for this reason, and also due to the intrinsic performance of each detector, there are many partially automated methods, in which detection is carried out in an offline approach. However, even partially automated methods are worth since the steps susceptible to automation are the most laborious, time-consuming and with a higher degree of manipulation.

Fully automated determination of radionuclides, i.e. with online detection, can be carried out using detectors such as ICP-MS, with more sensitivity for the determination of isotopes with very long lives [1,2]. Another possibility is the coupling with a radiometric detector, usually liquid scintillation counters (LSCs) which require counting-times of few minutes to measure samples with high activities, e.g. nuclear waste samples. However, LSCs require longer times, in the order of hours, to measure samples with environmental levels. Radiometric detectors are a very sensitive detection technique for radioisotopes with short lives ($T½ < 10$ years) but are also used for the determination of most long-lived radionuclides in environmental samples [3].

Thus, the grade of automation reached will be limited principally by the level of activity found in the sample and the detector selected. The latter will mainly depend on the half-life of the radionuclide of interest, in order to achieve the greatest sensitivity. Figure 8.1 shows different approaches used for the analysis of radionuclides, indicating the steps of the analytical protocol susceptible to be automated.

Since some detectors are unable to discriminate between radioisotopes, e.g. proportional counters, and others require a sample matrix clean-up, sample pretreatment is very often unavoidable. Thus, radionuclides are isolated before measurement to avoid the presence of other radionuclides and interfering elements, and therefore radiochemical methods usually include separation steps in their protocols. Apart from a limited number of radionuclides that can be analyzed by high-resolution gamma spectrometry, e.g. ^{54}Mn, ^{60}Co, ^{134}Cs, ^{137}Cs, ^{154}Eu, ^{155}Eu, and ^{125}Sb, direct analysis is impossible, that is the sample needs to be previously chemically processed. Furthermore, if samples to be analyzed have low activities, e.g. environmental samples, a preconcentration procedure is mandatory.

Besides specific radionuclide determinations, screening methods to determine radioactive elements levels are very useful tools prior to laborious and expensive analytical protocols allowing isotopic analysis. These screening tools could replace gross alpha and gross beta index providing information about total element concentration. In order to develop fully automated methodologies based on inexpensive instruments, and consequently affordable for any routine laboratory, spectrophotometric detection can be used for this purpose.

8.2 Online radiochemical separations

Most commonly used radiochemical protocols for radionuclides isolation are based on precipitation or coprecipitation processes, adsorption processes, ion exchange using anionic and cationic resins,

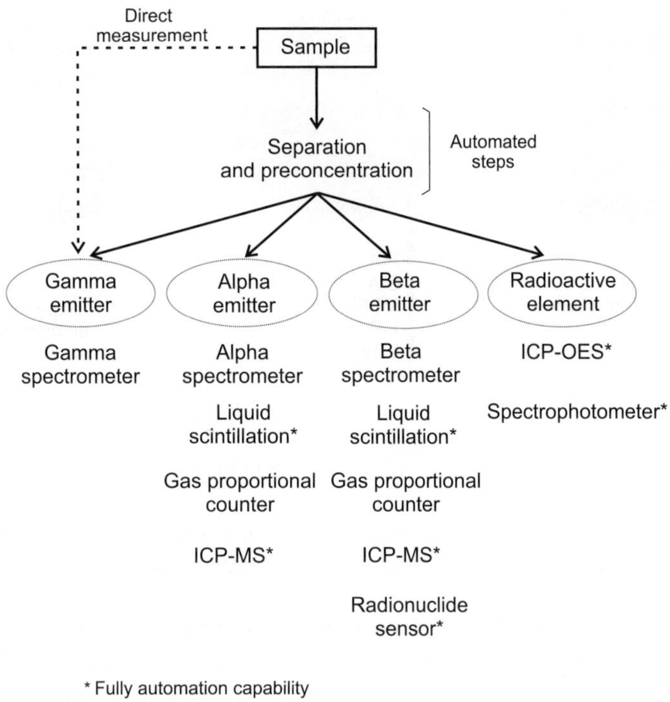

FIGURE 8.1

Different approaches for the analysis of radionuclides.

liquid–liquid extraction (LLE) and chromatographic separations. The combination of classical methods with modern instrumentation, such as flow techniques, makes possible the total or partial automation of many conventional analytical methods in the radiochemical field (Figure 8.2).

Separation by extraction chromatographic materials appears as the most suitable technique to develop automated radiochemical methodologies followed by LLE. There are a large number of selective resins for radionuclides determination, e.g. Ra, Ni, Pb, Th, U, Np, Pu, Am, Cm, Sr, Tc, ^{3}H, Fe and Pa [4], which have been largely included in protocols with online separation [5–8]. Automatic separation exploiting flow systems can also be carried out by LLE which involves the formation of complexes between several organic compounds and radionuclides [9–11].

Other separation strategies have also been used in flow systems, such as adsorption processes [12], capillary electrophoresis providing short separation times and high separating efficiencies [13], and electrochemically modulated separation coupled to ICP-MS for determination of radioisotope ratios [14].

8.2.1 Selective resins for radionuclide separations

Selective resins for radionuclide separations are commercially available in different forms, such as cartridges, packed columns, beads and disks. Moreover, there are available various pore size ranges,

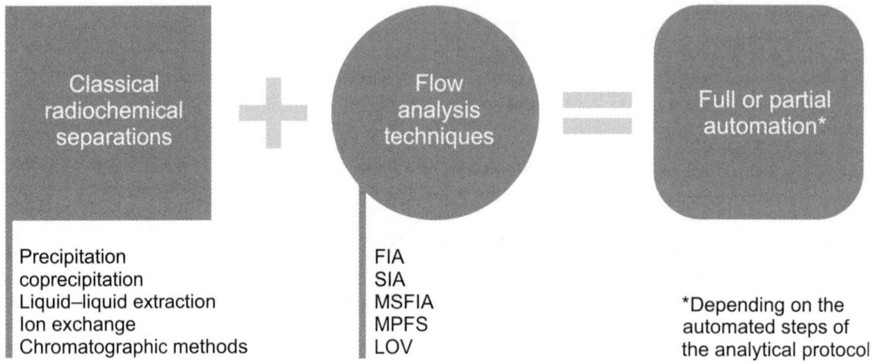

FIGURE 8.2

Combination of classical radiochemical methods with flow analysis techniques to automate radiochemical analysis. (For color version of this figure, the reader is referred to the online version of this book.)

between 20 and 800 μm. The range of 50−100 μm pore size is recommended in flow systems since these do not cause overpressure and beads have a high surface/volume ratio, exposing more active sites than larger diameter beads.

Beads packed in microcolumns is the most reported in literature, since users can customize the quantity of resin according to the capacity of it, to the volume of sample to be loaded, and to the minimum detectable activity (MDA) of the detector used. In general, packing is manually replaced in flow systems based on flow injection analysis (FIA), sequential injection analysis (SIA), multisyringe flow injection analysis (MSFIA) and multipumping flow systems (MPFS). By the contrary, lab on valve (LOV) allows the manipulation of heterogeneous solutions, i.e. bead injection, achieving the automated replacement of the resin. In Chapter 3 are described in detail the parts of the microcolumns and the way to fill them. Table 8.1 summarizes the variety of available resins from TrisKem International [4].

8.2.2 Radionuclide separation by liquid-liquid extraction

Separation of radionuclides exploiting flow systems can also be carried out by LLE which involves the formation of complexes between several organic compounds and radionuclides. LLE performed in a flow system can be carried out in a pipette tip, in-syringe or by a wetting film. The selection of the material, i.e. plastic, Teflon, C18 or glass, where LLE is carried out depends on the organic solvent used as extractant and the kind of LLE performed. Thus, if LLE is carried out by mixing aqueous and organic phases, it is preferred that no liquid remains adhered on the walls. Nonetheless, if a wetting film has to be used, the absorption of the extractant on the inert support should be taken into account.

LLE performed by using a wetting film has been satisfactorily applied to environmental and biological sample matrices in order to extract ^{90}Y and ^{90}Sr. In both, the extraction process is carried out in a pseudostationary phase or a coating film consisting in the extractant adhered on an inert support. In the case of yttrium, the extractant was adhered on C18 beads packed into a minicolumn [11]. The extraction LLE protocol for ^{90}Sr was based on the formation of a pseudostationary organic layer on the inner walls of PTFE tubing due to hydrophobic interactions [10]. This methodology allows the

Table 8.1 Commercially Available Resins for Radiochemical Separations

Resin Name	Radionuclide
Actinide resin	Actinide separation/total alpha/Be
BE resin	Beryllium
CL resin	^{36}Cl, ^{129}I
Cs resin	^{134}Cs, ^{137}Cs
DGA resin	Am, actinides, lanthanides, Y, Ra
MnO_2 resin	Ra
RE resin	Th, U, Np, Pu, Am, Cm and rare earth elements
Sr resin	Sr, Pb
TEVA® resin	Tc, Th, Np, Pu, Am/lanthanides
TRU resin	Fe, Th, Pa, U, Np, Pu, Am, Cm
UTEVA® resin	Th, U, Np, Pu
Tritium columns	^{3}H

reduction of sample and residual solvent volumes, an increase in sampling frequency, and avoids the release of organic vapors.

Online LLE can also encompass the mixture between the organic and the aqueous phases, i.e. the extractant and sample, respectively, in a pipette tip or in-syringe. After a contact time between the phases, usually assisted by flow movements to facilitate the mix, the mixture remains at rest allowing the phases separation.

Dispersive liquid–liquid microextraction (DLLME) enhances the effective surface area of the LLE, reaching the equilibrium in a shorter time [15,16]. DLLME is based on the use of an immiscible extractant solvent and a disperser solvent, which is miscible in both phases, aqueous and organic solvent. Extractant and disperser are mixed and injected rapidly into the sample, producing a turbulent mixture due to the formation of small droplets of the extractant throughout the aqueous sample.

The selection of the extractant and the disperser solvent if used will depend on the radionuclide of interest. Thus, 4,4′-(5′)-bis(tert-butylcyclohexano)-18-crown-6 solution in 1-octanol has been used as extractant for the selective isolation of strontium [10], di-2-ethylhexylphosphoric acid (HDEHP) for yttrium extraction [11], tri-*n*-octylphosphine oxide and tri-*n*-butyl phosphate (TBP) were used as complexing ligands of uranium [17], TBP in dodecane has been applied to the extraction of uranium, thorium and plutonium [18,19], and a mixture of aliquat® as extractant and 2-propanol as disperser solvent can be used to isolate technetium.

8.3 Detection of radionuclides

Usually, direct analysis of alpha, beta and some gamma emitters is impossible if the sample is not previously isolated and preconcentrated. There is little use of direct radiometric detection of beta emitters by beta spectroscopy or liquid scintillation because of its low capacity to discriminate between different beta particles. This is because beta particles from a single isotope are emitted with a range of energies, giving a spectrum that overlaps with those of other beta-emitting radioisotopes. So, beta emitters normally have to be isolated.

Alpha spectroscopy is a more selective technique, since alpha particles are emitted with characteristic energies, which facilitate simultaneous determination of alpha radionuclides with good-quality energy resolution. Therefore it could be used directly as detection technique for some alpha radionuclides. Nevertheless, various alpha emitters can be affected by the presence of spectral interferences, e.g. $^{241}Am/^{238}Pu$ and $^{237}Np/^{234}U$ and by auto-absorption effects, requiring careful sample preparation.

Gas proportional counters for alpha and beta emitters also require previous separation of radionuclides due to their low capacity to discriminate between particles with different energies.

ICP-MS is a good option for determining various radioisotopes with limits of detection comparable to radiometric methods in a shorter time, especially sensitive for the determination of isotopes with very long lives. However, low-resolution quadrupole ICP-MS is susceptible to isobaric interferences, e.g. $^{241}Pu/^{241}Am$, $^{99}Tc/^{99}Ru$, $^{238}U/^{238}Pu$, molecular interferences, e.g. $^{239}Pu/^{238}UH^{+}$, and spectral interferences, e.g. $^{237}Np/^{238}U$, also high concentration of dissolved solids should be avoided in order to enlarge the good performance of the instrument, making necessary matrix removal prior introduction into the ICP-MS.

Summarizing, most of the detectors used for radionuclide measurements require the isolation of the analyte of interest and the elimination of the sample matrix.

Regarding to the online connectivity of the detectors, LSCs [20,21] together with its modified versions for radionuclide sensors [22–24] and ICP-MS [25–27] have been successfully coupled to flow-based methods.

If the objective is the determination of a radioactive element, spectrophotometry appears as an alternative to the above-mentioned detection techniques, since it has good precision and accuracy, allows full automation of the system and a lower cost. It should be taken into account that the term radioactive element here is referred to elements whose isotopes are all radioactive, such as uranium, or have stable and radioactive isotopes, such as strontium. In any case, the screening method would be used to determine when an element exceeds a reference level, in such case an isotopic analysis should be carried out. Due to the low sensitivity and selectivity of spectrophotometric determinations of radionuclides, it is difficult to carry out a direct determination without preconcentrating and isolating the analyte. Furthermore, to enhance its sensitivity, it can be coupled with long path-length waveguide capillary cells, with 1–2 m optical lengths [6].

8.3.1 Radiometric detectors

8.3.1.1 Offline detection

When using radiometric detectors for environmental sample analyses, the detection is usually carried out offline due to the long counting-times required. As an example of a method with offline detection, we describe a system that combines two flow techniques, i.e. MSFIA and MPFS, for the separation and preconcentration of radium [12]. Radium is a naturally occurring element, whose isotopes are all radioactive. Among them it is worthwhile mentioning ^{226}Ra which is an alpha emitter well known for its long half-life ($T½$ 1600 years), widely spread in the environment and, therefore, considered one of the major contributors to the population internal dose. Since the analyte concentration in water samples is usually very low, preliminary steps of separation and above all of preconcentration will be needed in order to eliminate the possible interferences and to achieve the required MDA, respectively. The principle of the used extraction technique is based on the adsorption of radium on MnO_2 and subsequent elution and coprecipitation with Ba^{2+} (Figure 8.3). The analytical protocol was controlled automatically with the software AutoAnalysis.

First, 0.1 mol/l permanganate potassium solution was impelled towards the column, remaining in contact with the cotton fiber for 2 min to allow the formation of MnO_2. Then, permanganate solution excess was eluted with water. Micropump 1 loaded the sample into the column at 5 ml/min, and then micropump 2 supplied water to the column to remove possible interferences that had not been adsorbed on the MnO_2. Next, the multisyringe module dispensed 4 ml of 0.1 mol/l NH_2OH that reduced and dissolved completely the MnO_2 adsorbed onto the cotton fiber and eluted jointly the deposited Ra. In order to coprecipitate radium as $Ba(Ra)SO_4$, 4 ml of the acetate buffer/Ba^{2+} solution and 8 ml of 2 mol/l Na_2SO_4 were automatically added to the fraction previously collected in the autosampler. The formed slurry was left for aging for 30 min aiming to assure quantitative coprecipitation of Ra. Next, it was filtered through a membrane filter of 0.22 μm pore size. The retained precipitate was dried over a stainless steel planchet under an infrared lamp. Finally, the analyte alpha activity was measured with a low background proportional counter (LBPC). The cotton filling could be reused up to 30 times without affecting the recovery of radium. This method, with an MDA of 0.05 Bq/l, was satisfactorily applied to different types of water (tap, mineral and seawater).

Thus, coupling solenoid micropumps to MSFIA allowed the extraction and preconcentration of large sample volumes in a short time avoiding repetitive filling of the syringes. These facts are of special benefit in radioactivity environmental analysis where isotope concentrations are usually very low.

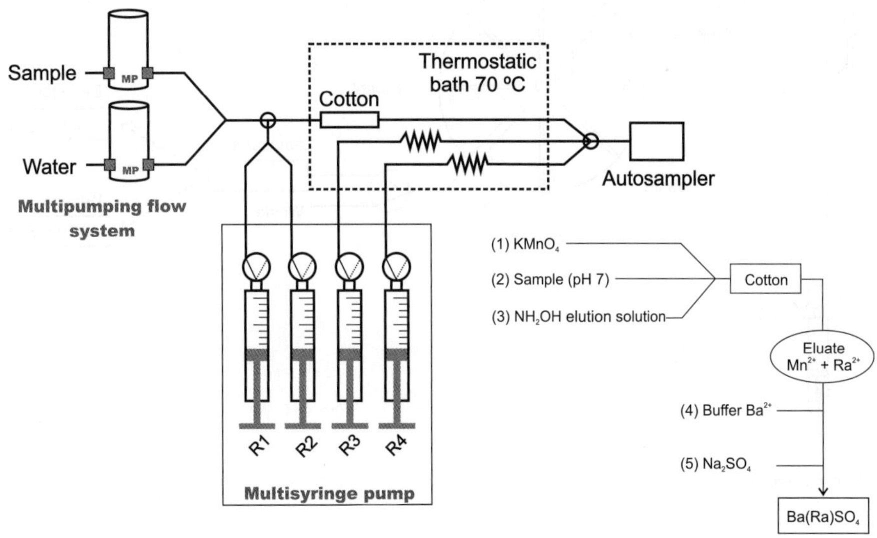

FIGURE 8.3

MSFIA-MPFS system for ^{226}Ra extraction and preconcentration. MP: micropump, R: reagent.

Other examples of automatic radiochemical methods with offline detection are automatic methods for stable and radioactive strontium determination [28,29]. In these works two detectors were proposed, ICP-OES and a LBPC, to measure stable and radioactive strontium isotopes, respectively.

Figure 8.4 shows a system based on LOV technique, exploiting Sr resin as selective extraction material [29].

The use of LOV allows the bead injection process, so that the replacement of the resin packing was performed automatically. Contrary to the longer manual protocol to isolate Sr which lasts days, the automatic separation and preconcentration system allowed analyzing up to five samples per hour depending on the preconcentrated volume. The analytical protocol started with the automatic filling of the microcolumn with Sr-resin. Then it was conditioned with 2 ml of 8 mol/l nitric acid. Once the column was ready, a variable volume of sample was dispensed toward the column. An additional volume of 0.5 ml of conditioning solution was propelled in order to eliminate interferences. After that, the elution of Sr was carried out dispensing 5.1 ml of 0.01 mol/l nitric acid. For the radiostrontium quantification, the eluate was dried and measured in a LBPC.

The Bateman equation was used at two different times for calculation of the initial $^{90}Sr/^{90}Y$ activities. That is to say, total beta activity, i.e. ^{90}Sr activity and its daughter ^{90}Y, was measured at two different times, one along the first day after the separation process and another when the secular equilibrium was reached. Main advantages of this method are the little amount of resin used and its long durability (up to 30 injections) achieving an MDA of 0.008 Bq ^{90}Sr.

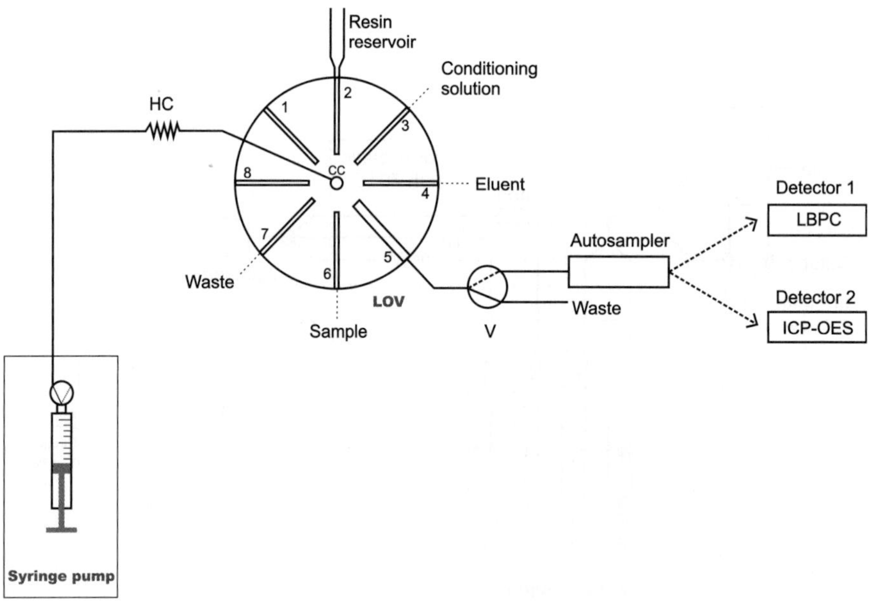

FIGURE 8.4

LOV system for stable and radiostrontium isolation exploiting ICP-OES and low background proportional counter (LBPC), respectively. CC: central conduit, HC: holding coil, V: valve.

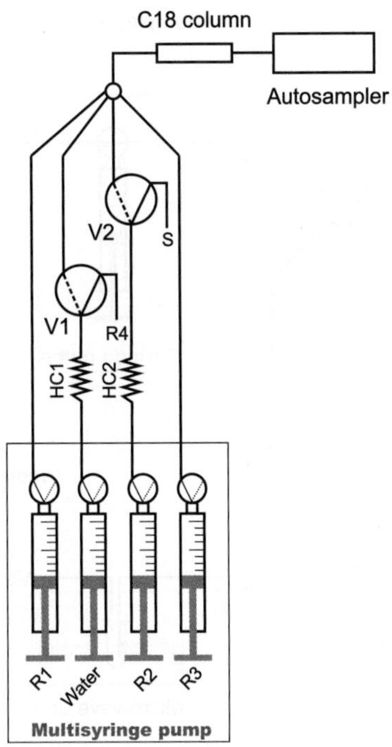

FIGURE 8.5

MSFIA system for separation and preconcentration of yttrium from water and biological samples. HC: holding coil, R: reagent, S: sample, V: valve.

An MSFIA method for the separation of both stable and radioactive yttrium exploiting LLE with HDEHP retained on a C18 support is shown in Figure 8.5 [11]. In this work, two offline detectors were proposed, an ICP-OES and a LBPC to measure stable and radioactive yttrium, respectively. The formation of an $Y(DEHP)_3$ complex is the basis of the extraction process. ^{90}Y (T½ 64.1 h) is generated from ^{90}Sr, and it is a pure beta emitter used in nuclear medicine owing to its properties, mainly for its high energy (2.27 MeV) and its relatively short half-life. The extraction system consisted of a column packed with preparative C18 of 55−105 μm mesh size. Once the adequate volume (0.3 ml) of HDEHP was retained on the C18 support into the column, the separation protocol was carried out. Thus, sample volume was dispensed through the column, and then yttrium was eluted with 10 ml of 4 mol/l HNO_3. The total beta activity was measured offline on the eluted fraction by using a LBPC. The elimination of the HDEHP film was performed with 5 ml of ethanol, after which a new coating film was created.

8.3.1.2 Online detection

As an example of a fully automated system for radionuclide determination, there is a method based on SIA technique developed for ^{99}Tc determination, including microwave assisted treatment, solid phase extraction (SPE) and online detection (Figure 8.6) [21].

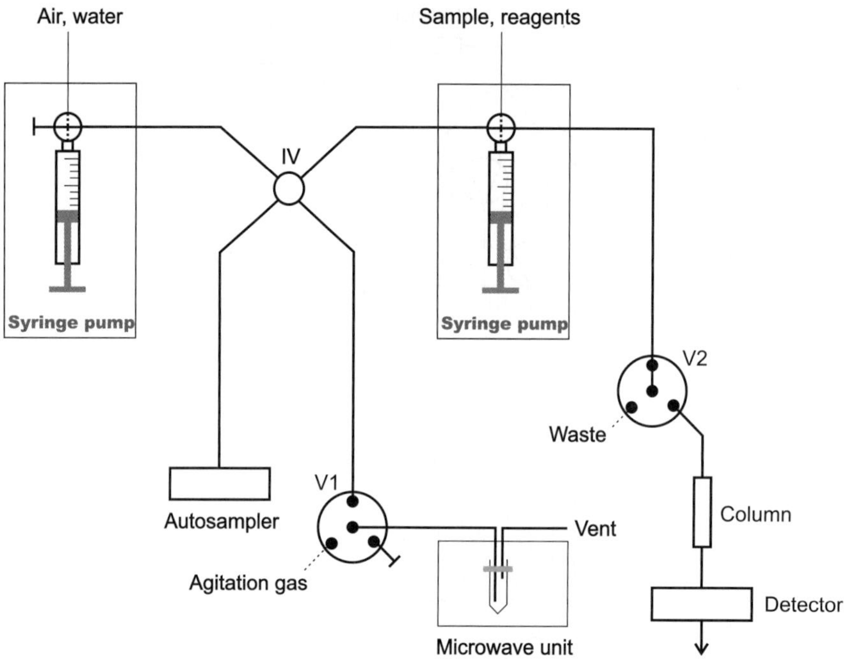

FIGURE 8.6

Fully automated system for ^{99}Tc determination. IV: injection valve, V: valve.

An oxidation procedure using peroxydisulfate in acidic solution was online performed to convert reduced technetium species into pertechnetate in samples with high content of reducing organics. The separation of $^{99}Tc(VII)$ from interfering radionuclides was accomplished using a microcolumn packed with strongly basic anion exchanger. Flow-through scintillation counting using a detector equipped with a 1 ml glass scintillator flow cell was employed for the detection of $^{99}Tc(VII)$ activity eluted from the separation column.

Analytical procedure started loading 1.4 ml of sample into the reaction vessel of the microwave, and mixing with 2.8 ml of 1.6 mol/l HNO_3 and 1 ml of 2 mol/l $Na_2S_2O_8$. Then, the separation was carried out by loading a 1.6 ml aliquot of the digested sample onto the column preconditioned with 0.2 mol/l nitric acid. Next, a sequence of column wash steps using 5 ml of 0.2 mol/l HNO_3, 6 ml of 1 mol/l NaOH, 6 ml of 0.2 mol/l HNO_3−0.5 mol/l $H_2C_2O_4$, and 5 ml of 2 mol/l HNO_3 was used to remove interfering radionuclide species. Finally, $^{99}Tc(VII)$ was eluted using 5.6 ml of 8 mol/l HNO_3 and the activity of the eluate was online measured by LSC. The total time required to execute the automated sample treatment protocol was 7.5 min. Here, it should be noted again that the possibility of carrying out an online measurement is due to the high activity of the analyzed samples, which were caustic aged nuclear waste matrixes from the Hanford site, a mostly decommissioned nuclear production complex.

Another example of system for radionuclides determination using online detection is an SIA system with bead injection (Figure 8.7) [20]. The use of a renewable column packing with selective

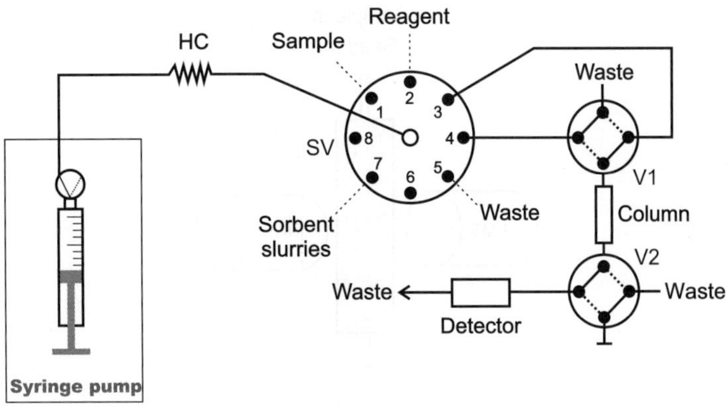

FIGURE 8.7

SIA system with online detection for separation and analysis of ^{90}Sr, ^{241}Am, and ^{99}Tc. HC: holding coil, SV: selection valve, V: valve.

resins, allowed the elution of each radionuclide separately eliminating carry over on the column from one separation to the next, and the same system could be used for multiple separations implementing different selective resins without requiring changes of the manifold. The system was applied to aged nuclear waste samples for the separation and analysis of ^{90}Sr, ^{241}Am, and ^{99}Tc employing Sr-resin, TRU resin, and TEVA resin, respectively.

The length of the column used in the analysis was 7.1 cm, with a calculated volume of 0.25 ml. The time required to automatically pack the microcolumn was 4.5 min, including washing steps. After elution, online detection was carried out by a flow-through liquid scintillation detector, configured with a 0.5 ml flow cell. The detector integration time to accumulate counts for each data point reported was 6 s.

Online detection of radionuclides is also integrated in sensors developed for environmental monitoring. Sensors are constituted by a preconcentrating minicolumn packed with scintillating microspheres, which integrates radiochemical separation and radiometric detection steps within a single device [24].

Scintillating microspheres with selective radionuclide uptake properties, serve to separate and to preconcentrate radionuclides within a detector of well-defined geometry as well as to emit a photometric signal. Since these sensors are built in flow mode, the sensing material can be regenerated with eluent or renewed online by fluidic replacement of the beads. The latter method allows the use of materials that bind analytes irreversibly or are unstable under regeneration conditions. Radionuclide selective scintillating microspheres can be prepared by coimmobilization of scintillating fluours and selective organic extractants within the pores of an inert polymeric support. Figure 8.8 shows an example of these sensors.

A sensor configured for ^{99}Tc(VII) analysis was successfully applied toward the analysis of groundwater samples from the Hanford site, using standard addition techniques for quantification (Figure 8.9) [22].

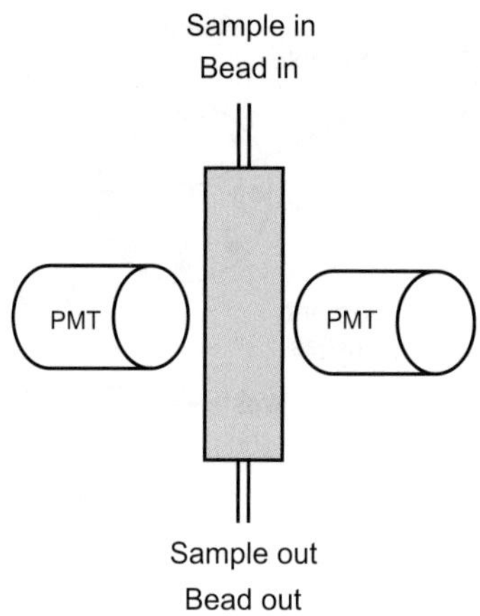

FIGURE 8.8

Depiction of a sensor for radionuclide separation, preconcentration and detection. PMT: photomultiplier.

This flow system is based on SIA technique and also uses injection valves. The sensor column flow cell was fabricated from a poly(methyl methacrylate) (PMMA) block, where the central column channel was 3 cm long and 3 mm internal diameter (i.d.), with a volume of 212 μl. A frit disk was placed directly into the threaded outlet of the column bed channel in order to retain scintillating sorbent beads within the sensor flow cell. The detection was performed using an online flow-through scintillation detector, in which the sensor flow cell assembly was positioned in the place of a conventional flow cell. To prevent

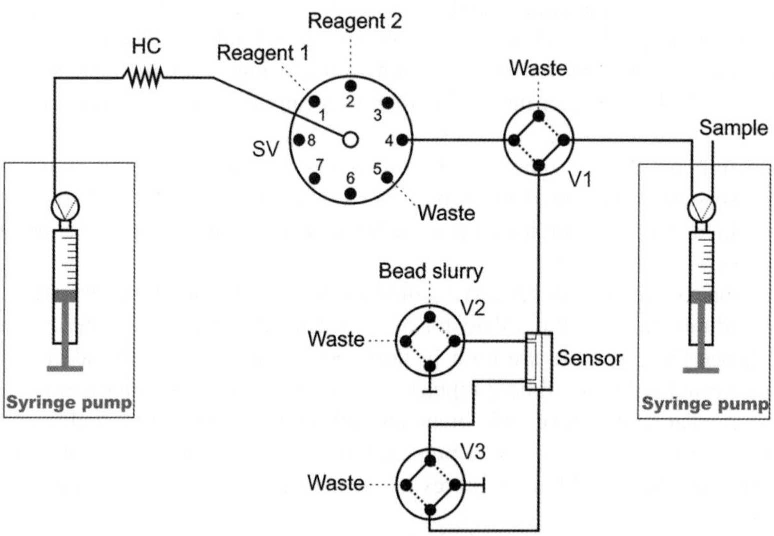

FIGURE 8.9

SIA system exploiting a sensor for ^{99}Tc determination. HC: holding coil, SV: selection valve, V: valve.

stray light from entering the sensor flow cell assembly, black FEP Teflon tubing was used to construct inlet and outlet lines of the sensor flow cell. The scintillating microspheres were prepared at the laboratory and stored in the dark for 24 h prior to use to eliminate sporadic chemiluminescence signals. Using a 50 ml sample volume and a counting time of 30 min, the detection limit for ^{99}Tc(VII) was 6 Bq/l.

Later, an equilibrium-based sensor using a preconcentrating minicolumn for ^{99}Tc and ^{90}Sr determinations was developed [23]. In this approach, a solution with a given analyte concentration was delivered through the minicolumn until resin breakthrough was complete. Nevertheless, if the analyte concentration is low, the column is not saturated, it is equilibrated, and the amount captured is proportional to the sample concentration, and the exit concentration equals the inlet concentration. Theory was developed to describe the signal at equilibration and to describe analyte uptake as a function of volume and concentration, using parameters and concepts from frontal chromatography. The main advantage lies in a long-term application, since the precise volume delivery of the pump over time is not a factor of maintaining calibration, as occurs when using the FIA technique, which requires periodic recalibration of flow rates.

The equilibrium-based sensor was applied for the detection of ^{99}Tc using anion-exchange beads mixed with scintillating beads and light detection, and for ^{90}Sr using beads of a selective resin mixed with scintillating beads. Radiometric measurements of the equilibration sensor were performed using a flow-through radioactivity detector. The detector integration time was typically 1 min, and the analysis required flow rates below 0.5 ml/min.

In order to reach the regulatory levels, a signal acquisition time of only 2 min was required for ^{99}Tc, while it required a counting time of 20 h for ^{90}Sr. Nevertheless, a larger size sensor column may be implemented to facilitate more rapid ^{90}Sr measurements.

8.3.2 ICP-MS for radioisotope determinations

Some considerations have to be taken into account about the advantages of an online coupling of ICP-MS. Even though the coupling involves benefits such as the decrease of analyte losses risk, and of the samples contamination together with the analyst safety improvement, coupling an ICP-MS for online detection requires the instrumentation setup in transient signal mode, leading to longer times for the reading of all the isotopes of interest. Moreover, the time needed to perform the extraction and preconcentration steps may take up to several minutes, requiring also additional time to other steps such as washes between different samples and column conditioning. Thus, a decrease of the injection frequency occurs and also it can result in an inconvenient due to the higher argon gas consumption, which is a critical concern when using ICP-MS. By contrast, with offline detection, the ICP-MS works in steady signal mode, requiring less time of reading. In consequence, the consumption of argon gas decreases at the same time that increases the analysis frequency. So, both approaches should be taken into consideration before proposing the ICP-MS coupling, establishing a compromise between the sampling frequency/sample contamination/analyst safety and argon gas consumption.

A fully automated LOV–MSFIA–ICP-MS system was proposed as a fast, sensitive and selective methodology for determination of thorium and uranium at environmental levels [27]. Both radionuclides have natural radioisotopes with long half-lives, e.g. ^{238}U $T½$ 4.47×10^9 years and ^{232}Th $T½$ 1.40×10^{10} years, and therefore the ICP-MS is a potential tool to measure them with a good sensitivity.

Online separation and preconcentration of thorium and uranium was carried out by means of UTEVA resin. LOV made possible the full automation of the system by the online regeneration of the column. Figure 8.10 shows a scheme of the flow system configuration.

The online coupling of the LOV–MSFIA system with the ICP-MS was performed through the digital output of the multisyringe burette which triggered the ICP-MS through a relay interface, using a connection cable described in Chapter 5.

Limits of detection (LODs) reached were 0.4 ng/l of uranium and 2.8 ng/l of thorium. The reproducibility of the LOV–MSFIA–ICP-MS was 1.7% expressed as relative standard deviation (RSD). Moreover, a high sensitivity, a wide working range, e.g. 0–200 μg/l for both ^{238}U and ^{232}Th, and an injection frequency up to 9/h (depending on the sample volume) should be highlighted. Different water sample matrices (seawater, well water, freshwater, tap water and mineral water), a phosphogypsum sample with natural uranium and thorium content and a channel sediment reference material were satisfactorily analyzed with the proposed method.

Other FIA systems have been coupled to ICP-MS for determination of uranium and thorium. As an example, an online SPE method, linked to ICP-MS was developed to overcome the problems of a high sample matrix concentration interfering with the determination of low levels of uranium and

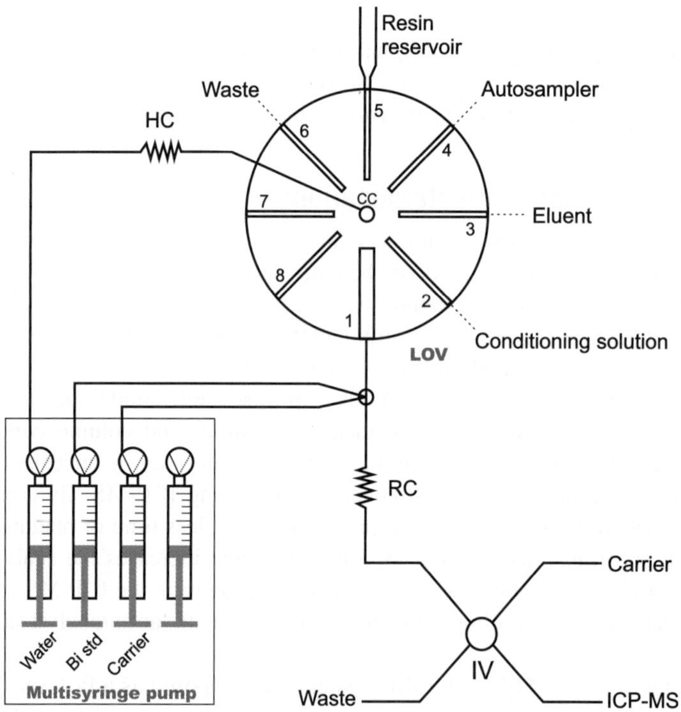

FIGURE 8.10

LOV–MSFIA–ICP-MS system for uranium and thorium determination. CC: central conduit, HC: holding coil, IV: injection valve, RC: reaction coil.

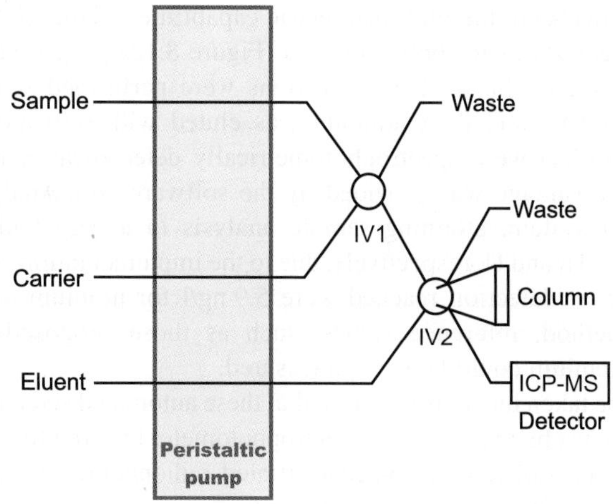

FIGURE 8.11

FIA–ICP-MS system for uranium and thorium determination. IV: injection valve.

thorium in aqueous samples [30]. Using TRU resin as a selective resin for actinide-specific extraction in a small column incorporated into the sample introduction manifold, good recoveries were obtained from seawater, ground water and samples containing significant levels of dissolved organic carbon. Samples were injected directly onto the column, without prior dilution. Two injection valves were used for sample treatment and introduction (Figure 8.11). A linear working range of 0–10 μg/l, with RSDs of 4.3 and 5.8% for uranium and thorium, respectively, and LODs of 15 ng/l for uranium and 6 ng/l for thorium were obtained. An injection frequency of 12/h was reached.

8.4 Screening methods for radioactive elements determination

New trends suggest the use of screening methodologies for individual radioactive elements determination in replacement of gross alpha activity or gross beta activity screening. Thus, if the reference concentration is exceeded, additional analyses of the specific radioisotopes should be developed. Since isotopic analyses usually require extensive protocols to separate the single radionuclides and expensive instrumentation, screening methods exploiting spectrophotometric detection are very handy, enabling fast, sensitive and low cost determinations.

As mentioned before, the lack of sensitivity of the spectrophotometric detection technique can be solved by implementing long path-length liquid core waveguide capillary cell (LWCC). Taking into account that the absorbance is directly proportional to the path length, exploiting 100 cm path length an improvement of two orders of magnitude in the signal regarding to a 1 cm path length cell can be accomplished.

A screening method for uranium and thorium was performed, in which the hyphenation of MSFIA and LOV coupled to an LWCC made of this system a powerful rapid screening tool for analyzing environmental samples, thanks to the multiparametric capabilities of the LOV, allowing the analysis in a wide range of concentrations for both elements (Figure 8.12) [31]. UTEVA resin was used for retention of both elements, in this system extractions were performed sequentially, first thorium was eluted with 5 mol/l HCl and then uranium was eluted with 0.01 mol/l HCl. After elution, thorium(IV) and uranium(VI) were spectrophotometrically detected after reaction with arsenazo-III. Another additional advantage was provided by the software AutoAnalysis which allowed the development of a smart system, attaining sample analysis in a very wide concentration range, 0–1200 and 0–2000 μg/l Th and U, respectively, due to the implementation of dilution or preconcentration protocols. Limits of detection reached were 5.9 ng/l for uranium and 60 ng/l for thorium. Therefore, with this method, reference values, such as those proposed by the World Health Organization [32], for uranium could be easily measured.

In addition, it has to be taken into consideration that these automated screening methods can also be used for isotopic analysis just by replacing the spectrophotometer by a fraction collector and taking the fractions which contain the isolated and preconcentrated radionuclides to a detector which allows isotopic analysis when the reference value is surpassed.

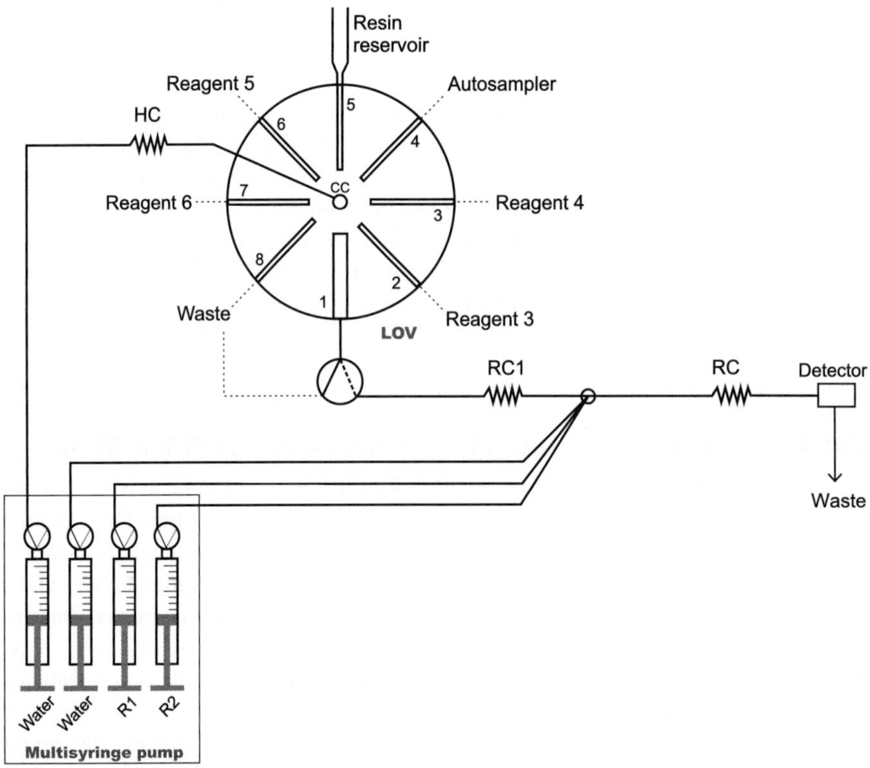

FIGURE 8.12

Smart system used as screening tool for uranium and thorium determination in a wide range of concentrations. CC: central conduit, HC: holding coil, R: Reagent, RC: reaction coil.

References

[1] J.S. Becker, Mass spectrometry of long-lived radionuclides, Spectrochim. Acta Part B At. Spectrosc. 58 (2003) 1757–1784.

[2] D. Lariviere, V.F. Taylor, R.D. Evans, R.J. Cornett, Radionuclide determination in environmental samples by inductively coupled plasma mass spectrometry, Spectrochim. Acta Part B At. Spectrosc. 61 (2006) 877–904.

[3] Y. Fajardo, J. Avivar, L. Ferrer, E. Gómez, M. Casas, V. Cerdà, Automation of radiochemical analysis by applying flow techniques to environmental samples, Trends Anal. Chem. 29 (2010) 1399–1408.

[4] TrisKem International: http://www.triskem-international.com.

[5] Y. Fajardo, L. Ferrer, E. Gómez, F. Garcias, M. Casas, V. Cerdà, Development of an automatic method for americium and plutonium separation and preconcentration using an multisyringe flow injection analysis-multipumping flow system, Anal. Chem. 80 (2008) 195–202.

[6] J. Avivar, L. Ferrer, M. Casas, V. Cerdà, Automated determination of uranium(VI) at ultra-trace levels exploiting flow techniques and spectrophotometric detection using a liquid waveguide capillary cell, Anal. Bioanal. Chem. 397 (2010) 871–878.

[7] J. Avivar, L. Ferrer, M. Casas, V. Cerdà, Lab on valve-multisyringe flow injection system (LOV–MSFIA) for fully automated uranium determination in environmental samples, Talanta 84 (2011) 1221–1227.

[8] M. Villar, J. Avivar, L. Ferrer, M. Galmés, F. Vega, V. Cerdà, Automatic and simple method for ^{99}Tc determination using a selective resin and liquid scintillation detection applied to urine samples, Anal. Chem. 85 (2013) 5491–5498.

[9] E.P. Horwitz, D.G. Kalina, H. Diamond, G.F. Vandegrift, W.W. Schulz, The TRUEX process – a process for the extraction of the transuranic elements from nitric acid wastes utilizing modified purex solvent, Solvent Extr. Ion Exch. 3 (1985) 75–109.

[10] M. Miró, E. Gómez, J.M. Estela, M. Casas, V. Cerdà, Sequential injection ^{90}Sr determination in environmental samples using wetting-film extraction method, Anal. Chem. 74 (2002) 826–833.

[11] Y. Fajardo, E. Gómez, F. Garcias, V. Cerdà, M. Casas, Multisyringe flow injection analysis of stable and radioactive yttrium in water and biological samples, Anal. Chim. Acta 539 (2005) 189–194.

[12] Y. Fajardo, E. Gómez, F. Garcias, V. Cerdà, M. Casas, Development of an MSFIA-MPFS pre-treatment method for radium determination in water samples, Talanta 71 (2007) 1172–1179.

[13] B. Kuczewski, C.M. Marquardt, A. Seibert, H. Geckeis, J.V. Kratz, N. Trautmann, Separation of plutonium and neptunium species by capillary electrophoresis-inductively coupled plasma-mass spectrometry and application to natural groundwater samples, Anal. Chem. 75 (2003) 6769–6774.

[14] M. Liezers, S.A. Lehn, K.B. Olsen, O.T. Farmer III, D.C. Duckworth, Determination of plutonium isotope ratios at very low levels by ICP-MS using on-line electrochemically modulated separations, J. Radioanal. Nucl. Chem. 282 (2009) 299–304.

[15] F. Maya, J.M. Estela, V. Cerdà, Completely automated in-syringe dispersive liquid–liquid microextraction using solvents lighter than water, Anal. Bioanal. Chem. 402 (2012) 1383–1388.

[16] F. Maya, B. Horstkotte, J.M. Estela, V. Cerdà, Lab in a syringe: fully automated dispersive liquid–liquid microextraction with integrated spectrophotometric detection, Anal. Bioanal. Chem. 404 (2012) 909–917.

[17] A.R. Ghiasvand, E. Mohagheghzadeh, Homogeneous liquid–liquid extraction of uranium(VI) using tri-n-octylphosphine oxide, Anal. Sci. 20 (2004) 917–919.

[18] T. Kiliari, I. Pashalidis, Alpha spectroscopic analysis of actinides (Th, U and Pu) after separation from aqueous solutions by cation-exchange and liquid extraction, J. Radioanal. Nucl. Chem. 284 (2010) 547–551.

[19] T. Kiliari, I. Pashalidis, Thorium determination in aqueous solutions after separation by ion-exchange and liquid extraction, J. Radioanal. Nucl. Chem. 288 (2011) 753–758.

[20] O. Egorov, M. O'Hara, J.W. Grate, J. Ruzicka, Sequential injection renewable separation column instrument for automated sorbent extraction separations of radionuclides, Anal. Chem. 71 (1999) 345–352.

[21] O. Egorov, M. O'Hara, J.W. Grate, Microwave-assisted sample treatment in a fully automated flow-based instrument: oxidation of reduced technetium species in the analysis of total technetium-99 in caustic aged nuclear waste samples, Anal. Chem. 76 (2004) 3869–3877.

[22] O. Egorov, S. Fiskum, M. O'Hara, J.W. Grate, Radionuclide sensors based on chemically selective scintillating microspheres: renewable column sensor for analysis of ^{99}Tc in water, Anal. Chem. 71 (1999) 5420–5429.

[23] O. Egorov, M. O'Hara, J.W. Grate, Equilibration-based preconcentrating minicolumn sensors for trace level monitoring of radionuclides and metal ions in water without consumable reagents, Anal. Chem. 78 (2006) 5480–5490.

[24] J.W. Grate, O. Egorov, M. O'Hara, T.A. DeVol, Radionuclide sensors for environmental monitoring: from flow injection solid-phase absorptiometry to equilibration-based preconcentrating minicolumn sensors with radiometric detection, Chem. Rev. 108 (2008) 543–562.

[25] J.H. Aldstadt, J.M. Kuo, L.L. Smith, M.D. Erickson, Determination of uranium by flow injection inductively coupled plasma mass spectrometry, Anal. Chim. Acta 319 (1996) 135–143.

[26] J.B. Truscott, L. Bromley, P. Jones, E.H. Evans, J. Turner, B. Fairman, Determination of natural uranium and thorium in environmental samples by ETV–ICP-MS after matrix removal by on-line solid phase extraction, J. Anal. At. Spectrom. 14 (1999) 627–631.

[27] J. Avivar, L. Ferrer, M. Casas, V. Cerdà, Fully automated lab-on-valve-multisyringe flow injection analysis-ICP-MS system: an effective tool for fast, sensitive and selective determination of thorium and uranium at environmental levels exploiting solid phase extraction, J. Anal. At. Spectrom. 27 (2012) 327–334.

[28] Y. Fajardo, E. Gómez, F. Mas, F. Garcias, V. Cerdà, M. Casas, Multisyringe flow injection analysis of stable and radioactive strontium in samples of environmental interest, Appl. Radiat. Isot. 61 (2004) 273–277.

[29] R. Rodríguez, J. Avivar, L. Ferrer, L.O. Leal, V. Cerdà, Automated total and radioactive strontium separation and pre-concentration in samples of environmental interest exploiting a lab-on-valve system, Talanta 96 (2012) 96–101.

[30] E. Unsworth, J. Cook, S. Hill, Determination of uranium and thorium in natural waters with a high matrix concentration using solid-phase extraction inductively coupled plasma mass spectrometry, Anal. Chim. Acta 442 (2001) 141–146.

[31] J. Avivar, L. Ferrer, M. Casas, V. Cerdà, Smart thorium and uranium determination exploiting renewable solid-phase extraction applied to environmental samples in a wide concentration range, Anal. Bioanal. Chem. 400 (2011) 3585–3594.

[32] World Health Organization, Guidelines for Drinking-Water Quality, WHO, Geneva vol. 1 (2004), 3rd edn.

Glossary

4-AAP	4-aminoantipyrine
A	agitator
AAS	atomic absorption spectroscopy
A/D	analogical / digital
AFS	atomic fluorescence spectroscopy
AP	aqueous phase
ASV	anodic stripping voltammetry
AU	absorbance units
BI	bead injection
BOD	biological oxygen demand
BP	back-propagation
BTB	bromothymol blue
BTEX	benzene, toluene, ethylbenzene and total xylenes
C4D	capacitively coupled contactless conductivity detection
CC	central conduit
CCD	charge coupled device
CE	capillary electrophoresis
ChOV	chip on valve
CIA	cross-injection analysis
CL	chemiluminescent or chemiluminescence
CNC	computer numerical control
CNT	carbon nanotube
CNF	carbon nanofiber
COD	chemical oxygen demand
CPE	cloud point extraction
CRM	certified reference material
D/A	digital / analogical
DC	direct current
DEP	differential electrolytic potentiometry
DIC	dissolved inorganic carbon
DLL	dynamic-link library
DLLME	dispersive liquid-liquid micro-extraction
DMA	dimethylarsinic
DMPD	N,N-dimethyl-p-phenylenediamine
DMSA	dimethylsulphonazo
DMSO	dimethylsulfoxide
DOC	dissolved organic carbon
DOP	disolved organic phosphorus
DPP	differential pulse polarography
DTNB	5,5'-dithiobis-2,2'-dinitrobenzoic acid
EC	electrochemical cell

ETAAS	electrothermal atomic absorption spectrometry
ETFE (Tefzel)	ethylene-tetrafluoroethylene copolymer
FAAS	flame atomic absorption spectroscopy
FIA	flow injection analysis
FIASS	FIA with sándwich sampling
FEP	fluorinated ethylene propylene
GC	gas chromatography
GD	gas diffusion
GR	general regression
HC	holding coil
HDEHP	di-2-ethylhexylphosphoric acid
HG	hydride generation
HLB	hydrophilic-lipophilic-balanced
HPLC	high-performance liquid chromatography
ICP-MS	inductively coupled plasma mass spectrometry
ICP-OES	inductively coupled plasma optical emission spectrometry
I/O	on-off
IR	infrared
ISAS	ionic strength adjustor solution
ISE	ion-selective electrode
IV	injection valve
KEL-F	poly-chlorotrifluoroethylene
LBPC	low background proportional counter
LC	liquid chromatography
LED	light-emitting diode
LIS	lab in a syringe
LLE	liquid-liquid extraction
LOD	limit of detection
LOV	lab on valve
LSC	liquid scintillation counter
LWCC	liquid waveguide capillary cell
MB	methylene blue
MBS	multicommutation and binary sampling
MCE	microcolumn extraction
MCFIA	multicommutated flow injection analysis
MDA	minimum detectable activity
MEMS	micro electro-mechanical system
MFE	mercury film electrode
MGD	membraneless gas diffusion
MLR	multiple linear regression
MP	micropump
MPFS	multipumping flow systems
MS	mass spectrometry
MSC	multisyringe chromatography

MSFIA	multisyringe flow injection analysis
NED	N-(1-naphthyl) ethylenediamine
NIR	near-infrared
NP	nitrophenol
OE	oxygen electrode
OP	organic phase
OPA	o-phthalaldehyde
P	product
PAH	polycyclic aromatic hydrocarbon
PAPS	polyfluoroalkyl phosphate esters
PAR	4-(2-pyridylazo) resorcinol
PCB	polychlorinated biphenyls
PE	processing element
PEDD	paired emitter-detector diode
PEEK	poly-ether ether ketone
PFA	perfluoroalkoxy
PM	pervaporation module
PM10	particulate matter <10 μm
PMMA	poly-methylmethacrylate
PMT	photomultiplier
PP	poly-propylene
PTFE	poly-tetrafluoroethylene
PTV	programmable temperature vaporizer
PVC	poly-vinylchloride
PVDF	poly-vinylidenefluoride
R	reagent
RC	reaction coil
RE	reference electrode
rFIA	reversed FIA
RGB	red-green-blue
RS232/485	serial interfaces
RSD	relative standard deviation
S	sample
SAM	sulfanilamide
SFA	segmented flow analysis
SIA	sequential injection analysis
SIC	sequential injection chromatography
SMA	sub-miniature version A connector
SPE	solid phase extraction
SPR	solid-phase reagent
SRP	soluble reactive phosphorus
SV	selection valve
T½	half-life
TA	total alkalinity

TBP	tri-n-butyl phosphate
TDC	total dissolved carbon
TDP	total dissolved phosphorus
TEA	triethanolamine
THF	tetrahydrofuran
TIC	total inorganic carbon
TISAB	total ionic strength adjustment buffer
TP	total phosphorous
TPS	total particulate solid
TSS	total suspended solids
TTL	transistor–transistor logic
μ-TAS	micro total analytical system
UHMW-PE	ultra high molecular weight poly-ethylene
USB	universal serial bus
UV-Vis	ultraviolet-visible
V	multicommutation solenoid valve
WAD	weak acid dissociable
WE	working electrode

Index

Note: Page numbers with "*f*" denote figures; "*t*" tables.

E

F

G

H

M

N

R

S

T

U

V

W

X

Y

Z